Advanced polymer composites for structural applications in construction

Proceedings of the first international conference, held at Southampton University, UK, on 15–17 April 2002

Edited by R A Shenoi, S S J Moy, L C Hollaway

Published for the Conference Organisers by Thomas Telford Publishing, Thomas Telford Ltd, 1 Heron Quay, London E14 4JD.
http://www.thomastelford.com

Distributors for Thomas Telford books are
USA: ASCE Press, 1801 Alexander Bell Drive, Reston, VA 20191-4400, USA
Japan: Maruzen Co. Ltd, Book Department, 3–10 Nihonbashi 2-chome, Chuo-ku, Tokyo 103
Australia: DA Books and Journals, 648 Whitehorse Road, Mitcham 3132, Victoria

First published 2002

Conference organised by
Centre For Advanced Composites In Construction (CACIC)
University of Southampton
University of Surrey
The Building Research Establishment

A catalogue record for this book is available from the British Library

ISBN: 0 7277 3122 X

Printed and bound in Great Britain by MPG Books, Bodmin, Cornwall, UK

Preface

Fibre reinforced polymer (FRP) based composite materials have a number of features that make them highly suitable for modern construction projects. FRP has been used for several decades in construction applications and its use is set to increase to meet demands for improvements in construction processes. The science and technology associated with this subject are advancing at a rapid rate. It is important for engineers and researchers to meet to exchange ideas in this area in order to promote the safe and cost-effective use of FRP for structural applications in construction. The theme of this conference is to review progress towards the greater exploitation of polymer composites and FRP materials in infrastructure.

The conference is being held under the auspices of a network dealing with Advanced Polymer Composites for Structural Applications in Construction (or CoSACNet). This initiative is funded by the Engineering and Physical Sciences Research Council of the UK and is intended to promote knowledge and expertise underpinning the potential use of polymeric composites for structural applications in civil construction. The conference is being organised by the University of Southampton, University of Surrey and the Building Research Establishment. It is intended to be the first of a series of such gatherings to be held in two-year intervals.

The year 2002 also marks the Golden Jubilee of the University of Southampton. A number of events are being held throughout this year to celebrate the occasion. The conference ACIC 2002 is part of the effort of the School of Engineering Sciences and the Department of Civil and Environmental Engineering towards the celebration.

Grateful thanks are due to Margaret Binnie, Hilary Smith, James Blake and Ajaya Nayak for their help in compiling these proceedings.

R A Shenoi, S S J Moy, L C Hollaway

April 2002

Contents

6   Analysis and design

7   All-composite structures and materials issues

Session 1 : **Keynote lectures**

1.1 The development and the future of advanced polymer composites in the civil infrastructure

L C Hollaway
University of Surrey, Guildford, Surrey, UK.

INTRODUCTION

A polymer composite is a material consisting of fibres bound into a polymer matrix to maximise specific performance properties of the final product. A composite structural system is one in which two dissimilar materials are combined to form an efficient component such that, when under load, the two materials act to maximise specific performance properties. Both of these definitions relate to composite structural systems relevant to the civil infrastructure.

Composites in structural/civil engineering are utilized in broadly three ways, these are:
(a) as load bearing infill panels,
(b) as all composite material structures,
(c) as one of two or more materials which are combined to form a structural system and to act compositely under load

The development of each of these systems will be discussed separately as observed from a UK perspective.

Glass fibre/polymer composites (GFRP) were first introduced into the UK during the Second World War as materials suitable for housing radar and electronic equipment. GFRP composites are non-conductive, consequently, no distortion of radio waves occurs as they are transmitted through the material. Similar housings are currently used on the north east coast of England and generally consist of eight or more segments bolted to a steel base ring which is attached to concrete foundations and to a top cap moulding made in GFRP. Adjacent panels are connected by return of flanges which, when correctly sized, can form an integral frame within the radome.

From the late forties to the late sixties the interest in and the use of polymer composites in the civil infrastructure was sporadic and any systems which were developed were expensive. In addition, because of the ease of manufacturing GFRP hand lay-up systems a number of unscrupulous small manufacturers produced cheap and inferior components which would generally have had no post-cure; these composites degraded very quickly and damaged the reputation of composites in the civil engineering industry.

ACIC 2002, Thomas Telford, London, 2002

In the early seventies serious research commenced into the utilization of polymer composites in the civil infrastructure with two major structures being conceived and developed in the UK and then shipped to and erected in North Africa and the Middle East. The first one was the dome structure at Benghazhi, Libya, in which a series of load bearing sandwich panels, were erected on to an aluminium skeletal dome structure, shown in Figure 1. The second, in 1972, was the Dubai airport roof canopy, shown in Figure 2, in which a series of inverted 'umbrella' type units encapsulating steel angle sections, to provide stiffness at the corner folds of the units, were used. This was the first known time that a dissimilar material, such as steel, had been used seriously in conjunction with polymer composites. Although the coefficient of thermal expansion of these two materials is vastly different, tests showed that no cracking or damage of the resin resulted from the relatively high differential coefficient of expansion of the two components.

Figure 1 The Bengazhi Dome

Figure 2 Dubai Airport Roof

From these early beginnings several 'all polymer' composite panels were conceived and developed in the UK during the 1970 and early 1980s to form prestigious buildings, notably the Morpeth School in London, Mondial House on the north side of the Thames at Blackfriars, London, the American Express Building, Brighton and the Covent Garden Flower Market at Nine Elms, London, which is illustrated in Figure 3. With the exception of Covent Garden Flower Market, these structures were constructed from a skeletal system manufactured from beams and columns which were of reinforced concrete or steel construction and the GFRP units were utilised as load bearing or semi-load bearing systems, in-filling the space between the beams and columns. The systems formed a lightweight building component but particular attention had to be paid to the fixing details of the panels because of the differential coefficient of expansion between the high composite value and the relatively low concrete or steel values. The GFRP components were generally manufactured by the hand lay-up method using the randomly orientated glass fibre arrays with a composite fibre/matrix ratio by weight of 30%. The Covent Garden Flower market was constructed as a double layer steel skeletal structure and the GFRP 'buckets' were placed into the top layer of the skeletal system; every bucket had to be separately drained.

Figure 3 Covent Garden Flower Market London - 1974

Figure 4 All composites building structure erected in Lancashire -1974

During 1974 the first all composite building structure, erected in Lancashire, was manufactured from a series of identical unit building blocks. The basic structural system was the icosahedron shape in which the flat surfaces of this geometrical form were folded into four unit pyramidal shapes (the unit building blocks) and these pyramids were all joined along their flanged edges. The structure is shown in Figure 4. It is extremely rigid, as a result of the geometric shape, and strong but a poor architectural detail associated with this system is that the drainage paths are along the joints of the pyramids. It must be noted that all applications up to this date were in the realm of demonstrator structures rather than in the commercial area; this is understandable when a completely new material is introduced, for commercial exploitation, into an industry. It is possible that it will take another twenty or so years to achieve the status of other structural engineering materials before it is fully accepted.

Figure 5 Double layer skeletal structure manufactured from FRP-1987 pultrusions

Figure 6 Opening ceremony of the Bonds Mill Lift-bridge, Gloucestershire UK - 1994

In the 1980s double layer skeletal structures were being developed. These systems were generally made by the pultrusion process and were manufactured using pultruded glass fibre reinforced polyester resin in which the greater percentage of the fibres was in the longitudinal direction with only a small percentage in the hoop direction. The tubes, therefore, had anisotropic material properties and were efficient in this situation; they should be compared with the isotropic property of steel which is currently the material used to make a skeletal structures' members; in this latter form of construction the steel is used inefficiently. If the GFRP members require additional stiffness, glass and carbon fibre hybrid composites or carbon fibre composites (CFRP) could be used to provide uniformity in the skeletal structure;

these would be manufactured to the same diameter as the GFRP members in other parts of the structure. These systems are lightweight and stiff, but special node joints are required to connect members of the skeletal system and one type, developed at the University of Surrey, is shown in Figure 5.

Deployable skeletal systems, originally for antennas in space, in which energy loaded nodal joints were used to deploy the skeletal systems automatically, were conceived and manufactured in the 1980s at the University of Surrey. A five metre deployable has been described in Fanning and Hollaway [1]. For more that two decades, the University of Cambridge has been investigating, deployable systems and are currently analysing the mechanics of bi-stable laminated composite tubes which are straight, thin walled tubes with a semi/full circular cross-section. They are configurable between two stable states, a coiled state in which the shell is folded along its length and an extended state in which the shell is unfolded. These shells can be folded without permanent deformation and are capable of self-deployment through the controlled release of elastic strain energy accumulated during folding. The relevant study is an investigation of a centre-fed deployable mesh reflector for use on spacecraft. Deployable systems have only very specialised use in the construction industry.

During the 1980s it was realised that if 'all polymer composite' structural systems were to become competitive in the construction industry they should be manufactured, as far as possible, by a building block system (c.f. the Lancashire classroom system) and these building blocks should be made by a completely automated process. With this thought in mind the Maunsell structural plank (the advanced Composite Construction System ACCS) was conceived, designed and manufactured by the pultrusion process in which a well controlled heating range and the rate of travel of the fibres through the die was maintained. This 'honeycomb' plank was 605 mm wide and 80 mm thick; only the factory and/or transportation limited the length. One steel die was needed for the manufacture of this plank. It was then possible to form a series of planks into box structures and flat plate structures. These latter structural systems could be fabricated into square or rectangular buildings. The three well-known structures, which were made using this system are the Aberfeldy footbridge (1992), the Bonds Mill bridge (1994), and the two storey Maunsell building, one example of the use of this latter structure was for offices at the site of the Severn second crossing. The Bonds Mill bridge is shown in Figure 6.

THE CURRENT SITUATION IN THE CONSTRUCTION INDUSTRY AND R AND D INVESTIGATIONS
All Polymer composites
At the end of the 1980s and into the 1990s the utilization of advanced polymer composite materials have made large advances in the civil engineering construction field particularly in the bridge area. The history of bridge engineering is allied to the development of structural materials. In the past when a new material, such as wrought iron or steel, was invented a prototype bridge was often built, such as Ironbridge, to demonstrate the capability of the materials. Various forms of fibre

reinforced polymer materials have been available for over thirty years and during this period GFRP has been used in prototype bridge structures. However, the composite materials did not show the immediate and obvious advantages that iron and steel offered over timber and stone in the 19th century, except perhaps, their potential for the construction of extremely long span bridges. As a consequence, fibre reinforced polymers were not seen as a material likely to make an impact on general bridge engineering, until the past ten years when the full implications of corrosion of steel in modern bridges and the specific weight of the material were appreciated.

The most recent composite bridge structure (although not completely of 'all composite' construction) to be built in the UK is the Halgavor Bridge which is erected over the A30 near Bodmin, Cornwall. The foot- and equestrian bridge is almost 50 m span and 4 m wide and is one of the longest composite profile structures in Europe. The construction made use of advanced bonding, fastening and fabrication technologies; the latter included the resin transfer moulding process (SCRIMP).

The conventional civil engineering materials for bridge construction are likely to continue to be cheaper, in the near future, than FRP material but savings in fabrication costs of the latter material may be considerable if highly automated production of advanced polymer composite materials is developed. Thus complete box girder structures could be pultruded in the future. Speed of construction, savings in erection costs and in foundation sizes will all contribute to economy.

Composite enclosures and aerodynamic fairings for bridges (Developed in the 1980s)
Regular inspection and maintenance of bridge structures are required and any closure of bridges, required to facilitate maintenance, will cause considerable disruption to travellers, and in addition, the cost of closure will be high. Furthermore, the costs of maintenance work, over or beside busy roads and railways, are increasing due to the stringent requirements imposed by the Highway Authorities. Many bridges, designed and built over the past 30 years, do not have good access for inspection . In addition, in many North European and North American countries deterioration, caused by de-icing salts, is creating an increasing maintenance workload. To overcome this problem and to protect the structure from further deterioration a 'Bridge Enclosure' can be fitted to the soffit of the bridge. The enclosure is a suspended floor beneath the girders of steel composite bridges to provide inspection and maintenance access. It is sealed onto the underside of the edge girders to enclose the steelwork and to protect it from further corrosion and ingress of moisture. The addition of the fibre polymer composite enclosures around long span bridges not only enables maintenance costs to be greatly reduced but also allows the shape of the cross-section to be optimised by extending the enclosure into a fairing to give minimum drag.

The bridge deck
An important part of the overall design of a long span bridge is the weight of the deck, in addition, its form and stiffness are important with respect to aerodynamic stability. Furthermore, there are likely to be significant advantages in the utilisation of advanced composites in bridge decks particularly as the likely trend, into the 21st century, is to increase bridge spans beyond their previous limits. The material characteristics of composites and their successful applications in other areas are sufficiently encouraging to show that there are certain to be important developments in using them in bridge deck structures.

As a result of the deterioration of the wearing surface and the degradation of the bridge deck system itself, the deck generally requires the maximum maintenance of all elements in a bridge superstructure. In addition, there may be a need (a) to increase the load ratings and number of lanes on the existing bridge, (b) to accommodate the constantly increasing traffic flow, (c) to upgrade the bridge to conform to new codes. Furthermore, to undertake work on an existing bridge, it is necessary to keep in mind the important consequences relating to the overall economics of time loss and resources, caused by delays and detours of traffic during construction. Consequently the development of bridge decks utilising materials that are durable, lightweight and easy to install have been encouraged as the result of the above restraints. In addition to the potential savings in life-cycle costs due to increased durability, decks fabricated from advanced polymer composites, would be significantly lighter thereby effecting savings in substructure costs. These deck structures can either be used for replacements for existing, but deteriorated or substandard concrete/conventional deck or used as new structural components on existing or new supporting structural elements.

Cost Comparison of replacement FRP decks and conventional material decks
The lower bound cost for FRP decks appears to be $700/m^2 which corresponds to about $7/kg of material. This cost is greater than the construction of a new bridge or deck replacement with conventional materials which is typically about $322/m^2 [2]. However, the higher cost of the FRP composite decks can be absorbed, for instance, if a complete reconstruction would become necessary in the absence of a lightweight deck alternative. Further analysis needs to be undertaken to determine if the higher initial cost of the FRP composite deck can be justified on whole life costing and other economic considerations.

An 'all composite' bridge deck has been analysed and develop by a European consortium of industrial firms; the project ASSET consists of seven partners[1]. The highway bridge has been designed to carry 40 tonnes vehicle wheel load. One of the tasks within the project was to design all the vital connection details to achieve a complete decking system. Conventional connecting parts such as parapets, lamp-posts, existing main girders etc. had to fit into the system.

---

[1] Mouchel (UK), Fiberline (Denmark), Skanska (Sweden), IETCC (Consej Superior de Investigaciones Cientificas (Spain), and HIM (Netherlands).

An ASSET designed highway bridge to carry 40 tonnes vehicles will be constructed for Oxford County Council in the current year. The bridge will be made entirely from GFRP consisting of the deck and main girders; the latter will be constructed from GFRP pultruded I-sections.

The future challenge for bridge decks
In the USA there exists a growing interest in the future of these structures. There are a large number of FRP decks which are already in service [3-5] and several others are scheduled for installation in the near future [6]. It remains to be seen whether this form of construction will be of interest to engineers in the UK. after the completion, in 2002, of the Oxfordshire County Council bridge deck, which will be manufactured from polymer composites including the support beams for the deck.

As with all new concepts there are several technical problems that must be addressed. These may be listed as:
(1) the development of design standards and guidelines,
(2) the design and characterisation of the jointing technology of panel to panel and decks to stringers,
(3) the fatigue behaviour of the panels,
(4) the durability characteristics of the panels,
(5) the failure mechanism and ultimate strength.

The combination of advanced polymer composites and the more conventional civil engineering materials

The various structural forms of the combination between APC and the more conventional materials are:
- Upgrading and retrofitting flexural and shear structural beams and columns using APCs.
- Non-metallic rebars to reinforce concrete.
- Combination of APC/concrete units; the materials are used to their best advantage and act compositely.

Upgrading and retrofitting non-metallic structures and structural units using CFRP laminates

For a variety of reasons the performance of reinforced concrete, steel and cast iron structures may be found to be unsatisfactory. Design strengths of structures may not be achieved in practice because of deficiencies at the design phase; these deficiencies include marginal design/design errors causing inadequate factors of safety, the use of inferior materials, or poor construction workmanship/management. In service, increased safety requirements, a change in use or modernisation causing redistribution of stresses, an increase in the management or intensity of the applied loads require to be supported, or an upgrading of design standards may render all or part of a structure inadequate. In addition, the load carrying capacity of a member may be compromised by deterioration of the material as a result of the corrosion of internal reinforcement, in the case of reinforced concrete beams, from carbonisation of the concrete or alkali-silica reaction, hostile marine and industrial environments and structural damage. On highway structures, corrosion of the internal reinforcement is exacerbated by the use of de-icing salts. For pre-stressed concrete beams, strengthening measures may be required to prevent further loss of pre-stress. These inadequacies may manifest themselves by poor performance under service loading in the form of excessive deflections and material failure, or through inadequate fatigue or ultimate strength. There are two possible alternatives to restore a deficient structure to the required standard, these are the complete or partial demolition and rebuild, or the commencement of a programme of strengthening.

At the end of the 1980s EMPA (Swiss Federal Institute for Material Testing and Research), undertook pioneering and subsequent work in the field of plate bonding together with Professor Rostasy at the Institute for Building Materials in Germany. In addition, in the UK in May 1994, the ROBUST project was initiated, under the UK Government's LINK Structural Composites programme, to investigate the technical and the commercial viability of the use of carbon and glass fibre-reinforced polymer composites as an alternative to steel plates in bridge strengthening applications. The project, which incorporated both un-stressed and pre-stressed plates bonded to the soffit of RC beams, aimed to address a range of short- and long-term issues, as well as theoretical evaluation of the system. The findings of the programme have been recorded by Hollaway and Leeming, [7]. Other research undertaken in the UK relating to upgrading is given by Barnes and Mays [8], Quantrill and Hollaway [9], Garden and Hollaway [10], Garden *et al* [11], Hutchinson and Rahimi [12], Mays and Hutchinson [13]. The behaviour and strengthening in terms of flexure and shear of RC beams, and units utilising FRP plates and wraps, has been reviewed by Teng *et al* [14].

Strengthening and up-grading metallic structures and structural units using CFRP laminates
Although there have been some studies which involve the strengthening of steel and C I structural members Mertz and Gillespie [15], Mosallam and Chakrabarti [16], Sen *et al* [17] and Moy *et al* [18, 19] this type of upgrading has not been so widespread as that for RC and timber members since it poses a different set of problems. Firstly, the likelihood of lateral buckling makes it necessary to fabricate composite steel sections where the compression flange is continuously supported by a reinforced concrete slab. Secondly, the high strength and stiffness of steel makes it more difficult to strengthen particularly as the high modulus CFRP composites, which are commonly used in construction, have a lower modulus of elasticity than that of steel. This means that substantial load transfer can only take place after the steel has yielding unless much thicker, or the ultra high modulus, carbon fibre composites are used to achieve comparable strength and stiffness gains, typically obtained in strengthened reinforced concrete or wood members. Furthermore, in this system the CFRP/adhesive bond is the weakest link and will always control the mode of failure. Thus, consideration of surface preparation alone will not always be sufficient, and consequently, attention must be given to augmenting the capacity of the adhesive through the use of appropriate fasteners to ensure load transfer past

yielding of the steel. Failure modes of strengthened sections are generally ductile and accompanied by considerable deformation but the CFRP composite must be adequately anchored to the members to prevent peeling failure at the ends of the members. Figure 7 illustrates the final placement of the carbon and glass fibre prepreg layers around the flanges and web of a steel beam using a low temperature curing epoxy resin developed by Advanced Composite Group, Heanor, Derbyshire.

Figure 7 Final placement of the carbon fibre prepreg around flanges and web of beam (By kind permission of Taywood Engineering London, UK and ACG Derbyshire UK.

Figure 8 Sycamore Lane Footbridge, Warrington, (2000) – Compacting Horizontally wound carbon fibre composite sheet (Kind permission of Tony Gee and Partners Consulting Engineers, London).

Wrapping of columns using advanced polymer composites
Throughout the world a wide variety of carbon, aramid and glass fibre composites have been and are being wrapped and bonded to structural columns to solve degradation problems, although there has not been much practical work undertaken in the UK as yet. The reasons that structural columns may require to be strengthened or retrofitted, may arise from environmental exposure, inadequate design procedures and increased loads due to a greater use of the existing structures or by seismic vulnerability. These materials also help to prevent moisture from entering concrete surfaces and corroding internal rebar reinforcement. In addition, composite materials are relatively easy to install compared to steel. Consequently, a fibre reinforced composite tube having a high proportion of helical reinforcement is an ideal material for encasing concrete, because the latter material takes the entire axial load. The Poisson expansion of the FRP in the circumferential direction is smaller than that of the concrete and the tensile strength of the former material in the circumferential direction is very high. Thus the GFRP, AFRP or CFRP casing counteracts lateral expansion of the concrete under load and when used in short columns the axial strength of the concrete increases over its uni-axial value and can reach a tri-axial failure strength in excess of four times the uni-axial value.

Retrofitting column structures to withstand earthquakes is a recent and wide-spread task and one of the more complex engineering challenges in terms of strengthening.

Seismic reinforcement of structural support columns typically addresses two problems, compressive failure and shear cracking. At the commencement of a seismic event, columns are known to dilate causing the first seismic failure mode, with concrete spalling, buckling of the rebars and a reduction of the compressive strength of the column. The uni-directionally aligned polymer composites wrapped around the column provide confinement of the concrete and provide an increased hoop strength without adding vertical (longitudinal) stiffness to the structure. In contrast, steel jackets add equal strength in both the hoop and the longitudinal directions, thus making the column less ductile in the vertical direction. Shear cracking is the second seismic failure mode for columns. Earthquakes will produce lateral forces on columns which often create 45^0 shear cracks, because of insufficient shear strength in the structure. Again the wrapped column with unidirectional composite will prevent shear cracking.

Another reason for wrapping advanced polymer composites around a column is to strengthen the column against impact in the longitudinal direction. The deck of the Sycamore Lane footbridge, Warrington, which is over a heavily trafficked road was struck by an abnormally high load carried by a truck. Design calculations on the central column to this bridge revealed an inadequate design against impact to support current UK design loads. To improve the design, external FRP reinforcement, which consisted of eleven layers of vertically aligned and two horizontal wound carbon fibre composite sheets were laid by the hand lay-up technique around the rectangular tapered cross-section. Each layer was impregnated by resin saturant and the final laminate was protected from UV light by a protective coating, Figure 8 shows the wrapping of carbon fibres around the column.

The future prospects of upgrading and retrofitting structural systems using APCs.
The upgrading and retrofitting of structures and bridges utilising advanced polymer composites is becoming well establishes. Although the cost of the material is greater than conventional materials, the number, cost of fabrication and erection benefits far outweighs the initial cost. There are strong indications that the FRP strengthening technique will continue to be the preferred choice for this type of application.

Non-metallic rebars to reinforce concrete
FRP composite rebar reinforcement to concrete
The steel rebars or pre-stressing tendons, in conventional reinforced concrete, are protected by the alkalinity of the concrete and this results in durable structures. However, the protection of the steel is overcome in aggressive environments, where chloride ingress or concrete carbonation can occur, and corrosion results leading to a reduction in mechanical strength. The more critical effect of corrosion is the formation of hydrated ferrous oxide and its associated volume expansion, which generates large internal stresses in the concrete. These are sufficient to lead to cracking and eventually spalling of the concrete cover, resulting finally in the structure becoming unserviceable and unsafe. In humid saline or other chemically aggressive environments the concrete spalling process is accelerated. If the steel

rebar is protected by an epoxy-coating, and that coating becomes damaged, corrosion can occur in an alkaline environment.

To improve the durability of reinforced concrete there are five main approaches. These are:

- to prevent or delay the aggressive environment from reaching the steel reinforcement by reducing the permeability of the concrete,
- to improve the quality of the concrete,
- to apply cathodic protection,
- to epoxy coat the steel rebars,
- to reduce the susceptibility of the rebar to attack by substituting the steel with an alternative and more durable material, such as non-metallic reinforcement.

Oxidation and corrosion from sea-water, de-icing salts and the caustic environment of concrete can be resisted by the FRP rebars. It has a better strength to weight ratio than steel. The additional advantage of magnetic transparency, makes GFRP rebars ideal, for instance, for reinforcing the foundations of magnetic resonance imaging (MRI) units and other magnetic energy applications. Furthermore, the natural non-conductivity of GFRP composites allows them to be used in electrical applications without interference or hazard. However, there has only been limited but specialised use of FRP reinforcement for concrete structures, particularly for hospital and the military installations, in the UK, USA, Japan and France; FRP bars and mats have been employed in deck replacement for bridges. The durability of GFRP rebars is an important design issue. When a GFRP rebar is exposed to the alkaline environment of the concrete, eventually it will be susceptible to degradation from the alkali attack and the design solution is to maintain a low stress level in the rebar. Beyond a particular strain level, micro-cracks will develop in the polymer of the rebars exposing the glass fibre to the alkali. The stress level of the rebar can be decreased by increasing the size of it or by minimising the spacing between the rebars

The fabrication of FRP rebars can take the form of one dimensional or multidimensional shapes. The majority of the multi-dimensional reinforcements are orthogonal two-dimensional grids although three dimensional grids, of various configurations, have been proposed for certain pre-cast structures.

FRP composite rebars were introduced into the civil engineering industry in the USA in the 1960s, [20], and into Europe [21] and Japan [22] in the 1970s. The FRP reinforcement has been used primarily in concrete structures requiring improved corrosion resistance or electromagnetic transparency. The Institution of Structural Engineers, London, Europe (FIB), USA (ACI), Canada and the Japan Society of Civil Engineers have all published design guidance for the use of structures reinforced with FRP. These have been referenced in the Bibliography section.

The future of FRP rebars
Smart structures sensor technology has been incorporated into the applications of FRP reinforcement in major innovative concrete structures. This has aided the performance data generation for FRP reinforced structures. With a view to producing long-term data, the Intelligent Sensing for Innovative Structures ISIS (Canada) has undertaken considerable research into the development of smart sensor technology in combination with research on FRP infrastructure applications. The continued development of design guides and codes throughout the world will aid the use of FRP reinforcements in construction practice.

Hybrid FRP/concrete sections.
Concrete has a high compressive strength but is very weak in tension. In a reinforced concrete beam the sole use of this material, below the neutral axis, is to position the reinforcing steel bars and to protect them from aggressive environments. This latter property, however, is not completely fulfilled as the concrete in this zone of the loaded beam will crack and allow aggressive substances to attack the steel, causing gradual degradation. Furthermore, advanced structural composites, (either glass or carbon fibre in vinyl ester or epoxy polymer matrix), possess excellent in-service properties, in particular, resistance to aggressive substances and with high fibre volume fractions have high specific strength and stiffness and excellent durability with minimum maintenance requirements over thirty years. Thus, advanced continuous fibre reinforced polymer composites and concrete would appear to be an example in which two very dissimilar materials can be joined to form a composite structure. If the high-compressive strength concrete is placed above, and the high tensile strength fibre/polymer composite is placed below the neutral axis of a duplex beam, a composite construction is formed and the two materials, if correctly designed, will be stressed to their ultimate limit. Recent research in this area, in particular, at the University of California, San Diego, USA, University of Surrey, UK., University of Warwick, UK. and EMPA., Switzerland, have focussed upon hybrid systems that combine advanced composites with conventional materials, in particular, concrete.

In Europe Triantafillou and Meier [23], Deskovic and Triantafillou [24] and Triantafillou [25] presented an innovative hybrid box section suited for simply supported span. The composite materials were combined with a low-cost construction material, namely concrete, to form the new system; this resulted in a new concept for the design of lightweight and corrosion free beams, with excellent damping and fatigue properties. In the UK, Hall and Mottram [26] presented a hybrid section combining GFRP pultruded sections with concrete, the GFRP sections used were commercially available as floor panels. The sections acted, simultaneously, as tension reinforcement and permanent formwork. Figure 9 give details of the development of composite/concrete duplex beams for a standard rectangular cross-section and for a Tee beam cross-section; the former is given by Canning, *et al* [27], and the latter is given by Hulatt, *et al* [28, 29]. The final versions of these beam sections consisted of either a sandwich construction, for the webs below the neutral axis, shown in Figure 9a, or hollow box beams with two vertical (webs) skins of polymer composite and web diaphragm stiffeners shown in Figure 9b.

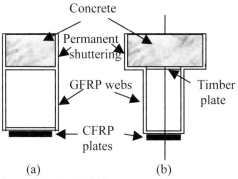

| (a) | (b) |

Figure 9   Rectangular and Tee Hybrid beams

Miscellaneous combinations of APC with concrete
Concrete filled tubular compression members
In new structural members, which consist of FRP tubes filled with concrete, the system utilises the best characteristics of the individual materials. The concrete core supports the FRP tube and prevents premature local buckling failure experienced in hollow tubes under compression. The tube confines the concrete and, in addition, the strength and ductility of the former material is increased. Furthermore, the system simplifies the construction procedures and reduces erection time. The concept of concrete filled FRP tubes was developed for two different uses. Firstly, to develop non-corrosive piles for marine environments; Mirmiran and Shahawy [30] and Parvathaneni *et al* [31] have discussed this, and secondly to enhance the ductility of the tubes. Seible *et al* [32-34] have discussed the second example in terms of bridge columns and piers in seismic zones. The filament wound process is used to manufacture the FRP tubular sections which are then filled with concrete., The jacket will have a low elastic modulus, in the longitudinal direction, due to it being lightly reinforced in this direction, consequently, the concrete will take the majority of the axial load. The FRP jacket will be loaded, essentially, in the hoop direction and, unlike a steel jacket, the FRP expansion in this direction, due to the Poissons ratio effect of the axial load, is less than that of the concrete; this provides a significant confinement The tube will, in this situation, be prevented from failing prematurely by local buckling as is the case with steel tubes. Shahawy and Mirmiran [35] and Mirmiran and Shahawy [36] have discussed a new piling system which was developed between 1994 and 1999, for the Florida, Department of Transport, and consisted of concrete filled GFRP tubes some of which were designed as under-reinforced and some as over-reinforced systems. A minimum sectional area of 0.0155 m^2 was required for corrosion resistance of the piles when they were driven into salt water or water containing a high chloride content and wet/dry cycles, regardless of the load level. The reason for encasing the concrete into GFRP composite tubes was to protect the concrete core and to alleviate the problems of moisture intrusion and permeability. The wall thickness of the FRP tubes was

14 mm and 6.6 mm for the over- and under-reinforced concrete-filled GFRP tubes, this resulted in 18.2% and 7.51% reinforcement ratios respectively

Applications in the Offshore Industry
The corrosion resistance and low weight of GFRP products is an advantage for offshore structural applications. They are particularly attractive for retrofitting applications because they are easy to handle and can be installed using only pneumatic tools. There are disadvantages, one is that GFRP composites can give off fumes under a fire and although it can be argued that it may actually be less risky to use than steel there is a reluctance to select GFRP access products in primary escape routes offshore. Newer resin systems (for instance the resin Modar) are now available which overcome this restriction. In addition phenolic resins are being investigated by Strongwells in the USA and by Fiberforce in Denmark. The phenolic material offers better fire resistance than the resin, Modar.

CONCLUSIONS – THE FUTURE FOR APC IN THE CIVIL ENGINEERING INDUSTRY
The future prospects for the utilisation of APC composites in the civil infrastructure are good. In this paper, the development of APC has been traced from its humble beginning to its present day uses in civil engineering and within a mere thirty years composites have been successfully implemented into the civil infrastructure and have become a major construction material for all designers, standing along side the more conventional civil engineering materials. Hollaway and Head [37] have illustrated that within a matter of 10 years the APC has become the preferred choice for many repair and rehabilitation projects involving buildings, bridges, historic monuments and other structures. However, there are many problems still to be solved, before optimum designs can be undertaken. In 2001, the Civil Engineering Research Federation (CERF) of the American Society of Civil Engineers undertook a Gap Analysis for the ductility of fibre reinforced polymer composites in the civil infrastructure. They identified seven areas where long-term durability and performance data are required, these are:
(1) environment (2) alkaline environment (3) thermal effects   (4) creep and relaxation effects (5) U.V. effect (6) fatigue performance (7) fire performance.

A future plan has been recommended by CERF for activities in the continuation of the study into polymer composites, these are:
* The development and implementation of an integrated knowledge system for durability of FRP composites in civil infrastructure.
* The development and implementation of a methodology for data generation, collection and assessment.
* The implementation of a plan for field assessment of durability.

A start had been made on these implementation items even before they were recommended. A number of monitoring programmes of structures in which FRP rebar reinforcement is the primary material have been set-up around the world and these should prove invaluable to designers in the coming years.

A number of design guides, standards and codes of practice (some of these have been given in the references) for the various constructional uses of FRP materials are being or have been written throughout the world. The most significant mechanical differences between the APC and the more conventional metallic materials are the higher strength, lower stiffness (although they have a greater specific stiffness and strength than steels) and the linear elastic behaviour to failure. Properties such as thermal expansion coefficient, moisture absorption and fire resistance must also be designed out.

REFERENCES

[1]   Fanning, P, and Hollaway, L.C. (1993) "A case study in the design and analysis for a 5.0m deployable composite antenna" Jnl. of 'Composites Engineering' Vol.3, N°11 1993, pp 1007-1023.

[2]   Lopez-Anido, R. (2001), 'Life-cycle cost survey of concrete decks – a benchmark for FRP bridge deck replacement', Proc. 80th Transportation Research Board Meeting, 7-11 January, Washington DC, CD-ROM.

[3]   GangaRao, H.V.S., Thippeswamy, H.K. Shekar, V. and Craigo, C. (1999), 'Development of glass fiber reinforced polymer composite bridge deck', SAMPE Journal, Vol 35 No. 4, pp 12-24.

[4]   Harik, I., Alagusundaramoorthy, P., Siddiqui, R., Lopez-Anido, R., Morton, S., Dutta, P. and Shahrooz, B. (1999), 'Static testing on FRP bridge decks panels', Proc. 44th International SAMPE Symposium and Exhibition, Vol. 2, Soc. Advancement of Materials and Process Engineering, Covina, CA, pp 1643-1654.

[5]   Temeles, A.B. Cousins, T.E. and Lesko, J.J., (2000) 'Composite plate and tube bridge deck design: evalution in the Toutville, Virginia weigh station test bed', Proc 3rd International Conference on Advanced Composite Materials in Bridges and Structures ACMBS-3 J.L.Humar and A.G. Razaqpur, Eds, CSCE, Montreal, Quebec, pp 801-808.

[6]   Davalos, J.F. Qiao, P. and Barbero, E.J. (1996), 'Multiobjective material architecture optimisation of pultruded FRP I-beams', International Journal Composite Structures, Vol 35 No. 3, pp 271-281.

[7]   Hollaway, L C and Leeming, M (eds) (1999) Strengthening of reinforced concrete structures using externally-bonded FRP composites in structural and civil engineering Woodhead Publishing Ltd. Cambridge.

[8]   Barnes, R.A. and Mays, G.C. (1999), 'Fatigue performance of concrete beams strengthened with CFRP plates', ASCE Journal of Composites for Construction, Vol. 3, No. 2 pp 63-72.

[9]   Quantrill, R.J. and Hollaway, L.C. (1998) 'The flexural rehabilitation of reinforced concrete beams by the use of prestressed advanced composite plates', Composites Science and Technology, vol. 58, pp. 1259-1275, 1998.

[10]   Garden, H.N. and Hollaway, L.C. (1997), 'An experimental study of the strengthening of reinforced concrete using prestressed carbon composite plates', Proc. 7 International Conference on Structural Faults and repair, University of Edinburgh, July 8-10 1997, Vol.2, pp 191-199.

[11]   Garden, H. N., Hollaway, L., and Thorne, A.M. (1997) "Strengthening of reinforced concrete members using bonded polymeric composite materials" Proc. of the Institution of Civil Engineers, Structures and Buildings Journal, Vol.123 May 1997 pp127-142.

[12]   Hutchinson, A.R. and Rahimi, H (1993), 'Behaviour of reinforced concrete beams with externally bonded fibre-reinforced plastics' Proc. 5th Int. Conf. 'On Structural Faults and Repair', ed. M.C. Forde, Engineering technics Press, Edinburgh, Vol. 3, pp 221- 228.

[13]   Mays, G.C. and Hutchinson, A.R. (1992), 'Adhesives in Civil Engineering', Cambridge University Press.

[14]   Teng, J.G., Chen, S.T., Smith, S.T. and Lam, L. (2002), Behaviour and strength of FRP-strengthened RC structures: a state-of-the-art to be published in Proc. ICE (Structure and Building)

[15]   Mertz, D.R. and Gillespie, J.W., (1996) 'Rehabilitation of steel bridge members through the application of advanced composites' Final report, NCHRP-93-ID01.

[16]   Mosallam, A.S. and Chakrabarti, P.R., (1999) 'Making the connection. Civil Engineering, ASCE, April, 1999, pp 56-59.

[17]   Sen, R., Liby, L. and Mullins, G., (2001), 'Strengthening steel bridge sections using CFRP laminates' Composites Part B Vol 32, pp 309-322.

[18]   Moy, S.S.J., Barnes, F., Moriarty, J., Dier, A.F., Kenchington, A., and Iverson, B., (2000a), ' Structural upgrade and life extension of cast iron struts using carbon fibre reinforced composites' FRC 2000: 8th International Conference on Fibre Reinforced Composites, University of Newcastle, September 2000.

[19]   Moy, S.S.J., Barnes, F., Dier, A.F., Kenchington, A., and Iverson, B. (2000b) 'Structural upgrade and life extension of cast iron struts and beams using carbon fibre reinforced composites', Proceedings of the Conference on Composites and Plastics in Construction Building Research Establishment, UK.

[20]   Dolan, C.W. (1999), 'FRP prestressing in the USA' Concrete International, ACI, Vol. 21, No. 19 pp 21-24.

[21]   Taerwe, L.R. and Matthys, S. (1999), 'FRP for concrete construction', Concrete International, ACI, Vol 21, No 10. pp 33-36

[22]   Fukuyama, H., (1999), 'FRP composites in Japan', Concrete International, ACI, Vol 21, No 1, Vol 21, No. 4 No. 10, pp 29-32.

[23]   Triantafillou, T.C. and Meier, U. (1992), 'Innovative design of FRP combined with concrete', Proceedings of the First International Conference on Advanced Composite Materials for Bridges and Structures (ACMBS), Sherbrooke, Quebec, 1992, pp 491-499.

[24]   Deskovic, N. and Triantafillou, T.C. (1995), 'Innovative design of FRP combined with concrete: short-term behaviour', Journal of Structural Engineering, Vol. 121, No. 7, July 1995, pp 1069-1078.

[25]   Triantafillou, T.C. (1995), 'Composite materials for civil engineering construction', Proceedings of the First Israeli Workshop on Composite Materials for Civil Engineering Construction, Haifa, Israel, May 1995, pp 17-20.

[26]   Hall, J.E. and Mottram, J.T. 'Combined FRP reinforcement and permanent formwork for concrete members', Jnl. of Composites for construction, ASCE, Vol.2, No. 2, May 1998, pp 78-86.

[27]   Canning, L., Hollaway, L. and Thorne, A.M. (1999) 'Manufacture, testing and numerical analysis of an innovative polymer composite/concrete structural unit'. Proc. Instn. Civ. Engrs. Structs & Bidgs. Vol.134, pp 231-241.

[28]   Hulatt, J., Hollaway, L. and Thorne, A. (2000) 'Characteristics of composite concrete beams' Bridge Management – 4 , Inspection, maintenance, assessment and repair. Eds. M J Ryall, G A R Parke and J E Harding. Pub. Thomas Telford. pp 483-491.

[29]   Hulatt, J., Hollaway, L.C. and Thorne, A.M., (2001), 'Developing the use of advanced composite materials in the construction industry', in Proceedings of the International Conference FRPRC-5, Cambridge, UK, July 2001.

[30]   Mirmiran, A. and Shahawy, M. (1996) 'A new concrete-filled hollow FRP composite column', Composite Part B: Engineering, Special Issue on Infrastructure, Elsevier Science Ltd. Vol. 27B, No. 3-4, June 1996, pp 263-268.

[31]   Parvathaneni, H. K., Iyer, S. and Greenwood, M. (1996), 'Design and construction of test mooring using superstressing', Proceedings of the Advanced Composite Materials in Bridges and Structures (ACMBS), Montreal, 1996, pp 313-324.

[32]   Seible, F., Burgueno, R., Abdalah, M.G. and Nuismer, R., (1995) 'Advanced composite carbon shell systems for bridge columns under seismic loads', in Proc. Of the National Seismic Conference on Bridges and Highways, San Diego, California, Dec. 1995.

[33]   Seible, F. (1996), 'Advanced composite materials for bridges in the 21st century', Proceedings of the First International Conference on Composites in Infrastructure (ICCI'96), Tucson, Arizona Jon. 1996, pp 17-30.

[34]   Seible, F., Karbhari, V.M. Burgueno, R. and Seaberg, E. (1998), 'Modular advanced composite bridge system for short and medium span bridges' Proceedings of the Fifth International Conference on Short and Medium Span Bridges', Calgary, Canada, July 1998, p 11.

[35]   Shahawy, M. and Mirmiran, A. (1998) 'Hybrid FRP-concrete beam-column', Proceedings of the 5th International Conference on Composites Engineering (ICCE/5), Las Vegas, July 1998, pp 619-620.

[36]   Mirmiran, A. and Shahawy, M. (1999) 'Comparison of over-and under-reinforced concrete-filled FRP tubes', Proceedings of the 13th ASCE Engineering Mechanics Division Conference, Baltimore, June 13-16, 1999.

[37]   Hollaway, L.C. and Head, P.R. (2001), 'Advanced Polymer Composites and Polymers in the Civil Infrastructure' Pub. Elsevier, pp 2001.

BIBLIOGRAPHY
Design guides:
The Institution of Structural Engineering Design Guide: the Structural Use of Adhesives.

Europe   International Federation of Structural Concrete (FIB) Task Group 9.3 – *FRP reinforcement for Concrete Structures (2001).*
Norway: *Eurocrete Modifications to NS3473 When Using FRP Reinforcement*
Report # STF 22 A98741 (1998).

USA   American Concrete Institute, 440H Committee Report – *Guide for the Design and Construction of Concrete Reinforced with FRP Bars (2000)*

Canada   *Canadian Highway Bridge Design Code – Section 16 Fibre Reinforced Structures (1998) CSA Standards S806 Design and Construction of Building Components with FRP (1990)*

Japan   Japan Society of Civil Engineers *Recommendations for Design and Construction of Concrete Structures Using Continuous Fibre Reinforcing Materials (1997).*

In Europe, FIB Task Group 9.3, (Lausanne) (2001), USA, - American Concrete Institute ACI 440 (1996), Guide for *the Design and Strengthening of Externally Bonded FRP Systems for Strengthening Concrete Structures*, (draft report), Canada Design Manual *'Strengthening Reinforced Concrete Structures with Externally-bonded Fibre Reinforced Polymesr, ISIS-MO5-00 (2001),* and Japan. In the UK, the Concrete Society established a technical committee to produce the Design Guidance for *Strengthening Concrete Structures using Fibre Composite Materials, 2000* (Technical Report No. 55).

1.2 The development of composites and its utilisation in Switzerland

U Meier

Director Swiss Federal Laboratories for Materials Testing and Research, EMPA, CH 8600 Dübendorf, Switzerland

INTRODUCTION

Since 1969, flexural and shear post-strengthening of reinforced concrete structures with externally bonded steel plates is well established in Switzerland. In a non-corrosive environment this technique shows very good long-term creep behaviour. However, for outdoor applications, it has limited use, because of corrosion of the steel plates at the concrete crack tips and technical problems in installing the heavy plates over long spans. Compared to conventional construction materials, carbon fibre reinforced polymers (CFRP) have superior properties with respect to strength, weight, durability, creep and fatigue.

DEVELOPMENT OF R&D
Post-strengthening with thin pultruded CFRP strips

From 1982, thin (0.7 to 1.5 mm) carbon fibre reinforced epoxy resin composites have been successfully employed at the EMPA laboratories in research and development (R&D) for the post-strengthening of reinforced concrete beams within several programs. Loading tests were performed on more than 100 flexural and 30 shear beams having spans of 2 to 7 metres. The research work shows the validity of the strain compatibility method in the analysis of various cross sections [1-5]. The work also shows that the possible occurrence of shear cracks may lead to peeling of the strengthening composite. Thus, the shear crack development represents an important design criterion.

When a change of temperature takes place the differences in the coefficient of thermal expansion of concrete and the carbon fibre reinforced epoxy resin composites result in thermal stresses at the joints between the two components. After 100 frost cycles ranging from +20°C to –25°C, no negative influence on the loading capacity of the three post-strengthened beams was found [1]. The situation is more critical in a hybrid structural component in which a principal Aluminium structure (Al) is reinforced with unidirectional CFRP strips. In a experimental study [6] an Al/CFRP box beam was subjected to a temperature change of, which is approximately the temperature change from initial room temperature (23°C) to the failure temperature of about -74.3°C. The dimensions of the beam in question were 30 mm, 30 mm, 210 mm. The thickness of the Aluminium box beam and the CFRP strips was 2 mm and 1.1 mm, respectively. Due to the large difference in thermal expansion coefficients between Aluminium (αAl = 23.4.10-6 °C-1) and high-modulus CFRP (in the fibre direction: αCFRP, 1 = - 1.46.10-6 °C, ECFRP, 1 = 305 GPa), stress concentration near the free edges caused delamination in the structure, however only at Δ T = -100°C.

As described above the static behaviour of flexural members strengthened with CFRP strips has been well documented in recent years, but there was a gap in knowledge regarding the effects of dynamic and impact loading of such members. Therefore tests [7] were arranged on four 8 m beams externally strengthened for flexure, two with CFRP strips and two with steel plates. The beams externally strengthened with CFRP strips performed well under impact loading, although they could not provide the same energy absorption as the beams externally strengthened with steel plates. Additional anchoring, at least at the ends of the CFRP strips, would improve the impact resistance of these beams.

EMPA performed fatigue tests on a T-beam with a span of 6 m under realistic loading conditions. The cross-section was 900 mm wide and 500 mm deep. The existing steel reinforcement consisted of 4 rebars of 26 mm diameter in the tension zone. The total load carrying capacity of this beam amounted to 610 kN for each of the 2 loading points without post-strengthening. When the beam was strengthened by bonding a CFRP strip, which had dimensions of 200 x 1 mm (Toray T300 fibre with a volume fraction of 70 %, a longitudinal strength of 2000 MPa and a longitudinal elastic modulus of 147.7 GPa), its load carrying capacity was increased by 32% to 815 kN per loading point. The fatigue loading was sinusoidal at a frequency of 4 Hz; the test set up corresponded to a four point flexure test with loading at the one third points. The calculated stresses in the CFRP strip and the steel reinforcement reached at the lower cyclic load of 125.8 kN 131 and at the upper load of 283.4 kN 262 respectively 210 MPa. The beam was subjected to this loading for 10.7 million cycles [6]. After 10.7 million cycles the tests were continued in an environmental condition where the temperature was raised from room temperature to 40 deg C and the relative humidity to a value of 95%. The aim of this test was to verify that the bonded CFRP strip could withstand very high humidity under fatigue loading. Initially the CFRP strip was soaked with water nearly to saturation. After a total of 12 million cycles the first steel rebar failed due to fretting fatigue. The joint between the CFRP strip and the concrete did not present any damage. In the next phase of the test programme, the external loads were held constant. After 14.09 million cycles the second steel rebar failed, also due to fretting fatigue. The cracks which were bridged by the CFRP strip rapidly grew and after failure of the third rebar, due to yielding of the remaining steel, the CFRP strip was sheared from the concrete. This fatigue test was a success exceeding all expectations.

STRENGTHENING WITH PRE-TENSIONED STRIPS

A further fatigue test was carried out analogous to that described in the previous section, but in this case the CFRP strip was pre-stressed to 1000 MPa. Thirty million cycles were performed under the same loading conditions without any evidence of damage what so ever. This demonstrates, that in certain cases it can be very advantageous to provide pre-stressing to the flexural-strengthening strips. In this way, the serviceability of the bridge structure can be improved and the shearing off of the strips due to shear failure of the concrete in the tension zone can be avoided. Therefore a method was developed to allow variation of the pre-stress in the strip [8]. This in turn permits application of the strip with a gradual reduction of pre-stress to zero at the ends, ruling out the risk of shearing in the concrete while obviating the need for elaborate end anchorage. The associated tensioning device comprises two tensioning units connected to each other in the required length via a reaction frame. Roughly three-quarters of the strip are carried around the tensioning rollers and the ends clamped in position. The strip may then be tensioned to the required level by turning one of the rollers. As the strip lies on top of the top of the tensioning device, it may be offered up directly to the structural element to be strengthened. Adhesive is applied to the tensioned strip, which is offered up to the prepared substrate. The pre-stress gradient is achieved by first bonding a fully pre-tensioned section in the middle of the strip, before slightly easing the tensioning force in the system and bonding the adjoining sections. The tensioning force is then again slightly reduced, and another section bonded at either side. The new method allows full exploitation of the high strength of the CFRP strips.

CFRP parallel wire bundles for stay cables and pre- and post-tensioning
The cables are built up as parallel wire bundles. The key problem facing the application of CFRP cables is how to anchor them. The outstanding mechanical properties of CFRP wires are only valid in a longitudinal direction. The lateral properties including interlaminar shear are relatively poor. This makes it very difficult to anchor CFRP wire bundles and obtain the full static and fatigue strength. EMPA researchers have been developing CFRP cables using a conical resin-cast termination. The conical shape inside the socket provides the necessary radial pressure to increase the interlaminar shear strength of the CFRP wires. If the load transfer media (LTM) over the whole length of the socket is a highly filled epoxy resin there will be a high shear stress concentration at the loadside of the termination on the surface of the CFRP wire. This peak causes pullout or tensile failure far below the strength of the CFRP wire. We could avoid this shear peak by the use of unfilled resin. However this would cause creep and an early stress-rupture. The best design is a gradient material. At the loadside of the termination the modulus of elasticity is low and continuously increases until reaching a maximum. This way a shear peak can be avoided. The LTM is composed of aluminium oxide ceramic granules with a typical diameter of 2 mm. To get a low modulus of the LTM, the granules are coated with a thick layer of epoxy resin and cured before application. To obtain a medium modulus the granules are coated with a thin layer. To reach a high modulus, the granules are filled into the socket without any coating. With this method the modulus of the LTM can be designed tailor-made. The holes between the granules are filled by vacuum-assisted resin transfer moulding with epoxy resin. Such terminations were tested in static and fatigue loading. The results prove that the anchorage system described is very reliable. The static load carrying capacity generally reaches 92% of the sum of the single wires. This result is very close to the theoretically determined capacity of 94% [9]. Fatigue tests performed on the above described 19-wire cables showed the superior performance of CFRP under cyclic loads [10]. The anchorage system is patented (EP 0710313, 03.05.2000).

A NEW TENSILE STRAP ELEMENT: NON-LAMINATED LAYERS OF THERMOPLASTIC CFRP TAPE

The use of non-laminated layers of thermoplastic tape overcomes several of the difficulties associated with laminated systems. In particular, when laminated CFRP elements are bent around a corner and loaded under tension, there is a differential state of stress through the thickness of the element. Since carbon fibres have a very low transverse strength, these stresses result in a premature failure in the bent portion of the laminate. In the non-laminated shear elements, several layers of tape are wrapped around pins or a beam. The outer layer is joined to the next outermost layer but all the other inner layers are free to slide relative to each other. When the strap is tensioned, the strain is evenly distributed amongst the inner layers and the premature failures which are characteristic of the laminated elements are avoided. Winistörfer [11] carried out on a comprehensive study of pin-loaded strap elements where the load was applied to non-laminated CFRP straps using two pins. The study verified the principle that, when tensioned, the tape layers would move relative to each other. Furthermore, using strain gauges which were attached to each layer, he determined that each non-laminated layer carried a similar amount of force.

In the case of shear test series performed by Stenger [12] and described below, each of the four applied strap elements consisted of twenty-five such layers. At the top and bottom edge of the membrane section the strap elements were led over semi-elliptical pads. After welding of the straps the upper steel pads were lifted by means of a hydraulic jack, thus generating the pre-stress in the straps. Finally, several distance plates were inserted in order to keep the pads in position and hence maintain the pre-stress level in the strap elements. The test series comprised experiments on five large-scale concrete coupling beams. With these tests the influence of a transverse pre-stressing on the load-carrying and deformational behaviour of both intact and predamaged beams could be investigated. The pre-stressing was provided by means of external strap elements made of carbon fibre reinforced plastic (CFRP). The test arrangement (Figure 1a) was identical for all tests and ensured well controlled boundary conditions at either end of the specimens; the load V was applied by raising one of the fixed ends, generating a constant transverse load and thus a moment distribution antisymmetrical to midspan. No normal forces were applied, and any normal forces were elimininated by adjusting the horizontal jack forces. All five specimens consisted of membrane sections with dimensions of 2240 mm x 1200 mm x 150 mm and two load transfer sections of 800 mm x 1200 mm x 800 mm. They had identical arrangements of the conventional internal steel reinforcement. The longitudinal reinforcement consisted of twelve 26-mm diameter rods concentrated in pairs of six at the top and bottom edges of the membrane section, and of six 8-mm diameter bars distributed over the middle section of the membrane. In the transverse direction the specimens were substantially under-reinforced in order to force shear failure of all specimens before yielding of the longitudinal reinforcement. The internal transverse reinforcement consisted of only six closed 6-mm diameter stirrups at a spacing of 370 mm. The total cross sectional area of the external strap elements was 384 mm^2; the initial pre-stress of the elements was 730 MPa, i.e. about 56 % of their ultimate tensile strength. The yield strength of the 6-mm diameter stirrups was 575 MPa, and their ultimate strength was 640 MPa; the uniaxial cylinder compressive strength of the concrete was 37 MPa. All specimens failed in shear when the stresses in the longitudinal reinforcement were still below the yield stress (Figure 1d). In summary, pre-stressing in the transverse direction resulted in higher cracking loads, higher reorientation of the principal compressive stress directions, reduced concrete straining, higher ultimate loads (+51%) and greater deformations (+138%) at ultimate. The CFRP strap elements have proven to be a promising device for effectively retrofitting concrete elements under shear.

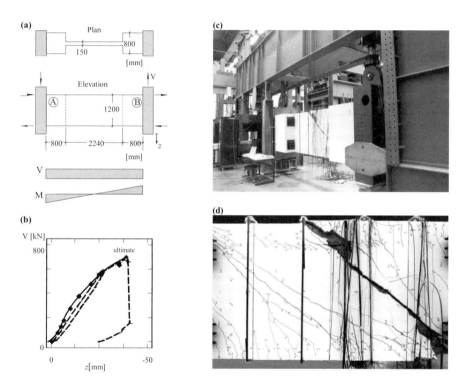

Figure 1: (a) dimensions of specimen, shear and moment diagrams; (b) vertical load V versus vertical displacements of fixed ends A and B; (c) Beam Element tester with specimen; (d) membrane section post-strengthened with strap elements after failure.

APPLICATIONS
Post-strengthening with thin pultruded CFRP strips
The Ibach Bridge near Lucerne was the scene of a major innovation in 1991. For the first time, very thin carbon fibre/epoxy strips were used to post-strengthen a bridge. In the following years this technique was applied to the historic covered wooden bridge near Sins, the City Hall in Gossau, the large multi-storey car park in Flims,

the tall chimney of the nuclear power plant in Leibstadt and the main railway station in Zürich. Apart from these projects, which are described elsewhere [13], approximately 800 structures of various sizes have been post-strengthened using thin CFRP strips in Switzerland since 1991. Despite the initially hesitant reaction of the Swiss market to the launch of CFRP strip for post-strengthening applications (1991: 6 kg, 1992: 126 kg), interest in the material has since risen sharply (approx. 130 t CFRP strip in 2001). Today, this technique has established a foothold in most parts of the world.

CFRP parallel wire bundles for stay cables and pre- and post-tensioning
The first suitable opportunity to test CFRP stay cables arose in 1996 in connection with the Stork Bridge, a 124-m-span, two-lane road bridge across 14 tracks at Winterthur railway station. Here, for the first time ever, two CFRP stay cables (Figure 2), each with a load capacity of 12 MN, were installed. As the main girders of this cable-stayed bridge are very flexible, the varying thermal expansion coefficients of the cable materials did not pose insuperable problems. Both the CFRP cables and the adjacent steel cables are being continually monitored by EMPA. So far results have met up with expectations and will become increasingly important in building confidence in years to come.

Figure 2: Stork Bridge, Winterthur

Figure 3: CFRP stay cables; each 241 wires of 5 mm diameter, load capacity of each cable: 12 MN.

A further three bridge projects in Switzerland using CFRP cable followed in quick succession in autumn 1998. The aim was to gather experience from practical CFRP cable applications over a longer period of time in order to promote confidence in the material. Two CFRP cables produced by a BBR/EMPA team were used as bottom chord for the bridge over the Kleine Emme River described in detail in [13]. This project saw the first use of CFRP wires with integrated optical fibre sensors for this kind of application.

The refurbishment of the Verdasio Bridge (Figure 4 and 5) represented the first attempt to make practical use of the results of extensive experiments performed in recent years on continuous two-span beams pre-stressed with CFRP cable [15]. Four cables each comprising 19 5-mm-diameter CFRP wires were used for external post-tensioning to replace corroded steel cross-sections. The main problem in this application was to carry the cable round relatively tight bending radii of 4.5 m, as CFRP wires are sensitive to transverse pressure due to their composition. However, a series of EMPA experiments with 2-span girders 0f 18 m with 3.0 m radii at the centre support suggested that a satisfactory solution was found to this problem.

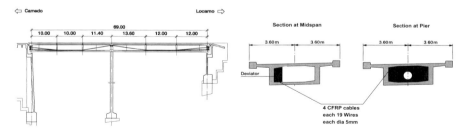

Figure 4: Longitudinal section of Verdasio bridge. 4 external CFRP cables inside the box girder are replacing corroded steel cables.

Figure 5: Cross sections of Verdasio bridge. The sustained stress on the CFRP cables is 1610 Mpa.

CONCLUSIONS
Despite the initially hesitant reaction of the market to the launch of CFRP strips for post-strengthening applications, interest in the material and applications have strongly increased within the last 10 years. Today, this technique has established a foothold in most countries and is continuing to gain ground. Similar trends cannot yet be anticipated for CFRP cables. Initial costs are too high but costs over the life cycle are comparatively low. More patience is needed. As soon as enough confidence for CFRP cables is built up it can be assumed, that decisions will be more and more based on life cycle considerations. This will be an advantage for CFRP cables.

REFERENCES
[1]   Kaiser, H.P., Strengthening of Reinforced Concrete with Epoxy-Bonded Carbon-Fibre Plastics. Doctoral Thesis, Diss. ETH Nr. 8918, 1989 ETH Zürich, CH-8092 Zürich/Switzerland (in German).

[2]   Meier, U., Bridge Repair with High Performance Composite Materials, Material und Technik, **15**, 1987, 225-128 (in German and in French).

[3]   Meier, U., Carbon Fibre-Reinforced Polymers: Modern Materials in Bridge Engineering, Structural Engineering International, 2, 1992, 7-12.

[4]   Meier, U., Deuring, M., Meier, H. and Schwegler, G., Strengthening of structures with advanced composites, Alternative Materials for the Reinforcement and Prestressing of concrete, edited by John L. Clarke, Chapman & Hall, 1993, 153-171.

[5]   Deuring,   M.,   Verstärken   von   Stahlbeton   mit   gespannten Faserverbundwerkstof-fen, EMPA-Bericht Nr. 224, 1993, (Post-strengthening of Concrete Structures with Pre-stressed Advanced Composites), published by the   EMPA   in   German   as   Research   Report   No.   224,   CH-8600 Duebendorf/Switzerland.

[6]   Motavalli, M., Terrasi, G. P. and Meier, U., On the behaviour of hybrid aluminium / CFRP box beams at low temperatures, Composites Part A, 28 A 1997, 121-129.

[7]   Erki, M. A. and Meier, U., Impact Loading of Concrete Beams Externally Strengthened with CFRP Laminates, Journal of Composites for Construction, August 1999, Pages 117-124.

[8]   Stöcklin, I. and Meier, U., Strengthening of concrete structures with prestressed and gradually anchored CFRP strips, Proceedings of FRPRCS-5 fibre-reonforced plastics for reinforced concrete structures, edited by Chris Burgoyne, Thomas Telford Ltd., 2001, p. 291-296.

[9]   Rackwitz R., Kabel für Schrägseilbrücken - Theoretische Überlegungen, VMPA-Tagung, Qualität und Zuverlässigkeit durch Materialprüfung im Bauwesen und Maschinenbau, TU München, 1990, 189.

[10] Meier U., Carbon Fiber-Reinforced Polymers: Modern Materials in Bridge Engineering, Structural Engineering International, 2, 1992,7-12.

[11] Winistörfer, A., CARBOSTRAP – An advanced composite tendon system, Proceedings of FRPRCS-5 fibre-reinforced plastics for reinforced concrete structures, edited by Chris Burgoyne, Thomas Telford Ltd., 2001, p. 231-238.

[12] Stenger, F., Tragverhalten von Stahlbetonscheiben mit vorgespannter externer Kohlenstofffaser-Schubbewehrung, Dissertation Nr. 13991, ETH Zürich, Dezember 2000.

[13] Meier, U. and Erki, M.-A., Advantages of composite materials in the post-strengthening technique for developing countries, Proceedings of the Sixth International Colloquium on Concrete in Developing Countries, January 4-6, 1997, Lahore, Pakistan, ISBN 0-921303-67-X, editors: S. A. Sheikh and S. A. Rizwan, p. 1-13.

[14] Nellen, Ph. M., Frank, A., Meier, U., and Sennhauser, U. Fiber Optical Bragg Grating   Embedded   in   CFRP   Wires,   SPIE's   6th   Annual   International Symposium on Smart Structures and Materials, Conference on Sensory Phenomena and Measurement Instrumentation for Smart Structures and Materials, Newport Beach, California, USA, March 1-5, 1999.

[15] Maissen, A., Concrete Beams Prestressed with CFRP Strands, Structural Engineering International (SEI) 4/97, p. 284-287, 1997.

1.3 Durability of FRP composites for civil infrastructure –myth, mystery or reality

V M Karbhari
University of California San Diego, La Jolla, CA, USA

INTRODUCTION

Since FRP composites are still relatively unknown to the practicing civil engineer and infrastructure systems planner, there are heightened concerns related to the overall durability of these materials, especially as related to their capacity for sustained performance under harsh and changing environmental conditions under load. Although FRP composites have been successfully used in the industrial, automotive, marine and aerospace sectors, there are critical differences in loading, environment and even the types of materials and processes used in these applications as compared to the materials-process-load combinations likely to be used in civil infrastructure. These materials have also a fairly successful record of use in pipelines, underground storage tanks, building facades, and architectural components. Anecdotal evidence provides substantial reason to believe that if appropriately designed and fabricated, these systems can provide longer lifetimes and lower maintenance costs than equivalent structures fabricated from conventional materials. However, actual data on durability is sparse, not well documented, and in cases where available – not easily accessible to the civil engineer.

The long-term durability of fiber reinforced polymer (FRP) composites is often stated as being the main reason for the use of these materials. However, their durability depends intrinsically on the choice of constituent materials, method and conditions of processing, and surrounding environmental conditions through their service lives. Although FRP composites do not corrode (as in rusting of steel) they do undergo physical and chemical changes, including oxidation. There is, today, a plethora of conflicting evidence related to the durability of composites resulting in an array of perceptions ranging from the absolute invincibility of these materials under terrestrial environments such that no maintenance would be required through their life to the extreme sensitivity of these same materials to environmental influences resulting in their rapid degradation within the first few months of use in a civil environment. Faced with these diverse and conflicting aspects it is no surprise that civil engineers, often with little knowledge of composites, treat the material and the entire technology with a great deal of caution. This paper attempts to address a few of the myths surrounding composites, demystify at least in part the question as to how in some applications composites can provide long life-times with very little maintenance while the same material combinations may degrade very rapidly in other situations, and then provide some data as a reality check.

THE MYTH

It is important to point out, at the outset, that there are a number of myths regarding FRP composites, chief among which are that:

- Composites are a panacea for the deterioration of civil infrastructure;
- Composites do not degrade and are not affected by environmental influences;
- Performance attributes of FRP composites used in civil infrastructure, at present, are comparable to those of aerospace grade prepreg based autoclave cured composites;
- Due to stress-rupture considerations glass fiber reinforced composites should only be used in architectural applications;
- Carbon-fiber reinforced composites are not degradable and have infinite life;
- All composites have well documented data bases; and
- Variation in field conditions, incoming raw materials, or manufacturing processes do not cause any changes in properties and performance.

Rather than being a panacea for the current deterioration in civil infrastructure, FRP composites do provide alternatives for rehabilitation and renewal not possible with conventional materials. However, these materials can degrade at the level of the fiber, polymer, and/or interface under specific circumstances. It must be remembered that the resin matrix allows moisture adsorption and this can lead to a variety of mechanisms, some of which result in deterioration of the polymer and, in some cases, of the reinforcing fiber. Although glass fiber reinforced composites are susceptible to creep- and stress-rupture, in general their use under sustained loads is possible as long as the stress levels do not exceed 25-30% of the ultimate. This has been shown to advantage in the resiliency of glass fiber reinforced pipes and underground storage tanks which have withstood very aggressive environments through the use of appropriate factors of safety [1]. Although carbon fibers are generally considered to be inert to most environmental influences likely to be faced in civil infrastructure applications the inertness does not apply to the fiber-matrix bond and the matrix itself, both of which can in fact be significantly deteriorated by environmental exposure. As emphasized in a recent study [2] significant gaps exist in the durability data base for FRP composites used in civil infrastructure even for environmental conditions such as moisture and alkalinity due to reasons ranging from actual lack of data on specific systems to difficulty in accessing data from investigations conducted in industry and government laboratories. Further, it is important to understand that since FRP composites are formed through the combination of micron sized fibers in polymer matrices the actual performance of the materials is intrinsically controlled by the microstructure, which is in turn controlled by the choice of constituent materials and form, interphase development, and the processing route taken (Figure 1).

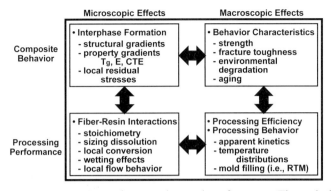

Figure 1: Interactions of Processing and Performance Through the Interphase

Minor changes in any of these aspects can result in significant changes, not just in short-term performance, but also in the overall durability of the FRP composite which depends on factors such as bonding, void content, degree of cure (Figure 2), and level of process induced residual stresses.

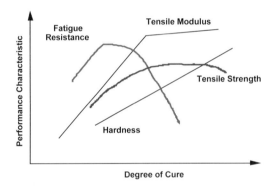

Figure 2: Schematic Showing Effect of Degree of Cure on Vinylester FRP Characteristics

Results from the aerospace industry are often used to show that if composites could be accepted for use in such a high precision application with very low tolerance for failure, then their use in civil infrastructure (often considered "low-tech" in comparison) should pose absolutely no problem.   It should, however, be remembered that the use of composites in aerospace applications has been predicated on extensive materials testing for the purposes of qualification followed by strict adherence to prescribed specifications for autoclave based fabrication in highly controlled factory environments. These materials and processes are unlikely to find significant application in civil infrastructure due to cost and processing specific aspects. Civil applications are more likely to (a) use processes such as wet lay-up, pultrusion and resin infusion than autoclave molding, (b) fiber and resin as separate constituents rather than in the form of preimpregnated material, and (c) resin systems such as polyesters, vinylesters, phenolics and lower temperature cure epoxies rather than the higher temperature curable epoxies and thermoplastics. Further, in the cases of rehabilitation, there is likely to be extensive use of processes under ambient conditions in the field, rather than fabrication in factory controlled environments. Even in the case of prefabricated elements, adhesive bonding to substrates has to be conducted under field conditions with little control, if any, of humidity and/or temperature. Thus, the civil engineering environment not only brings with it new challenges for the control of quality and uniformity of composites, but also makes it difficult (if not impossible) to use the well established databases generated by DoD (Department of Defense) sponsored research (such as those for AS4/3501-6 or T300/5208 based systems) for more than comparative baseline and trend analysis purposes. Further aerospace composites, are to a large degree extensively inspected at routine intervals within carefully controlled environments, whereas inspection and maintenance requirements in civil infrastructure are neither as regulated nor thorough.

THE MYSTERY
Despite the significant changes in material specifications and process methods, FRP composites have been shown to have very good long-term response when used in applications such as underground storage tanks, pressure vessels, and in corrosion resistant equipment. In addition their successful use in marine vessels over extended periods of time has been extensively documented. Yet, data from laboratory studies consistently shows significant levels of degradation in the same classes of materials when subjected to moisture/humidity, temperature, and stress. The mystery in the apparent dichotomy between "real-world" applications and laboratory data which is often forgotten in the discussion related to the use of composites in civil infrastructure can be explained in terms of the use of appropriate "knock-down" or safety factors.

Although FRP composites have been used extensively in naval and marine applications there is still a lack of design rules [3] and incomplete understanding of the complex scaling laws for anisotropic materials resulting in the use of very conservative safety factors that are often far higher than those used in similar applications using metals [4]. For the most part their use is predicated through the adoption of a factor of safety of 4-6 for strength [5-7] based on combined effects of stress-rupture and fatigue. It should be noted that although hulls face a greater degree of loading than a number of thin skin composite elements in other application areas, hulls in general are not subjected to sustained loading, but rather to impact and slamming loads of short and intermittent duration [5]. If the structure has to be designed to carry impact loads safety factors as high as 10 are often used [8]. In almost all cases, durability for long-term use is dependent on the use of an appropriate gel-coat without which there is a realistic fear of premature degradation through surface blistering which can lead to further deterioration and delamination of the FRP composites, especially those fabricated using orthophthalic polyesters.

FRP composites, primarily using glass fiber reinforcements, have a history of successful use in underground storage tanks, corrosion resistant equipment and pressure vessels. Because of a high level of regulatory activity this area has, perhaps, the highest development of standards for design and use. The U.S. standard – ASME RTP-1, for reinforced thermoset plastic corrosion resistant equipment provides for the use of factors based on whether design is conducted through the application of rules (Subpart 3A) or through stress analysis (Subpart 3B). When design rules are used a design factor of 10 is specified for the determination of minimum thickness of shells subject to internal pressure, and a factor of 5 is specified for determination of minimum moment of inertia for stiffening rings. In the case wherein design is done through the use of specific methods of stress analysis, minimum strength ratios between 8 and 10 for the inner surface and interior layer depending on the design method and type of non-destructive testing used [9]. For vessels designated for critical service the factor is required to be multiplied by 1.25. For all other layers in the laminate the minimum strength ratio is required to be 1.6 for vessels not in critical service and 2 for vessels in critical service. The British standard, BS 4994 for the design and construction of vessels and tanks in reinforced plastics [10] requires a minimum design factor of 8. It is noted that the forward explains that the factor was increased to 8 from the previous value of 6 specified in the 1973 edition "because experience has shown that the overall strain limitation does not permit lower values than 8." Under the provisions of the ASME Boiler and Pressure Vessel Code [11] two methods are allowable for qualification of design adequacy. Under Class-I design, which specifies qualification through pressure testing of a prototype the minimum qualification pressure is set at 6 times the design pressure. The maximum design pressure is limited to 1034 kPa (150 psi) for vessels fabricated using bag molding, centrifugal casting or contact molding, 10342 kPa (10342 psi) for filament wound vessels, and 20685 kPa (20685 psi) for filament wound vessels with polar boss openings. For the latter case only a 5:1 factor of safety is allowable. Specific conditions are also set for cycling and temperature conditions. Under Class-II design, stress analysis is required using either specific design rules or discontinuity analysis. If the specified design rules are used the membrane strains are restricted to 0.001 and the thickness is required to exceed 6 mm (0.25 in) with a design factor of 5 applied to the modulus parameter. In the case of discontinuity analysis the minimum buckling pressure is required to be at least 5 times the design external pressure and the average shear stress between the vessel and overlays/joints is restricted to 200 psi. Table 1 presents levels of factors of safety as prescribed by various codes for FRP pressure vessels and compares the values based on fiber type.

Table 1: Factors of Safety for Pressure Vessels

Standard	Glass	Aramid	Carbon
FRP-1	3.33	3.33	NA
FRP-3	3.5	3.0	2.25
ANSI AGA NGV-2	3.5	3.0	2.25
ASME-Boiler & Pressure Vessel-X	5.0	5.0	5.0
Aerospace	3-4	1.5-3	1.5-2

In all these examples of successful design it should be noted that stresses are kept to a very low level through the use of fairly high factors of safety, even in cases where inner metallic liners are used. This ensures a large part thickness with sufficient tolerance for some degradation and a very long period of time for moisture/solvent saturation or even through-thickness infiltration. The use of very small factors of safety as espoused by some for civil infrastructure negates this very concept.

THE REALITY

FRP composites have a long history of use in marine vessels, piping, corrosion equipment, and underground storage tanks, and anecdotal evidence and limited testing shows that they can be successfully engineered to have long service lives in contact with moisture and aqueous solutions. It is, however, a misnomer that FRP composites and polymers are "water-proof" since moisture diffuses into all organic polymers, leading to changes in thermophysical, mechanical and chemical characteristics. The primary effect of absorption is on the resin itself through hydrolysis, plasticization and saponification, which cause both reversible and irreversible changes in the polymer structure. At the minimum the absorption of moisture causes a depression in the glass transition temperature and hence it is critical that the FRP composite has a Tg a minimum of 30 degrees above the maximum service temperature to allow for this potential depression. In some cases the moisture wicks along the fiber-matrix interphase and has been shown to cause deleterious effects to the fiber-matrix bond resulting in loss of integrity at that level. Moisture and chemical solutions have also been shown to cause deterioration at the fiber level. In the case of glass fibers, degradation is initiated by the extraction of ions from the fiber by the water. These ions combine with water to form bases (alkaline solutions), which etch and pit the fiber surface, resulting in flaws that significantly degrade strength and can result in premature fracture and failure. Basic solutions, when in contact with glass fibers, can cause significant pitting and leaching, often resulting in the fiber loosing its core over a short period of time with the outer sheath being reactive, accelerating the leaching process.

FRP composites are increasingly being accepted for use in the seismic retrofit and strengthening of columns. Due to lower costs and greater overall familiarity E-glass reinforced epoxy systems are often used in preference to the others. Whereas both sustained load levels and environmental attack can be expected to be less severe in the case of seismic retrofit of columns as compared to storage tanks for corrosive media, the design of wraps/jackets still needs to consider aspects related to long-term durability. A number of studies conducted on flat coupon specimens of commercial systems have shown the degradation of these materials when exposed to water, as would be possible when columns are in a flood plain, raising significant concern related to the life-expectancy of the systems. Whereas moisture related degradation, especially in cases where there is either no protective coating/gel coat on the system, or the coating has been abraded off through accidental damage or scour, is likely it is possible to reduce the level of degradation through appropriate selection of materials and processes. An effective means of enhancing durability of these systems is through the use of appropriate sizings/finishes on the fibers themselves.

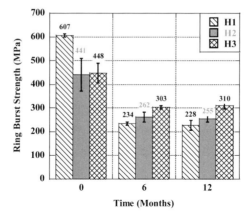

Figure 3: Change in Tensile Strength in the Hoop Direction as a Function of Exposure Time for the Three Differently Sized E-Glass/Epoxy Systems

Figure 3 shows the effect of using 3 different sizings on E-glass fibers for a commercially available E-glass/epoxy material system, for seismic retrofit, subjected to an accelerated exposure to water at 140°F (60°C). The different sizes/finishes used were the 111A from Owens Corning, the 2022 from PPG, and the 61H11 from Vertrotex Certainteed, designated in composite form for the purposes of this study as H1, H2 and H3, respectively. All three use 250 yield direct wound E-glass roving supplied with bushing applied commercial finishes for multi-resin compatibility. The specimens were fabricated in the form of ring blanks following the protocol described by Reynaud *et al* [12] and were then tested using a ring burst test to evaluate hoop direction properties. It is clearly seen that although the unexposed value of strength attained by set H1 is significantly higher than that of H2 and H3 respectively, the level and rate of degradation of strength are also significantly higher. At the end of the 12-month immersion periods, rings from set H1 show a 62.5% reduction from the original unexposed value whereas those from H2 and H3, both having different sizings, show a 42.19% and 30.77% reduction, respectively. It is noted that in all cases most of the loss in strength occurred in the first six months with subsequent deterioration being significantly less. Besides emphasizing the efficacy of tailoring of sizing to enhance durability of FRP composite systems, the results of the study suggest that it may be better to trade higher initial performance attributes for better long-term results. Further, based on the results presented it is also reasonable to conclude that the use of a tougher and higher elongation resin system would yield both higher initial and post-exposure results since a greater "stretch" or global strain would be required before local stress concentrations and notch effects would dominate. The combination of this with optimally tailored sizings is suggested as a means for achieving a durable and damage tolerant FRP system without sacrificing overall cost effectiveness.

Although there is a substantial body of literature related to aspects of moisture absorption in composites and there have been extensive studies of effects due to saltwater and marine environments there is a lack of information related to sea water effects on vinylesters and E-glass/Vinylester composites [13], with most previous efforts being restricted to epoxy and unsaturated polyester based systems.

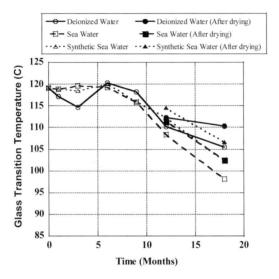

Figure 4: Change in T_g as a Function of Aqueous Solution and Period of Immersion

Table 2: Levels of Tensile Strength (MPa) as a Function of Exposure Type and Time (Unexposed Strength: 363 MPa, Std. Deviation: 4.21 MPa)

EXPOSURE TYPE	Time Period (Months)				
	1	3	6	9	12
	-	-	-	-	-
Deionized Water	355 [9.65]	349 [4.48]	341 [1.31]	335 [2.83]	322 [12.48]
Sea Water	354 [14.82]	360 [9.10]	341 [7.65]	322 [11.10]	314 [11.45]
Sea Water (Sealed Edges)	360 [16.62]	352 [6.76]	350 [11.03]	345 [3.72]	315 [8.34]
Sea Water (Cycling)	338 [7.93]	363 [25.86]	339 [19.03]	336 [18.55]	333 [18.13]
Synthetic Sea Water	343 [11.03]	351 [5.31]	340 [15.03]	341 [16.13]	316 [15.44]

The increasing consideration of E-glass/Vinylester composites in civil infrastructure systems such as piling, decking for piers and wharves, and fender systems, and in offshore applications such as risers, downhole tubing, and subsea flowlines, makes the lack of this information a severe drawback. It is also known that although moisture absorption in itself causes mechanisms of degradation, the actual effects are largely dependent on the type of aqueous solution being considered with the presence of salts such as NaCl reducing saturation moisture content in some cases [14, 15] and even causing substantial deviation from Fickian behavior in others including early onset of irreversibility. Results of comparative tests on a quadriaxial E-glass/vinylester FRP composite fabricated using wet layup with the application of vacuum throughout cure are shown in Figure 4 and in Table 2.

SUMMARY

Although FRP composites have made significant inroads in the civil engineering area, they have not been used to their full potential to date. With the increasing number of projects, especially in the area of rehabilitation, being completed there has been a marked increase in the level of comfort of civil engineers in using these materials. Although the durability of these materials can be significantly better than that of conventional materials it is emphasized that this is only possible through the selection of appropriate constituent materials, use of appropriate processes for manufacturing and fabrication, and incorporation of appropriate measures for environmental and damage protection. A schematic of needs from the constituent to the applications levels is shown in Figure 5.

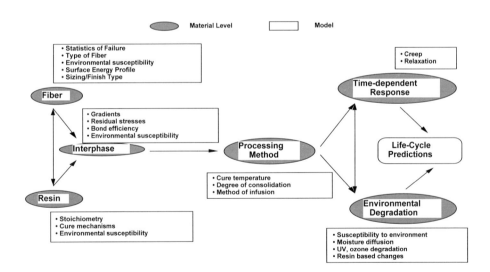

Figure 5: Schematic for Overall Evaluation of Durability

It is, important to emphasize that further research is needed at specific levels of materials, systems, and field implementation focusing on combined effects of stress and environmental exposure. The3 development of validated data bases for materials sets aimed at application in a civil infrastructure environment and the achievement of high quality products through control of processes along with the development of appropriate safety factors will result in the attainment of the potential of long-life with low maintenance – without these the durability of FRP composites in civil infrastructure will remain both a myth and a mystery.

REFERENCES

[1]   Karbhari, V. M., and Li, Y (2001), "Safety Factors and Materials Design Allowables for FRP Composites – Application to Civil Infrastructure," Proceedings of the 33rd International SAMPE Technical Conference, Seattle, 14 pp.

[2]   Karbhari, V.M., Chin, J.W. and Reynaud, R. (2001), "Gap Analysis for Durability of Fiber Reinforced Polymer Composites in Civil Infrastructure," CERF Report, 127 pp.

[3]   Mouritz, A.P., Gellert, E., Burchill, P. and Challis, K. (2001), "Review of Advanced Composite Structures for Naval Ships and Vessels," Composite Structures, Vol. 53, pp 21-41.

[4]   Evans, T.W., Swann, R.F. and Troffer, M.A. (1991), "The Submarine Perspective," Proceedings of the National Conference on the Use of Composite Materials in Load-Bearing Marine Structures, National Academy Press, Washington, DC, pp 11-17.

[5]   Puccini, G. Environmental Aspects in Composite Materials in Maritime Structures, Vol. 1, Eds. R.A. Shenoi and J.F. Wellicome, 1993, pp 44-85 (Cambridge University Press, Cambridge, UK).

[6]   Gibbs and Cox, Inc. Marine Design Manual for Fiberglass Reinforced Plastics (McGraw Hill, New York), 1960.

[7]   Cable, C.W. (1991), "The Effect of Defects in Glass-Reinforced Plastics," Marine Technology, Vol. 28[2], 91.

[8]   Chalmers, D.W. (1988), "The Properties and Uses of Marine Structural Materials," Marine Structures, Vol. 1, pp 47-70.

[9]   ASME-RTP-1d-1998 (1998), Addenda to ASME-RTP-1-1995 edition, Reinforced Thermoset Plastic Corrosion Resistant Equipment, ASME, New York, NY.

[10]   BS 4994 (1987), British Standard Specification for Design and Construction of Vessels and Tanks in Reinforced Plastics," British Standards Institute.

[11]   ASME Boiler and Pressure Vessel Code X (2001), FRP Plastic Pressure Vessels, ASME, New York.

[12]   Reynaud, D., Karbhari, V.M. and Seible, F. (1999), "The HITEC Evaluation Program for Composite Column Wrap Systems for Seismic Retrofit," Proceedings of the International Composites Exposition, Cincinnati, OH, pp 4A/1-6.

[13]   Sagi-mani, D., Narkis, M., Siegmann, A., Joseph, R. and Dodivk, H. (1998), "The Effect of Marine Environment on a Vinylester Resin and Its Highly Filled Particulate Quartz Composite," Journal of Applied Polymer Science, Vol. 69, pp 2229-2234.

[14]   Jones, C.J. and Dickson, R.F. (1984), "The Environmental Fatigue Behavior of Reinforced Plastics," Proceedings of the Royal Society of London, A396, pp 315-338.

[15]   Soulier, J.P. (1988), "Interaction of Fiber Epoxy Composites With Different Saltwater Solutions Including Isotonic Liquid," Polymer Communications, 29, pp 243-246.

1.4 Beyond rehabilitation - use of FRP composites in new structural systems: laboratory and field implementation

L Zhao, Y D Hose and V M Karbhari
University of California San Diego, La Jolla, CA, USA

INTRODUCTION

Fiber reinforced polymer (FRP) composite materials are increasingly being considered for use in the renewal of civil infrastructure. Over the past decade their use in applications such as the seismic retrofit of columns, strengthening of slabs and girders, and shear strengthening of walls has been proven to be an effective means of both increasing load capacity and serviceability levels, and in increasing the useful life of the structure. Despite similar levels of efforts being made for the introduction of composites into new construction applications there has been only limited success with most projects being at the level of demonstrations, rather than bid competitively with conventional materials, or being specified because of the inherent and essential advantages of FRP composites over conventional materials. In part their application in new construction has been limited due to high costs in comparison with components fabricated from conventional materials such as concrete and steel. Further there is a reluctance to use these materials for primary structural elements in new construction without sufficient data on long-term structural response and durability, and in the absence of appropriate guidelines, codes and standards.

There is however, no doubt, that these materials have significant advantages for use in new construction ranging from lighter weight which would translate to greater ease in construction without heavy construction equipment and use of smaller sub-structural elements, to their greater capacity to meld form and function thereby providing for ease in integration of aesthetics (especially to blend in with the environment) with functionality. The lighter weight also can be translated to longer unsupported span in bridges and larger clear areas in buildings and other structures. Their use in seismic zones is also advantageous not just due to weight, but also due to the capability of designing stiffness attributes and through specific location of load paths as defined through fiber architecture. Potentially life-cycle costs would also be lower due to the expected increased durability of these materials in most environments likely to be seen by civil structures. In most cases FRP composite systems can also be erected faster than conventional systems providing a significant savings in time and use of resources. A major disadvantage, to date, however, has been the tendency to use these materials in one-to-one replacement of existing steel or concrete components, thereby negating the inherent advantages of the anisotropy and nature of FRP composites.

To most effectively utilize the advantages of FRP materials, which are lightweight and improve constructability, innovative structural designs are needed which either incorporate "composites-efficient" forms or which combine these "new" materials with conventional ones. Modular systems using the latter approach have been under development at the University of California San Diego and have been implemented in bridge systems in collaboration with the California Department of Transportation and the Federal Highway Administration. This paper describes the development and implementation of a class of systems that uses principles of stability and confinement to optimize the use of both FRP composites and concrete. The intrinsic system consists of prefabricated filament-wound carbon/epoxy thin shells that can be used as columns or girders after being filled on-site with concrete. The shell serves the dual function of reinforcement and stay-in-place formwork for the concrete core. The concrete provides compression force transfer, stabilizes the thin shell against buckling, and allows the anchorage of connection elements. It should be noted that confinement of the concrete within the shell provides for use of shorter development lengths and better connections. Transverse ribs are provided on the inside of the carbon shell for full force transfer between the concrete infill and the shell. For the development of superstructure components, the concrete filled carbon shells are combined with a structural deck system, which consist of either a conventional cast-in-place reinforced concrete (RC) slab, or a FRP modular deck system. A schematic of the concept is shown in Figure 1.

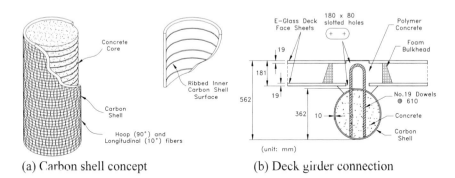

(a) Carbon shell concept   (b) Deck girder connection

Figure 1: Composite bridge component and connection concepts

SYSTEM DEVELOPMENT FOR SHORT- AND MEDIUM-SPAN BRIDGES
The development of the carbon shell system followed a systematic approach by performing analytical and experimental studies on small-scale components, followed by full-scale components, connection, and system subassemblages. The small-scale components included 15-cm-diameter circular and 15-cm-size "conrec" (square with rounded corners) carbon tubes, hollow and filled with concrete [1]. Full-scale

components included 36-cm-diameter carbon/epoxy tubes and a trapezoidal core based E-glass reinforced FRP composite deck. Connections that were evaluated included the girder-to-deck, girder-to-abutment, deck-to-deck and barrier-to-deck. At the system level, the structural characterization of carbon shell beam-and-slab assemblies was investigated based on two full-scale four-point flexure tests.

The geometry and dimensions for the test units were determined from a design study performed for the Kings Stormwater Channel Bridge as shown in Figure 2a, a two span continuous system with a beam-slab superstructure (shown in Figure 2b) and a multi-column intermediate pier. The bridge was constructed by the California Department of Transportation (Caltrans) on the new State Route 86. Its superstructure is composed of 6 longitudinal carbon shell girders, with 10 mm wall thickness, filled on-site with concrete and connected across their tops with a FRP composite deck system. A brief description of some of the tests is presented in the following sections.

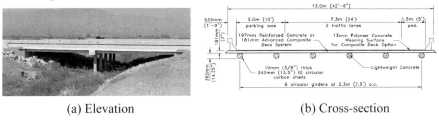

(a) Elevation   (b) Cross-section

Figure 2: Photograph of the Kings Stormwater Channel Bridge

Component tests
A concrete-filled carbon shell girder was tested in a four-point bending configuration as shown in Figure 3a. The span length of the girder specimen, as well as those of other superstructural subassemblages, represents the distance between inflection points in the superstructure of the Kings Stormwater Channel Bridge. This test served to verify the analytical and design models developed to predict the stress and strain states and failure mechanisms.

A 4.6 m by 2.3 m deck panel with trapezoidal cellular core and face sheets was tested simply supported as shown in Figure 3b. Two patch loads were applied at the mid-span simulating the AASHTO specified wheel loads. A linear-elastic load-deflection response, which was accurately predicted by an analytical model, was observed up to failure.

Connection tests
Shear connections at the girder-deck interface provide interaction between the two components to achieve composite action. Behavior of the shear connectors was characterized and validated by push tests [2].

Connection of a standard reinforced concrete crash barrier to the FRP composite deck was made by locally grouting "starter" steel reinforcement into the cell of the deck with polymer concrete, before the concrete section was formed. A hydraulic actuator was used to apply a horizontal force to determine the force resistance capacity of a 1.5-m-long section of the barrier and to validate the design.

"1-Girder" Subassemblage Tests
The test unit was composed of a carbon shell girder filled with lightweight concrete affixed to a 2.3-m-wide, 0.2-m-deep concrete slab (shown in Figure 3c) through steel shear dowels as shear connectors. The objectives of the test were to evaluate the flexural behavior and assess the capacity of the shear connectors.

(a) Girder test

(b) Deck panel test

(c) Girder-concrete slab subassemblage test

(d) "3-girder" subassemblage test

Figure 3: Photographs of component and subassemblage test setups

"3-Girder" Subassemblage Tests
A photograph of the full-scale "3-girder" superstructural subassemblage test setup is shown in Figure 3d. The subassemblage consisted of three concrete-filled carbon shell girders spaced 2.3m on center connected along their tops with a 0.18-m-deep FRP composite deck system, which was oriented such that its longitudinal direction (pultruded tube alignment) was perpendicular to the carbon shell girders. The webs of the deck had a thickness of 13 mm and an equivalent longitudinal and transverse elastic modulus of 17 GPa and 12 GPa, respectively. The compounded flanges (pultruded tube and face sheets) have a thickness of 19 mm and equivalent longitudinal and transverse elastic modulus of 30 GPa and 13 GPa, respectively. The shear connectors consist of two D19 G60 bars spaced at every 0.6 m along the length of the girder. The girders were connected to two continuous reinforced concrete end diaphragms, which were then simply supported on six rocking load cells, one under each girder end, as shown in Figure 4. Loading was applied

vertically to the top of the deck system by means of four servo-controlled hydraulic actuators attached to a load frame.

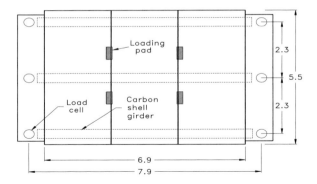

Figure 4: 3-girder test setup

The subassemblage was first subjected to two million cycles of fatigue loading up to the design service demand, before final loading was applied. Load-deflection response shown in Figure 5a indicated that the stiffness of the system does not decrease due to fatigue and permanent deflection stabilizes after approximately 1 million cycles.

The load versus deflection response from the final loading, as shown in Figure 5b, indicated that the system maintained largely linear-elastic up to failure. After failure occurred, the system still possessed approximately 85% of its maximum capacity. When loading resumed, it was able to achieve up to approximately 95% of its original maximum capacity, with a decreased stiffness. The upper and lower bounds obtained from a finite element model predicted the load-deflection behavior of the system. More significantly, the failure criteria established based on the classical lamination theory and the first ply failure criteria of a laminate accurately predicted the failure load. It also pointed to failure of an internal layer due to fiber compression, which could potentially cause delamination.

Failure was observed to have occurred near the location of the loading points, traversing the entire width as shown in Figure 5c, which coincided with where compressive stress was maximum from the analytical results. The observed failure mode was a delimination in the middle of the top flange of the deck panel as shown in Figure 5d, which agrees with the fiber compressive failure of an interior ply predicted by the analysis.

The load-deflection response also indicated that the capacity of the system was significantly higher than that of the demand levels (shown in Figure 5) for the design of the Kings Stormwater Channel Bridge.

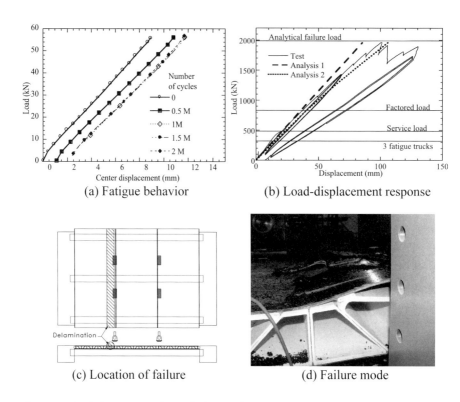

(a) Fatigue behavior

(b) Load-displacement response

(c) Location of failure

(d) Failure mode

Figure 5: 3-girder test results and observations

FIELD IMPLEMENTATION

Following the analytical studies, design, and full-scale laboratory validation, the carbon shell system was implemented in the actual construction of the Kings Stormwater Channel Bridge. Construction began with the center pier cap, which was made of conventional reinforced concrete. Then the hollow girders were installed with local steel reinforcement end cages to be cast with regular reinforced concrete in the pier cap and the abutments (shown in Figure 6). The steel reinforcement was not used through the entire length of the girders, but was rather only used at ends to develop a connection to the conventional abutment and to provide continuity and connection to the ceter cap. Then U-shaped dowels are installed in the pre-drilled holes at the top of the girders and high-slump concrete was pumped from one end of the girder to the other (shown in Figure 6b). The dowels provide a means of connecting the girder to the deck. After the concrete has set, the FRP composite deck panels were installed, with pre-drilled hole patterns matching those of the dowels (shown in Figure 6c). Polymer concrete was then poured manually from the top of the deck to locally grout the dowel in the core of the deck and form a saddle between the deck and the girder for shear connections. Finally, a polymer concrete

(a) Placing girders   (b) Dowels grouted in girders

(c) Dowels matching cut-outs   (d) Grouting connection

Figure 6: Construction of the Kings Stormwater Channel Bridge

wear surface was installed and conventional barriers were cast at ends with reinforcement being encapsulated in the deck locally through the filling of polymer concrete. It should be noted that in essence the deck system has hollow cores with only local areas being filled as needed by polymer concrete.

Field tests
Prior to the opening of the Kings Stormwater Channel Bridge, field-loading tests were conducted on the bridge to verify the performance of the system – see Figure 7. Three fully loaded trucks were used to provide specified loading, and the mid-span deflection was recorded.

The first peak (downward deflection of the bridge is defined as negative) in the figure occurred when the second axles of the three trucks travelled to the south mid-span, while the second peak occurred when the same axles travelled to the north mid-span. Both peaks reached approximately 9 mm, or $L/1118$. The maximum measured strain was approximately 450 $\mu\varepsilon$, equivalent to 4.5% of the ultimate strain capacity of the carbon shell (0.01) and 18% of the allowable strain for service loads (0.025). Based on an extended series of tests it was concluded that the all design criteria were satisfied. A long-term instrumentation program to continuously monitor the performance of the bridge is being installed [3] and currently the bridge is being continuously monitored through wireless instrumentation, thereby enabling remote monitoring of the condition of the bridge. It is expected that such remote monitoring systems will provide increased capacity for the continuous monitoring of life-lines enabling a higher degree of inspection.

 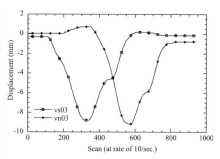

Figure 7 Field loading test

Pylon for cable stayed bridge systems
The carbon shell concept is currently being studied, at a much larger scale, for use as pylon legs in a cable-stayed bridge as shown in Figure 8a. Each pylon leg is composed of 10 m long filament-wound cylindrical carbon composite tubes filled with concrete that are spliced together as shown in Figure 8b. For the starter splice, steel reinforcement is placed continuously from the foundation through the first section into the second. Only a local steel reinforcement cage (30 D29 G60 rebars, 1.8 m each direction) is used for typical splices.

A full-scale test of the most critical typical splice section was conducted to evaluate the effectiveness of the splice design and the adequacy of the design splice length. A specimen as shown in Figure 9a was tested in a cantilever configuration, with a reinforced concrete footing fixed to a strong floor and a top loading stub loaded horizontally with a hydraulic actuator in both directions. A total constant axial load of approximately 18 MN was applied to the specimen by four external post-tensioning tendons. A photograph of the specimen while at extreme drift is depicted in Figure 9b. The force versus displacement hysteresis, which is shown in Figure 9c, indicated that despite the linear-elastic characteristics of the carbon composite, the pylon setup possessed significant ductility due to the design at it splice connections [4].

The Pylon Splice Connection Test successfully demonstrated that the design satisfied the performance objectives. The compressive and tensile strain levels in the composite shell were well below the limit states set forth in the *Design Criteria* [5]. The development length provided in the splice region was sufficient in transferring the forces to the composite shell. The amount of transverse reinforcement provided by the composite shell and steel hoops in the splice region was able to resist pull-out failure of the lap splice longitudinal bars and sufficiently confine the plastic hinge region for adequate ductility capacity. Preliminary assessments of the strain levels as well as acoustic observations during testing indicate that slip of the longitudinal splice reinforcement was avoided.

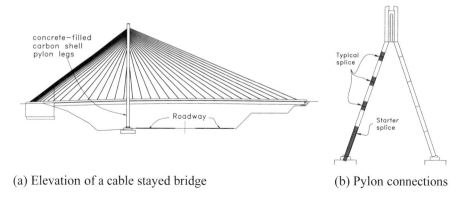

(a) Elevation of a cable stayed bridge   (b) Pylon connections

Figure 8: Schematic of Pylon and Bridge System

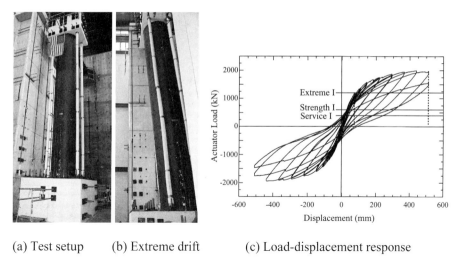

(a) Test setup   (b) Extreme drift   (c) Load-displacement response

Figure 9: Pylon Splice Connection Test

SUMMARY

An innovative construction concept that integrates advanced composite materials with conventional ones such as concrete and steel has been developed. Preliminary design studies have shown that new modular short- to medium-span bridge structures are feasible by using modular prefabricated FRP components taking advantage of their light weight, low maintenance and increased durability. The concept has been successfully applied to the development of a beam-and-slab type bridge structure, and is currently being used in a long-span cable stayed bridge.

Such systems, comprising of composite tube and plate assemblies used synergistically with concrete infill and selective use of conventional embedded connection details can provide design enhancements that were envisaged earlier with the use of concrete filled steel tubes but did not find widespread use due to difficulties associated with steel yielding, corrosion related degradation, and weight. Such concepts could also conceivably be extended for use in rapid fabrication of large industrial facilities without the use of heavy equipment in areas where labor is either unskilled or not readily available.

Recent advances in materials, manufacturing, mechanics, and design point to the continued development of innovative materials- and structural- systems that will (a) enable synthesis of form and function, (b) enable greater emphasis to be paid to aesthetics and blending of the structure into the environment, (c) allow construction of true long-span bridges, (d) enable use of intelligent, self-monitoring structural systems and, (e) development of very large scale structural systems (VLSSS) such as large integrated city-structural complexes built as independent environments on land, in the air/space, or underground. It is opined that through such developments FRP composites will transition from use as a material for rehabilitation (band-aids so to speak) to actual materials of construction, with an equal footing for use in projects as steel, concrete, masonry and timber.

REFERENCES

[1]   Karbhari, V.M., Seible, F., Burgueño, R., Davol, A. and Zhao, L., "Damage Tolerance and Durability of an Advanced Composite Bridge System," 1st International Conference on Durability of Composites for Construction, Sherbrook, Canada, August 5-7, 1998.

[2]   Zhao, L., "Characterization of deck-to-girder connections for FRP composite superstructures," Ph. D. dissertation, University of California, San Diego, 1999.

[3]   Zhao, L, Sikorsky, C.S., Karbhari, V.M. and Seible, F., "Field Monitoring of the Kings Stormwater Channel Advanced Composite Bridge," 6th Caltrans Seismic Research Workshop, Sacramento, June 2001

[4]   Hose, Y.D., Zhao, L., Karbhari, V.M and Seible, F., "Experimental Characterization of the I-5/Gilman Advanced Technology Bridge," 6th Caltrans Seismic Research Workshop, Sacramento, June 2001.

[5]   "I-5/Gilman Drive Advanced Technology Bridge Design Criteria," Draft version 2.2, University of California, San Diego, September 2001.

1.5 GRP offshore – applications and design for fire

J M Davies and P Currie
University of Manchester, Manchester, UK

INTRODUCTION
In many respects, advanced composites appear to be excellent, well-researched
construction materials looking for applications which are economically practical.
However, this simplistic overview is not valid for offshore installations where the
combination of high strength, light weight and supreme resistance to corrosion has
ensured their economic viability. In many potential practical applications, the main
hindrance to their widespread acceptance is cost. However, in the demanding
offshore environment, it is more likely to be their performance in fire.

Offshore, the main applications that have been considered to date are pipes, tanks
and vessels, including fire deluge systems, fire protection systems, cladding panels,
gratings and stairways. The most successful practical applications have been in
pipework and fluid handling. More recently, a feasibility study has been carried out
which considers the use of composites for the primary structural members of
offshore platforms. This paper will briefly review these applications before
proceeding to summarise some of the research that has been carried out at the
University of Manchester in order to demonstrate that advanced composites can
meet the design requirements for offshore applications. Particular attention will be
given to the fire performance of composites and it will be shown that,
notwithstanding the fact that all of the plastic materials used in the manufacture of
composites are flammable, it is not necessary to be pessimistic regarding their
performance in hydrocarbon fires, even in certain fire-critical situations.

APPLICATIONS - PAST, PRESENT AND FUTURE
The potential benefits of using composites offshore are clear. In the 1980's, a
number of barriers to their more widespread use were identified [1], namely:

1. Regulatory requirements, especially on combustibility
2. Lack of relevant performance information, especially in hostile environments
3. Lack of efficient design procedures and lack of familiarity by designers
4. The fragmented structure of the composites industry
5. Difficulty of scaling up small-scale processes to make large structures.

Since then, much has been achieved in removing the first three barriers, especially
where 'prescriptive' requirements have been replaced by performance-based

standards. There has been less progress with 4 and 5. Table 1 [1] summarises the current offshore applications of composites together with some expected in the near future. The future applications reflect the current change of emphasis from shallow to deep water and the increasing interest in sub-sea applications.

Table 1. Applications of composites offshore

(a) Recent applications:

Fire protection	Walkways and flooring	Lifeboats
Blast protection	Handrails	Buoys and floats
Corrosion protection	Sub-sea anti-trawl structures	ESDV protection
Partition walls	Casings	Boxes and housings
Aqueous pipe systems	J-tubes	Loading gantries
Tanks and vessels	Caissons	Pipe refurbishment
Firewater systems	Cable trays and ladders	Riser protection
Pipe liners	Accumulator bottles	Bend restrictors
Separator internals	Well intervention	Sub-sea instrument housings

(b) Future applications

Rigid risers	Coilable tubing	Flexible risers
Tendons	Primary structure	Separators

Evidently, in a short paper, it is not possible to discuss all of these applications in any depth. The remainder of this paper will, therefore, concentrate on three where the research group at the University of Manchester have made a significant contribution. A good source of further information concerning the wider picture will be found in the Proceedings of the series of Conferences "Composite Materials for Offshore Operations" at the University of Houston [eg 2].

PIPES FOR FIREWATER SYSTEMS [3]

The fire water system of an offshore installation generally incorporates a ring main which is kept full of stagnant water at all times. This ring main supplies dry riser pipes which, in turn, feed the sprinkler heads. In a fire situation, when activated by an automatic alarm system, the deluge system would normally be fully running within 30 seconds of the start of an emergency. If the automatic systems fail, this time may possibly extend to 3-5 minutes following manual activation. These "hand activation" times of 3 or 5 minutes are very often specified as a design requirement for the empty pipe work.

There are, therefore, three distinct conditions that need to be considered when using polymer composite materials in a fire deluge system. These relate to the internal conditions of the pipes, namely: empty and dry, filled with stagnant water, or filled with flowing water. The empty and dry internal condition is by far the most critical.

The empty and dry internal condition

The work reported in this paper concerns furnace-based fire tests in which the pipe is placed within a computer-controlled furnace which is programmed to follow a specified time-temperature regime, in this case the DoE hydrocarbon fire curve. The initial testing was carried out using a gas-fired furnace lined with stack bonded ceramic wool which provided a working volume of $1.5 \times 1.5 \times 1.5$m. After a certain amount of trial and error, the following method was developed as a reasonable method for the preliminary tests on glass reinforced plastic pipes:

1 metre long lengths of pipe were inserted horizontally into the furnace through an opening in the front face fixture leaving a short length protruding and any gaps were plugged with ceramic wool. The end of the pipe inside the furnace was also plugged and sealed with ceramic wool. The pipes were tested for the desired period to the DoE hydrocarbon time-temperature curve after which the burner was turned off. The pipe under test was then withdrawn from the furnace and any flames extinguished with a carbon dioxide fire extinguisher. The pipe was then stood on its end and filled with water, which was allowed to wet only the inside of the pipe. After cooling in this way the ends of the pipes were trimmed to length and the pipe was tested under static pressure applied by means of a hand pump. The maintainable pressure over three minutes was then recorded. A satisfactory outcome after testing in this way was considered to be a residual pressure capacity of 16 bar.

The tested pipes were all proprietary products which, in the majority of cases, were filament wound glass reinforced epoxy (GRE) pipes. It was found that the deterioration of empty unprotected pipes can be very rapid, with a total loss of strength after about 90 seconds, and that uneconomical wall thicknesses or some form of fire protection would be required if dry riser pipes are to survive for up to 5 minutes before the deluge system is activated

Monitoring the internal temperature of the pipe, and determining the internal wall temperature at which the pipe may be deemed to have failed suggested that a limiting internal wall temperature of 200°C could be assumed to ensure that the pipe wall maintained its functionality. This temperature was used in future tests to remove the risk of withdrawing burning GRE pipes from the furnace.

The use of ceramic fibre as a novel and durable form of fire protection that could be applied externally at the manufacturing stage was then investigated [4].   Four alternative fire protection arrangements were investigated by adding them to Ameron Bondstrand 2000M GRE pipes. These were in the form of either single or double layers of 3 mm thick ceramic wool blanket in either the dry or resin-wetted condition. In all cases the ceramic fibre was then covered with a double layer of resin-wetted 600 gsm glass woven roving. The initial tests with a single layer of protection showed that they were able to survive a 3 minute test but were not capable of withstanding a five minute hydrocarbon fire exposure. Table 2 summarises the results for the 5 minute test series.

Table 2 - Results of pressure testing pipes after 5 minutes hydrocarbon exposure

Test specimen	Pressure maintained for 3 minutes (bar)
Pipe + GRP wrap	0
Pipe + 2 × 3 mm dry ceramic wool	8
Pipe + 2 × 3 mm wetted ceramic wool	16

It can be seen that the inclusion of thin layers of ceramic wool has increased the fire resistance of the epoxy pipes substantially. Pipes with the added protection of two 3 mm resin-wetted layers of ceramic wool with a woven roving cover can survive a five minute hydrocarbon fire exposure whilst maintaining the pressure rating of the pipe.

A widely used and effective method of protecting empty GRP pipes against the rapid initial deterioration in a hydrocarbon fire is to use a thick intumescent external coating. Two tests were performed to determine the effectiveness and relative performance of intumescent systems. For the first test, a 100 mm Ameron Bondstrand 2000M GRE pipe was used with a 7.5 mm unreinforced Pitt-Char coating. The internal wall did not reach the 200°C critical temperature until about 11 minutes into the test. This form of protection is therefore more than adequate to protect the pipe until the water flow is established. However, without reinforcement, the intumescent tended to crack and split and did not appear to be very secure.

The second sample tested was similar to the first. However, in this case the pipe was protected by an 11.5 mm thick intumescent layer which was reinforced by a glass fibre mesh at about mid depth. The results for this test were similar to those for the thinner unreinforced layer of intumescent, again with the critical temperature being reached after approximately 11 minutes. The reason for the similarity of the results is that, although the reinforced intumescent is thicker, the glass layer tends to restrain the expansion during the fire test. It was noted that the reinforced system had much greater integrity after the fire exposure and could be easily handled without causing further damage.

These results show that the very rapid deterioration of empty GRE pipes in fire can be prevented using coatings of either ceramic fibre or intumescents.

The stagnant and low-velocity water-filled condition
When steel piping is used for the deluge system on offshore platforms, it is beneficial to maintain the pipes in the empty and dry condition in order to control corrosion of the internal walls of the pipes. Otherwise, it is possible that the corrosion products may block the sprinkler heads thus rendering the deluge system useless. It is then possible that the heating effect of the fire could cause the water in the pipe to boil and hence create very high pressures. In the worst case scenario, this may cause the pipe to explode, or burst violently. When using GRP materials, there is no internal corrosion in the pipes so that this is unlikely to happen. It is not, therefore, necessary to maintain the riser pipes in the empty dry condition and this leads to a simple alternative to the

provision of passive fire protection. If the pipes are maintained filled with stagnant water, this provides them with a substantial heat sink which may well be sufficient for them to survive for the "hand activation" time until the water starts to flow.

In order to test GRE pipes in the stagnant and flowing water conditions, an external furnace enclosure was built so that the personnel could be safely inside the laboratory building whilst the test was proceeding outside. After various exploratory tests, the first formal test showed that even a low rate of flowing water soon produced a stable condition with little, if any, deterioration of a 50 mm diameter GRE pipe. The thermocouple readings quickly settled down to a fairly steady state and there was little increase in internal temperature during the subsequent 30 minutes of testing. This same pipe was then subject to a further test of 2 hours duration during which the flow rate was reduced in stages from 18 to 7 litres per minute and a final test of 30 minutes with water flowing at only 3.5 litres per minute. During the latter test, the pipe wall temperature exceeded $100^{\circ}C$ for most of the test indicating that the wall was isolated from the full cooling effect of the water due to the formation of steam. After all of this abuse, the pipe was pressure tested to 16 bar and it maintained its functionality, though with slight leakage, losing only 30 cc of water over a period of one minute.

A similar fire test was carried out on a l00 mm internal diameter GRE pipe subject to a 10 minute period of stagnant water followed by 20 minutes of flowing water at 18 litres/min. The water inside the pipe began to boil after approximately 5 minutes of testing to the hydrocarbon curve. During the remainder of the stagnant period, the boiling of the water inside the pipe became more violent. After approximately 8 minutes of testing, the water inside the pipe was being forced out of the stand pipes and hence the pipe wall was isolated intermittently from the cooling effect of the water. After cooling, this pipe then survived the pressure test.

Table 3 shows the results of pressure tests up to 16 bar on several pipes after fire testing using different internal regimes. It should be noted that the flow rates used in these tests were much lower than would be typical of a deluge system. The performance with a realistic flow rate would be expected to be even better.

Medium velocity water-filled condition
It has been that the heat sink provided by even a modest amount of water is sufficient to cool the pipe wall to the point where, once the water has started to flow, further breakdown of the resin matrix does not occur. Experiments conducted under conditions of higher water flow further highlight the benefits of maintaining GRE pipes in the stagnant/flowing water filled condition. With the confidence gained by testing outdoors with low flow rates, a more sophisticated facility was constructed indoors by connecting an existing large furnace to a water supply with a much larger head and a cooling facility. This facility, located in the fire testing laboratory of the University of Manchester, allowed the testing of water-filled pipes to be continued with more complex pipe configurations and the possibility of much higher flow rates.

Table 3 - Summarised results of fire tests on water-filled pipes

Test No.	Dia. Mm	Test conditions	Result
1.	50	2 minutes stagnant then flowing	No leakage
2.	50	5 minutes stagnant then flowing	Leak at 10 bar
3.	50	2 minutes dry then flowing	Leak at atmosph. pressure
4.	100	2 minutes stagnant then flowing	Leak at 10 bar
5.	100	5 minutes stagnant then flowing	No leakage
6.	100	repeat of test 5 (low TC outputs)	No leakage
7.	100	Pitt-Char coated. 50 minutes stagnant only	No leakage
8.	100	10 minutes stagnant then flowing	No leakage
9.	100	Pitt-Char coated. 5 mins stagnant then flowing	No leakage
10.	100	2 minutes stagnant then flowing for 58 mins.	No leakage

In a typical test on a 100 mm diameter Ameron Bondstrand 2000M GRE pipe containing flowing water from a 50 mm diameter, 3 bar supply giving a flow rate in the tested pipe of approximately 240 litres per minute, there was little rise in temperature of the internal wall of the GRE pipe during the fire exposure. As in previous tests, after the first 10-15 minutes of testing, the temperature appeared to have settled into a quasi-steady state condition with an internal temperature of approximately 30°C. It is anticipated that, under these fire and flowing water conditions, the pipes would be able to survive indefinitely.

GRATINGS
Another investigation considered the performance at elevated temperature of GRP grating systems formed from pultruded sections, which were to be used for walkways on offshore installations. Tests were performed, in conjunction with Petrobras and Reverse Engineering Limited, to determine the performance of a proprietary GRP grating system at elevated temperatures. The test procedures developed during the course of this investigation were subsequently incorporated into a Petrobras Standard [5].

The scenario considered was that, in the event of a fire, although they would not be directly exposed to burning products, the gratings may experience a significant rise in temperature due to radiation. In these circumstances, they would still be required to remain functional in order to provide means of escape and to provide access routes for fire fighters and their fire fighting equipment. Various tests including full grating unit tests and small-scale constituent element tests were performed at a range of temperatures up to 120°C. The full-scale tests were performed to simulate the loading of a fire fighter and fire fighting equipment whilst the small-scale tests were performed to determine the mechanical properties of the constituent elements as a prelude to more detailed numerical analysis.

Figure 1 shows a section of grating 2.0 m wide × 1.2 m span after testing in a heated chamber. For convenience, the load, here a single load over an area 200 mm × 50 mm, was applied upwards, rather than downwards as in practice. The pultruded primary members are linked by interlocking secondary members. It can be seen from Figure 1 that these secondary members are effective in spreading a point load over a significant number of primary members.

Figure 1: Grating after testing

The component tests showed that the decrease in bending strength of the primary members was about 28% at 90°C, increasing to about 39% at 120°C. The stiffness decrease was only 10-12% in this temperature range.

WALL PANELS
Another recent project is studying the effects of elevated temperatures on the behaviour of GRP stringer panel construction. The aim of this research is to assess the feasibility of using pultruded GRP stringers (wall studs) for wall construction in low rise modular building construction both on-shore and off-shore.

Small-scale cellulosic fire tests were performed on several different panel constructions in order to determine their fire performance and to accurately monitor the development of temperatures throughout the panel. An example of the type of construction tested is shown in Figure 2. This panel consisted of 12.5 mm thick gypsum board faces, a 100 mm thick mineral wool core and a 4 mm thick GRP pultruded channel stringer.

Figure 3 shows the temperatures recorded in the GRP stringer after 30 minutes testing in a cellulosic fire. It can be seen that, although the fire-exposed surface attained a high temperature, the majority of the GRP section remained below 80°C.

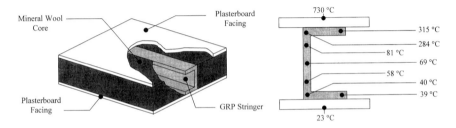

Figure 2: Stringer Panel Construction   **Figure 3**: Temperatures after 30 mins

Identical sections to those used to form the GRP stringers within the panels were then tested as stub columns in a custom-built compression rig capable of applying axial load under controlled conditions of elevated temperature. Tests were carried out to determine the properties of the pultruded sections at elevated temperatures up to 250°C. The failure loads obtained give an indication of the performance decrease with rise in temperature. The results are summarised in Figure 4. This research is continuing.

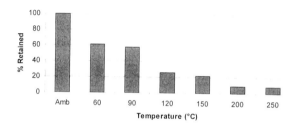

Figure 4: Results of compression tests at elevated temperature

NUMERICAL MODELLING OF FIRE PERFORMANCE

Due to the considerable complexity of the phenomena that take place, it is not practical to form a complete numerical model for all of the chemical and physical changes that occur as a polymer composite degrades in fire. There is a general absence of public domain information concerning the majority of the processes involved. A promising approach is to form a mathematically viable but relatively simple model [6,7] that can capture the main features of the pyrolysis process and the consequent heat transfer behaviour. In order to simplify the model, several idealisations have to be made:

a)   The GRP material is assumed to be homogenous, and the transfer of heat and mass only takes place in a direction perpendicular to the face of the material.

b)   There is thermal equilibrium between the decomposition gasses and the solid material and there is no accumulation of these volatile gasses within the solid material.

c)   The feedback of energy from the flaming of volatiles released by the GRP is ignored due to its small contribution in comparison to the high heat flux created by the furnace.

In addition to these assumptions, which can be used for all GRP materials, pipes may be subject to the following additional assumptions:

d)   The helically wound nature of FRP pipes ensures that full scale delamination does not occur.   Consequently, the delamination model developed at the University of Manchester [8] is not directly relevant to pipes.   However, delamination should be considered for all FRP components subject to long fire exposures and can be a particularly important consideration in the case of phenolic resin laminate panels.

e)   The loss of heat from the cold surface (i.e. the inside of the pipe) may be restricted due to the closed nature of the pipe. This may lead to a more rapid increase in the temperature than would be observed, for example, in the case of a panel.

For one-dimensional problems, the finite difference method proves to be the most convenient [7] and this can be used with polar coordinates for pipes or Cartesian coordinates for (sandwich) panels. For two- and three-dimensional problems, the finite element method has been used. The Author's research group at the University of Manchester has developed a full suite of such programs and further details of the material properties assumed in the analysis are given in reference [7].

The boundary conditions on the exposed and unexposed sides may be either a prescribed temperature or a combined radiative and convective boundary condition. In order to exclude the uncertainties of heat transfer from the furnace to the specimens under test, the measured temperature was generally used as the boundary condition on the exposed surface. On the unexposed inside surface of empty pipes, an adiabatic surface was assumed on the basis that the heat capacity of the inside air is negligible and the temperature distribution inside is uniform. For the water-filled condition, an empirical formula for turbulent flow is used to calculate the internal heat transfer from the pipe to the water.

The validity and accuracy of the models described above can be seen from Figure 5 which shows a typical comparison of the prediction given by the numerical model and the actual test result for a fire-exposed 100 mm diameter empty Ameron Bondstrand 2000M GRE pipe.

CONCLUSIONS

Glass reinforced plastic members have a number of attractions as structural members, particularly in offshore applications. A primary hindrance to their more widespread use is their perceived poor performance in fire and at elevated temperature. This paper summarises some recent research and provides some indicative results for the performance of pipes, gratings and stringer panel walls at

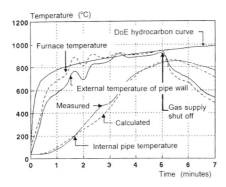

Figure 5: Typical comparison of results

elevated temperature. These give grounds for optimism regarding the use of GRP pultrusions as gratings and members of wall panels. It may be noted here that the examples given all involved relatively thin composite material. Thicker composites generally show a significantly better performance in fire because they tend to form a protective char and then burn away rather slowly as a result of the endothermic reaction.

In practice, the most realistic scenarios for fire water systems involve pipes under conditions of empty or stagnant water followed by flowing water. The stagnant water condition is advantageous and unprotected pipes of appropriate wall thickness are clearly viable when they are maintained in this state. It has been shown that, in medium or high water flow conditions, GRE pipes can last for extended periods in fire with little or no further damage after the quasi-steady state condition has been achieved.

REFERENCES
[1] A G Gibson "Composite materials in the offshore industry", Chap. 6.23, Comprehensive Composite Materials, Eds. A Kelly and K Zweben, Elsevier, 2000.

[2] S S Wang et al (Eds) "Composite Materials for Offshore Operations – 3" Proc. Conf. CMOO-3, University of Houston – CEAC, Oct. 31 – Nov. 2, 2000.

[3] J M Davies, D Dewhurst and H-B Wang "FRP pipe fire performance: Modelling and testing", Ibid pp 269-285.

[4] J M Davies, J B McNicholas and D Dewhurst "Ceramic fibre fire protection of GRP pipes and panels", The Inst. of Materials 6th Int. Conf. on Fibre Reinforced Composites, University of Newcastle upon Tyne, 29-31 March 1994, pp 20/1-20/10.

[5]   "Composite floor grating", Petrobras Technical Standard N-2614, Comissao de Normas Tecnicas, SC-05 Marine Installations and Operations, Aug 2000.

[6]   J M Davies, H-B Wang and J B McNicholas "A numerical and Experimental Heat Transfer Study of GRP Materials Subject to Standard Cellulosic and Hydrocarbon Fire Tests", Interflam '96. Proc. 7th Int. Fire Science and Engineering Conf., Cambridge, March 1996. Interscience Communications Ltd, pp 123-132.

[7]   J M Davies and H-B Wang "Heat transfer analysis of GRP pipes in fire-water systems exposed to fire", Proc. 17th Int. Conf. on Offshore Mechanics and Arctic Engineering, OMAE'98, Lisbon, 5-9 July 1998.

[8]   J M Davies and H-B Wang, "Glass reinforced phenolic laminates subject to hydrocarbon fire tests", 7th Int. Conf. on Fibre Reinforced Composites, FRC'98, Univ. of Newcastle upon Tyne, 15-17 April 1998, Woodhead Publishing Ltd., pp 377-385.

Session 2 : **Application of FRP to concrete structures**

2.1   Effect of curvature on the bond between concrete and CFRP sheets

M A Aiello, *N Galati*, *and A La Tegola*,
Department of Innovation Engineering, University of Lecce, Italy

ABSTRACT
FRP sheets or laminates are widely utilized for flexural and/or shear strengthening of beams, for confining concrete columns and improving joints performance, mostly in seismic areas. The success of retrofitting techniques depends on more extent on the bond performance between FRP reinforcement and concrete. In fact, under both service and ultimate conditions the interface behaviour plays a fundamental role; cracking and deformations are influenced from bond laws as well as, at the ultimate state, the bond failure can cause an unexpected collapse of the concrete elements.

In particular, when curved concrete members are strengthened with FRP sheets, a combination of shear and normal stresses is transferred at the interface. Such combination involves, in many cases, a premature failure of bond.

In this paper an experimental investigation, aimed to analyse the bond between FRP sheets and curved concrete elements, is reported. At this aim a bending test (modified RILEM test) has been performed on curved elements reinforced with CFS (Carbon Fiber Sheet). In particular the state of stress and strain at the interface is analyzed varying the curvature of the elements and the stiffness of the reinforcement.

Obtained results are discussed and compared with those regarding concrete members without curvature.

INTRODUCTION

In the recent years there is an increasing need to repair and upgrade reinforced concrete structures, due to several reasons. An important phenomenon is the degradation caused by environmental conditions; in particular very dangerous effects are due to carbonation, chloride attack, freeze and thaw, reinforcement corrosion, etc. On the other hand, added needs of the today's society impose to comply with new standards; a continuous improvement of the bearing capacity should be attained, mostly as regard infrastructures because of the increasing live loads. Finally a structure must be repaired because of the occurrence of an accident or human errors made at the design stage and improper construction. Several structural deficiencies have been registered worldwide and in many cases the whole replacement of damaged structural elements has been the adopted solution. However because of the high cost or the necessity of providing uninterrupted service strengthening and retrofitting techniques are more appropriate.

The use of FRP (Fiber Reinforced Polymers) for repair and strengthening reinforced concrete structures seems to be a very promising technique. Composite materials are characterized from many advantageous properties with respect to traditional materials as steel, in particular the low weight, the good mechanical properties, the absence of corrosion that improves the durability of concrete structures reducing the maintenance costs, the easier handle on site, the opportunity of optimising the reinforcement as a function of the specific design requirements.

The effectiveness of the repair and strengthening technique is strictly depending on the bond between the reinforcing materials and the substrate. In fact bond performance affects both the serviceability and the ultimate conditions. As kwon, cracking and deformations are influenced from bond laws, on the other hand structural elements may undergo a premature collapse because of the crisis at the interface reinforcement-substrate. On the basis of these considerations many theoretical and experimental studies have been carried out to analyse the bond behaviour between reinforcement and substrate [1-10]. Therefore numerous experimental data are available in the literature and some theoretical models have been proposed to analyse the bond between concrete and reinforcement at the service condition and up to the collapse.

Conversely few researches have been addressed to the analysis of the bond when curved concrete members are repaired using FRP laminates or sheets to improve structural behavior; typically is the case of old bridges made by arches. Because of the presence of the curvature the state of stress at the interface is characterized from a combination of radial and shear stresses; for this reason a reduction of the bond strength may take place, increasing the probability of a premature collapse due to the bond failure.

A research work about the interface performance of curved structural elements strengthened with CFS (Carbon Fiber Sheets) has been undertaken from authors. At this aim an experimental investigation has been planned on curved elements strengthened with CFS and the influence of curvature on bond performance has been investigated, varying some key parameters as the stiffness and the width of the reinforcement. First results have been reported in other studies [11, 12] where also some theoretical considerations have been made and an analytical relationship has been proposed in order to evaluate the bond strength of curved elements. In the present paper the influence of both curvature and stiffness variations are discussed.

The progress of the experimental investigation will give further information's about the role of other substrate and reinforcement parameters on the interface behavior. In addition the analysis of a wider number of specimens will be more effective in order to define a more reliable evaluation of the bond strength referred to curved elements.

EXPERIMENTAL INVESTIGATION

The bond performance between curved concrete elements and FRP sheets, glued at the intrados of the curved elements for repairing and/or upgrading, has been analysed by an experimental investigation carried out on six groups of beams characterized from different curvature values and different reinforcement stiffness. Geometrical details of tested specimens are reported in Table 1 and Figure 1.

Table 1. Tested Specimens

SPECIMEN	Width (mm)	Height (mm)	Number of layers	Curvature radius (mm)
A1–50, A2-50, A3-50	200	250	1	500
A1–100, A2-100, A3-100	200	250	1	1000
A1– 00, A2-00	200	300	1	∞
B1–00, B2-00	200	300	2	∞
B1–100, B2-100	200	250	2	1000
B1–50, B2-50	200	250	2	500

Carbon Fiber Sheets (CFS's), utilized as reinforcement, have been glued at the concrete substrate for 200 mm length and 50 mm width. One layer (thickness of the layer=0,12 mm) and two layers of reinforcement have been utilized to give different stiffness. Tensile strength and elastic modulus values of the reinforcement are 3400 MPa and 275 GPa respectively. Standard compression test have been made to determine the compression strength of the concrete, the corresponding mean value has been 30.22 MPa.

The bond test performed has been a flexural test, in particular a modified RILEM bond-test, where CFS's have been attached on tension face of the concrete beams. Specimens have been subjected to a four points load with a distance of 400 mm between the support and the point load. Beams have been precracked in correspondence of the hinge; to assure that debonding takes place in the curved zone an additional CFS, placed in the direction normal to the tested CFS has been provided in correspondence of the straight part of the specimens in order to avoid the crisis in that zone.

Strain gauges have been placed on CFS in many locations along the some line parallel to the loading direction to measure the strain values and the strain path along the sheets. The uniformity of the strain distribution in the transverse direction has also been checked by strain gauges glued on the width of the reinforcement at fixed positions. Details of strain gauges positions have been already reported in [12]. The scheme of the test apparatus is reported in Figure 2.

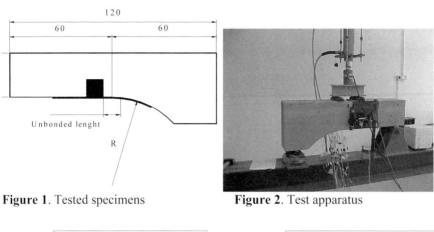

Figure 1. Tested specimens   **Figure 2**. Test apparatus

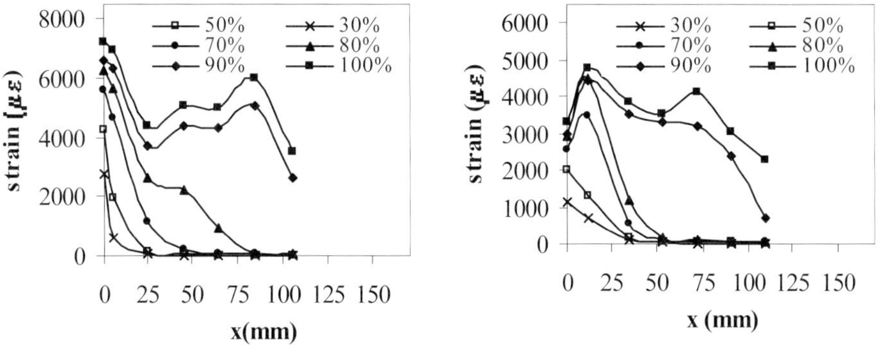

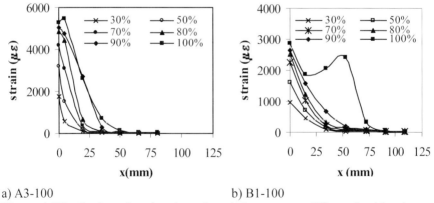

a) A2-00   b) B1-00
Figure 3. Distribution of strains along the reinforcement at different load levels

a) A3-100   b) B1-100
Figure 4. Distribution of strains along the reinforcement at different load levels

In the Figures 3-5 strain distribution along the reinforcement, is reported referring for some tested specimens. In particular Figures 3a, 4a and 5a refer to specimens with different curvature values reinforced with one layer of CFS; Figures 3b, 4b and 5b show the strains pattern when two layers of reinforcement are employed.

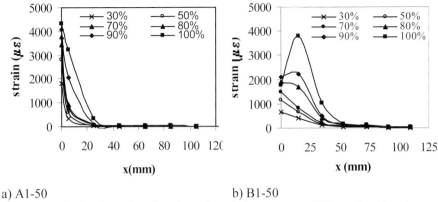

a) A1-50   b) B1-50

Figure 5. Distribution of strains along the reinforcement at different load levels

Analysing the figures it can be observed as the strains distribution along the reinforcement is influenced both from curvature of the specimens and stiffness of the reinforcement itself. Referring to straight specimens, the strains pattern is almost exponential at low load level while when load increases the peak of the strain shifts along the reinforcement, till an almost uniform trend is recorded at high load level. For curved beams the transfer mechanism of stresses is localized near the loaded end of the reinforcement up to the crisis of the interface. However for specimen reinforced with one layer sheet an exponential trend of strains is evidenced almost at any load level, while for specimen with higher stiffness the exponential trend is not evidenced and the peak of strain shift increasing load along the sheet. Therefore the reinforcement length involved for stresses transfer decreases increasing the specimen curvature but it is higher when stiffness arises.

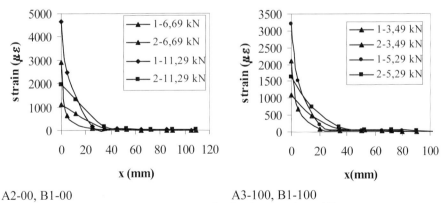

A2-00, B1-00   A3-100, B1-100

Figure 6. Comparisons between strains distribution varying stiffness

In the Figure 6 comparisons between strains distribution along the reinforcement, for specimens with one and two layers, are reported in correspondence of two load levels. In particular, comparisons are made for specimens without curvature and for specimens with R=1000 mm. It can be observed as the stiffness affects the strain gradient, being the strain gradient lower for higher stiffness. In addition it can be observed as maximum strain values are lower for specimens reinforced with two layers sheet.

On the basis of measured strains and imposing equilibrium conditions the shear stresses and the radial stresses along the FRP sheets can be evaluated. In the Figure 7 the shear stresses distribution within the reinforcement is drawn for specimens at different curvature and stiffness.

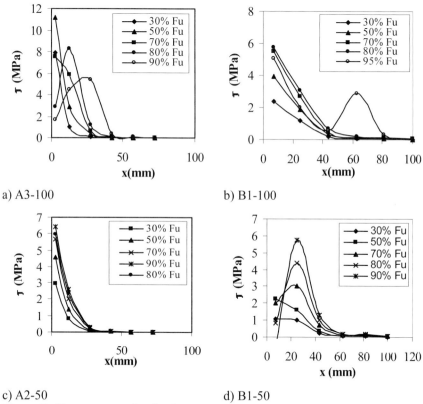

c) A2-50   d) B1-50

Figure 7. Shear stresses distribution along the reinforcement at different load levels

Analysing results it appears as increasing the stiffness the transfer length of the reinforcement increases and at the same time the peak shear stress generally decreases. In addition, increasing curvature the radial stresses arise, therefore debonding starts in correspondence of lower shear stresses values.

In the Figure 8 the ultimate load is drawn versus the curvature value for all tested specimens. As one can see a relevant decrease of the ultimate load value is recorded passing from straight to curved specimens. On the other hand a higher stiffness, obtained utilizing two layers of reinforcement, leads in all cases a higher ultimate load.

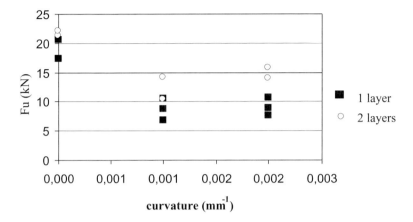

Figure 8. Ultimate load versus curvature

CONCLUSIONS
In the paper results of an ongoing analysis on bond performance of concrete curved elements strengthened with CFS have been reported and discussed. Some interesting remarks can be drawn:
- the presence of curvature, involving a combination of radial and shear stresses at the interface, leads a decrease of the bond strength and a variation of the transfer mechanism between concrete a reinforcement;
- when stiffness increases the ultimate load increases as well. In addition a higher stiffness causes a reduction of strain gradient and a greater transfer length involved.
- further theoretical and experimental investigations are suggested in order to define analytical relationships able to determine the bond strength of curved elements taking into account the presence of radial stresses.

REFERENCES
[1]   Bizindavy, L., and Neale, K. W. (1999), "Transfer Length and Bond strength for Composites bonded to Concrete". Journal of Composites for Construction, Vol. 3, No. 4, 153-160.

[2]   Brosens, K., and Van Gemert, D. (1997), " Anchoring stresses between concrete and carbon fibre reinforced laminates". Proc., Third Int. Symposium on Non Metallic (FRP) Reinforcement for concrete structures, JCI , Vol.1, 271-278.

[3] Brosens, K., and Van Gemert, D. (1999), " Anchorage Design for externally Bonded carbon Fiber Reinforced Polymer Laminates". Proc., Fourth Int. Symposium on FRP Reinforcement for concrete structures, Baltimore, 635-645.

[4] Maeda, T., Asano, Y., Sato, Y., Ueda, T., and Kakuta, Y. (1997), " A study on bond mechanism of carbon fiber sheet". Proc., Third Int. Symposium on Non Metallic (FRP) Reinforcement for concrete structures, Vol.1, 279-286.

[5] Nanni, A., Bakis, C., Boothby, T., (1997), " Externally bonded FRP composites for repair of RC structures". Proc., Third Int. Symposium on Non Metallic (FRP) Reinforcement for concrete structures, Vol.1, 303-310.

[6] Täljsten, B. (1994), "Plate Bonding. Strengthening of existing concrete structures with epoxy bonded plates of steel or fibre reinforced plastics". Doctoral Thesis, Luleå University of technology, Sweden.

[7] Ueda, T., Sato, Y., Asano, Y., (1999), "Experimental Study on Bond Strength of Continuous Carbon Fiber Sheet", Proc., Fourth Int. Symposium on FRP Reinforcement for concrete structures, Baltimore, 407-416.

[8] Bisby, L. A., Green, M. F., Beaudoin, Y., Laboissière, P., (2000)," FRP Plates and Sheets Bonded to Reinforced Concrete Beams", Proc., Advanced composite materials in Bridges and Structures, J. Humar, A. Ghani Razaqpur, ed., Canadian Society for Civil Engineering, Montreal, Quebec, 209-216.

[9] Aiello, M.A., Pecce, M., (2001), "Experimental Bond Behaviour between FRP Sheets and Concrete", Proc., 9th Int. Conference Structural Faults + Repair 2001, London, UK .

[10] Brosens, K., and Van Gemert, D. (2001), " Anchorage analysis of externally bonded bending reinforcement". Proc., Int. Conference on Composites in Construction, Porto, Portugal, 211-216.

[11] Aiello, M.A., La Tegola, A. (2001), Strengthening of Structural Arches using FRP Composites: Analysis and Design" Proc.,Tenth Int. Congress on Polymers in Concrete, Honolulu, Hawaii

[12] Aiello, M.A., Galati, N., La Tegola, A. (2001), "Bond Analysis of Curved Structural Concrete Elements Strengthened using FRP Materials", Proc.,fifth Int. Conference on Fibre-reinforced plastics for reinforced concrete structures, Cambridge, UK, 399-408.

2.2 Cracking analysis of FRP reinforced concrete tension members

M A Aiello, M Leone and L Ombres
Innovation Engineering Department, University Of Lecce, Lecce, Italy

ABSTRACT

The paper is devoted to the analysis of cracking evolution of Fiber Reinforced Polymers (FRP) reinforced concrete tension members and, as consequence, their deformability in terms of load-elongation relationships. Such analysis allows to determine the tension stiffening effects. The stiffness of the reinforced tension members, in fact, can be evaluated and compared with that corresponding to the naked rebars; the obtained scatter gives a measure of the reduced deformability due to the contribution of the tensile concrete.

Results of an experimental investigation, carried out on cylindrical concrete specimens in order to analyse the influence of geometrical properties on cracking and deformability of the members, are reported in the paper. Comparison with results obtained from a theoretical analysis and from Codes relationship are also made, aiming to give useful information from a design point of view.

INTRODUCTION

The low elastic modulus of Fiber Reinforced Polymers (FRP) reinforcement involves high deformability of concrete structures reinforced with non-metallic rebars. As a consequence the design of FRP reinforced members is mainly governed by the deformability check instead of the ultimate strength. For this reason an accurate analysis of cracking and deformations is needed, taking into account the contribution of tension stiffening and the bond between concrete and reinforcement.

The tension stiffening, that is the contribution of concrete between cracks in bearing tensile stresses, is strongly dependant on the bond between the concrete and the reinforcement.

For traditional steel reinforced concrete structures, the bond behaviour and the tension stiffening effects are well defined and considered, by using Codes relationships [1, 2] in the structural analysis. In the other hand, for concrete structures reinforced with FRP rebar the bond behaviour and the tension stiffening effects aren't completely defined. In this case, in fact, parameters involved are numerous and their influence on the physical phenomenon is not clearly evidenced.

ACIC 2002, Thomas Telford, London, 2002

Many researches, both theoretical and experimental, are still in progress on this topic.

The evaluation of tension stiffening contribution needs to analyse the cracking behaviour of concrete members considering the influence of parameters influencing the bond and the tension stiffening. The shape of the reinforcement, the size and the outer surface treatment of FRP rebars, the concrete strength and the concrete cover thickness are the main parameters governing the bond behaviour and the tension stiffening effects [3].

Useful information on the tension stiffening of FRP reinforced concrete structures can be obtained by means of a cracking analysis of tension members as described in this paper. Results of an ongoing investigation devoted to the study of the tension stiffening of FRP reinforced concrete are reported and discussed.

The analysis of tension concrete members reinforced with commercial Carbon FRP rebars is performed both theoretically and experimentally. The experimental investigation is carried out on cylindrical specimens varying the concrete cover thickness; the theoretical analysis refers to a model founded on the hypothesis of perfect bond between the concrete and the reinforcement. Code relationships are also used in the analysis. A theoretical and experimental comparison has been made; results of the comparison allow to evaluate the tension stiffening effects and to give useful information from a design point of view.

EXPERIMENTAL INVESTIGATION
FRP reinforced concrete members subjected to axial tension have been tested. Specimens have been made with ordinary concrete and reinforced with commercial Carbon FRP rebars; a variation of the confining action is considered, given from the thickness of the concrete embedding the FRP reinforcement.

MATERIALS
The compression strength of concrete has been determined by standard compression tests on cubes 150 mm high; the mean value has been f_c= 46.82 MPa. The average tensile strength, determined by splitting test on cylinders with diameter of 150 mm and height of 300 mm, has been f_{ct}= 4.35 MPa.

The CFRP rebars, produced from the Sireg Co., Itay, have a diameter of 8 mm and are sanded and spiral wounded with carbon fibers.
The mean values of strength and Young modulus, determined by standard tensile tests have been f_r= 2401 MPa and E_r= 109 790 MPa, respectively.

SPECIMENS AND TEST SET-UP
Cylindrical concrete specimens have been made with a CFRP rebar embedded in a concrete block. Geometrical details of specimens are reported in the Table 1.

Table 1. Dimension of specimens

Number of specimens	Specimen No	Height (mm)	Diameter (mm)
2	T40A, T40B	450	40
2	T50A, T50B	450	50
2	T60A, T60B	450	60
2	T80A, T80B	450	80

A tensile force has been applied by means a testing machine, under displacement control. Electrical strain gauges have been glued to the FRP rebar before the casting to measure the strain of the reinforcement during the test. To record the strain evolution within the concrete, electrical strain gauges have been put on the concrete, at the same positions of those glued to the rebar, and on both side of the block in order to evidence any load eccentricity. In addition, two LVDTs have been used to measure the elongation of the rebar embedded within the concrete block and that of the concrete.

EXPERIMENTAL RESULTS
Results of tests, in terms of applied tensile load versus crack width, crack spacing and elongation have been recorded. In the Figure 1, the diagrams tensile applied force versus specimen strain, ε, are drawn together with the curve corresponding to the naked rebar. The strain is determined as ratio between the elongation ΔL and the initial length, L, of the specimen.

Analysing obtained results, it appears as the rebar embedded in the concrete cylinder presents a higher stiffness at all stress levels, with respect to the naked rebar. The tension stiffening effects can be measured as difference between each curve and that corresponding to the naked rebar. The tension stiffening contribution increases with the diameter of the concrete surrounding the rebar. However, the Figure 1a shows that the scatter between curves reduces above a certain amount of reinforcement; relevant difference has been recorded between strain of the T40 and T50 specimens while small differences have been evidenced between T50, T60 and T80 specimens.

On the basis of obtained results, it seems that the effective area of the CFRP rebars, is equal to the area of the T50 specimens that is the area of specimen having a 50 mm of diameter. As a consequence the value c/d= 3 seems to be the optimal value of the concrete cover thickness, c, to the rebar diameter, d, ratio.

THEORETICAL ANALYSIS
Theoretical analysis has been carried out by using two models; an analytical model founded on equilibrium and compatibility conditions and the model adopted in the Eurocode EC2 for steel reinforced concrete members.

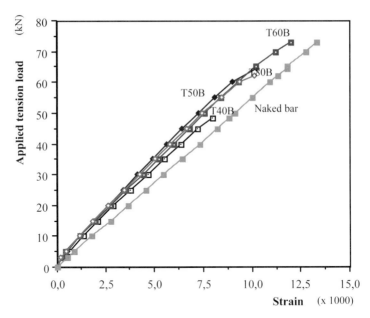

Figure 1. Experimental results

The analytical model
This model is founded on the validity, up to failure, of the Bernouilli hypothesis (that is the section remain plane after deformation) supposing a uniform distribution along the reinforcement of the bond between the concrete and the FRP rebar. Furthermore the constitutive laws of the FRP and the tensile concrete have been assumed as linear with the Young's modulus.

The tensile force corresponding to the first cracking of the specimen is expressed as:

$$F_{cr} = f_{ct} \, A_c \left(1 + n\,\mu\right) \tag{1}$$

being f_{ct} the tensile strength of the concrete, A_c the concrete specimen area, $\mu = A_r/A_c$ the percentage of tensile reinforcement, A_r the FRP reinforcement area and $n = E_r/E_{ct}$ the ratio between the elastic moduli of the FRP and the tensile concrete, respectively.

In the cracking stage, $F > F_{cr}$, the equilibrium conditions involved in the solution of the static problem are:

Equilibrium conditions for cross section

$$F = \sigma_c \, A_c + \sigma_r \, A_r \tag{2}$$

being σ_c *and* σ_r the stress values for the concrete and the reinforcement, respectively,

Axial equilibrium for FRP rebars

$$d\sigma_r(z) = \tau(z)\,p\,dz \qquad (3)$$

being $\tau(z)$ the bond stress; σ_r (z) the tensile stress and p the perimeter of the rebar; z the axis of specimen. Assuming τ (z)= τ_{ad}, we obtain the following expression for the minimum crack spacing, l_{min}

$$l_{min} = \frac{f_{ct}\,A_c}{\tau_{ad}\,p} \qquad (4)$$

and for the mean value of the reinforcement strain

$$\varepsilon_m = \frac{1}{E_r}\left[\frac{F}{A_r} - \frac{2\tau_{ad}}{p\,l_{min}}\right] \qquad (5)$$

THE DESIGN MODEL
According to the Eurocode EC2 [2], for steel reinforced concrete structures, the average strain in the reinforcement is expressed as

$$\varepsilon_{sm} = \frac{\sigma_s}{E_r}\left[1 - \beta_1\,\beta_2\left(\frac{\sigma_{sr}}{\sigma_s}\right)^2\right] \qquad (6)$$

where σ_s is the tension in the rebar, σ_{sr} the tension in the rebar in the cracked section when the first crack occurs (σ_{sr}= F_{cr}/A_r); β_1 is a bond quality coefficient; β_2 is a coefficient dependent on the load duration (β_2 =1 for short term loads). Even if the Equation 6 is validated for steel reinforced concrete structures, it has been used for the analysis of FRP reinforced concrete structures, assuming an adequate value for the bond coefficient β_1.

COMPARISON BETWEEN MODELS
By using the relationships of the above-described models, a comparison has been made with experimental results. The comparison has been made in terms of applied tension load versus strain of specimens.

In the comparison, theoretical predictions furnished by the Equation 5 have been calculated using τ_{ad}=15 MPa; this value has been obtained by pull-out tests carried out on the same CFRP rebars used in this analysis [4].

In the Figure 2 curves for the T40B specimen are drawn together with the curve corresponding to the naked bar. Analysing results it appears as the theoretical model and the Code model predict very well the experimental results.

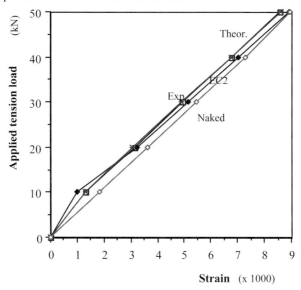

Figure 2. Theoretical-experimental comparison for the T40 specimen

In the Figure 3 curves for the T60 specimen are drawn. The analysis of results shows as theoretical and Code predictions overestimate experimental results for high values of applied forces while they are in good agreement with experimental results for low level of applied forces.

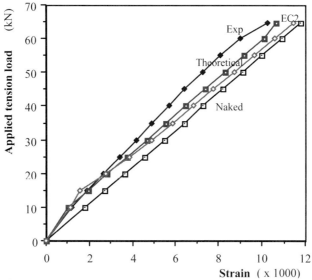

Figure 3. Theoretical-experimental comparison for the T60 specimen

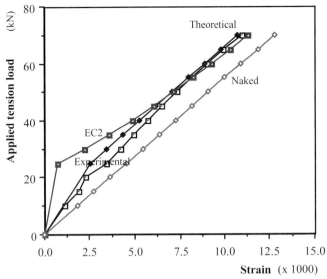

Figure 4. Theoretical-experimental comparison for the T80 specimen

For the T80 specimen results of comparison are drawn in the Figure 4. It is evident as theoretical predictions are in very good agreement with experiments while the Code predictions underestimate experimental values for high level of applied forces.

CONCLUSIONS
Even if the number of experimental tests is limited and, consequently, does not allow to make general conclusions, on the basis of the comparisons between theoretical and experimental results, it is possible to evidence the following:
-    the tension stiffening contribution is very influential and effective in reducing the strain of FRP reinforcing rebars;
-    the ratio concrete cover-to-rebar diameter is the main parameter influencing the tension stiffening;
-    models founded on the Bernouilli hypothesis (plane sections remain plane) are able to predict the behaviour of tension FRP reinforced concrete elements;
-    the Eurocode model, calibrated for the analysis of steel reinforced concrete members can be used for the analysis of FRP reinforced concrete members by introducing an adequate value of bond coefficient.

REFERENCES
[1]   CEB-FIP Model Code 1990. Bulletin d'information n. 203-205, 1991

[2]   Commission of the European Communities. Eurocode n. 2- Design of Concrete Structures, 1994

[3] Aiello M.A. and Ombres L., Load-deflection analysis of FRP reinforced concrete flexural members, Journal of Composites for Constructions, No 4, 2000, pp. 164-171.

[4] Aiello M.A., Leone M. and Pecce M., Experimental analysis on bond between FRP rebars and concrete, Proceedings of the International Conference Composites in Construction-CCC 2001, Porto, October 2001, pp. 199-204

2.3 Guidance on installing, inspecting and monitoring FRP strengthened concrete structures

J L Clarke
The Concrete Society, Crowthorne, UK

INTRODUCTION

There is a growing interest in the use of fibre reinforced polymer composites (commonly known as FRPs), using carbon, glass or aramid fibres, for strengthening concrete structures. Table 1 indicates some applications in the UK.

Table 1. FRP strengthening applications

Structure	Type of strengthening and reason
Buildings	Floor slabs to carry additional imposed loads or following structural alterations
	Columns to carry additional load or to replace missing links.
Bridges	Slabs to carry additional traffic loadings
	Edge of slab in shear
	Arched support (using carbon sheets).
	Columns to improve impact resistance
Industrial buildings	Power station cooling towers to resist wind loading
	Walls to resist accidental loading
	Lighthouse

The Concrete Society carried out a project to develop design guidance in line with British design codes, which was published as Technical Report 55 [1] in 2000.

It was clear from the various discussions that were held during the Project that one of the major concerns of Owners was the long-term monitoring and inspection of the structures once they have been strengthened. There are a number of different aspects that need to be addressed:

- What is the long-term durability of the materials used in the environments likely to be encountered in practice?
- What inspection techniques are available and how suited are they for use on site both during installation and for routine inspection?
- What instrumentation should be installed?
- What testing (i.e. type and frequency) should be carried out on control specimens, both at the time of strengthening and subsequently?

ACIC 2002, Thomas Telford, London, 2002

The Concrete Society is currently carrying out a Project with a view to addressing these issues.

TYPES OF INSPECTION

Inspection should take place at all stages of the strengthening process, as part of the agreed Quality Control procedure. Immediately after installation of the FRP, a full inspection should be undertaken to determine the 'As installed' condition. This will act as a reference against which the results of other inspections should be compared. Depending on the criteria for acceptance, there may be some initial imperfections, such as local delamination. The extent to which imperfections may be acceptable will be very dependent on the type of strengthening and the location in the structure. For example, the consequences of delaminations in the wrapping of a column will be significantly less than the delamination of a plate on the soffit of a beam. The important factor in subsequent inspections will be the identification of new areas of delamination or increases in existing areas.

All strengthened structures should be subjected to routine visual inspections, to check on the condition of the composite, and more detailed inspections as shown in Table 2.

Table 2. Inspection intervals

Structures	Inspection interval	
	Routine	Detailed
Bridges	One year	Six years
Buildings	One year	Change of occupancy or use but not more than 10 years
Others	Depending on use	Depending on use but not more than 10 years

For buildings it will probably only be practical to inspect a portion of the strengthened area. For industrial structures inspections should be carried out when the plant is shut down for operating reasons. For example, the cooling towers of power stations are shut down annually for inspection and servicing. It is recommended that detailed inspection of the FRP is carried out at shorter intervals in the first few years after installation.

DAMAGE TO COMPOSITE MATERIALS IN SERVICE

The most significant risks to the FRP in service are fire and mechanical damage. These should have been considered at the design stage and, where possible, steps taken to minimise the chances of occurrence.

Fire

Fire presents a serious risk to structures strengthened with fibre composites. The fibres themselves are unlikely to be affected by the elevated temperature, but the resin binders on the surface layers will rapidly soften at elevated temperatures, as will the adhesive. Hence composite action between the FRP and the concrete will be lost. Following a fire, a complete reappraisal of the structure will be required. It is likely that all the FRP in the affected area will have to be removed and replaced with new material. Special attention will have to be paid to the condition of the surface of the concrete to which the new FRP will be bonded, as this may have been affected by the fire.

Accidental mechanical damage
The most likely cause of mechanical damage will be impact. Vulnerable locations include strengthened soffits of bridges, which are prone to impact from over-height vehicles. Similar damage may occur to strengthened columns in car parks. A second form of damage is that caused by subsequent work on the structure, for example drilling through the composite to install fixings into the concrete. To avoid this type of damage, suitable plates or other markings warning of the presence of the FRP, should be fixed on or adjacent to the composite.

Anticipated mechanical damage
In some cases the FRP will be covered by a surfacing that has a limited life and hence will be removed at some stage, which may result in damage to the FRP. An example of this is FRP bonded to the top surface of a bridge deck, which is covered by the waterproofing and the blacktop. The latter will be removed after 15 to 20 years. There is obviously a very high risk that the FRP will be damaged by the planing machine. In a building, the FRP may be covered by a decorative coating, that will be removed when the building is refurbished. In both cases the owner must be made aware of the consequences of damage to the FRP and appropriate remedial actions.

INSPECTION OF STRENGTHENED STRUCTURE
Any inspection must also consider the concrete structure itself. Reference should be made to Concrete Society Technical Report 54 [2].

Visual inspection
The nature of FRP materials means that they should need little or no maintenance while in service, provided the composite and adhesive have been correctly specified and installed. However, the fibre composite should be inspected for signs of crazing, cracking or delamination which would indicate some level of overall deterioration. In addition, the composite should be inspected for local damage, for example caused by impact or abrasion. Particular attention should be paid to anchorage regions, locally at cracks, and locations at which the thickness of the adhesive changes, for example due to steps in the concrete or where pultruded laminates overlap.

The nature of many buildings makes visual inspection difficult, with structural elements hidden behind false ceilings and cladding panels. Ceiling voids may be filled with air conditioning ducts, wiring conduits etc. Inspection will be made easier by the provision of readily removable sections of fire protection and cladding. It is important that representative parts of the strengthening are visually inspected and not just those that are easily accessible. Where the composite has been covered with over-coating, for protection or decorative purposes, any inspection of the composite will have to be limited to uncoated control samples.

Identification/warning labels, where fitted, should be checked and missing ones should be replaced. This is particularly important where there is the likelihood of future work that could damage the fibre composite material, such as the installation of fixings for services.

The fibre composite may have been covered by paint or other form of protective layer, e.g. for protection from ultraviolet light, which will have a limited life. It will be necessary to check the condition of this layer and to replace it when required, in accordance with the suppliers recommendations, with a material that is compatible with the fibre composite.

Details of each inspection should be recorded, noting which parts of the FRP have been NOT been inspected as well as those that have. The reasons for not inspecting certain parts should be recorded. It is particular important to record any significant changes since the previous inspection. Ideally these should be in a standard format for easy reference. The information should be checked by a competent engineer who, with reference to the Health and Safety File, can determine what remedial action, if any, should be taken.

In situ tests on FRP
There are only limited methods for testing the FRP once it has been installed. Tapping the structure gently with a light hammer or coin is a simple, established method and relies on a change in sound when different areas of bond quality are tapped. This method is very dependent on the skill of the operator but can be used to locate areas of large adhesive voiding or significant debonding. Other methods being developed include thermography and acoustic emission.

Tests of sacrificial test zones of strengthened structures
It is strongly recommended that additional samples of the fibre composite material should be bonded to the structure away from the region to be strengthened. Alternatively, FRP can be bonded to concrete samples, such as short beams, which can be stored on, or adjacent to, the structure so that they are in the same environment. Samples can be inspected and tested as part of the inspection regime.

Mechanical measurements
The most likely area for degradation will be at the adhesive/concrete interface. Partially cored pull-off tests may be carried out on the additional samples to identify any changes. The removed samples of FRP may be tested further to determine whether there have been any changes in the material.

Instrumentation and structural testing

It may be appropriate in some situations to install instrumentation to monitor the performance of the strengthened structure. The decision whether or not to install instrumentation will depend on the relative costs and benefits. It will probably most appropriate for highly critical structures or ones that are not readily accessible for other forms of inspection. It will obviously also be appropriate when the strengthening is being used in novel applications. Load tests may be appropriate in some cases. However, these should be considered as proof loads; the ability to carry the load does not give any indication of the ultimate capacity of the structure.

Where instrumentation has been installed prior to the structure being strengthened this should be used to check that the change in response of the structure following strengthening is in accordance with predictions. Again the 'As installed' readings from the instrumentation will form the basis for comparison with readings taken at subsequent inspections. However, depending on the type of instrumentation, any change in the results will generally indicate a change in the overall response of the structure. This may not necessarily be due to a change in the FRP.

ROUTINE MAINTENANCE

Moisture is one of the main causes of deterioration of FRP strengthening systems. Hence it is important that all gutters, drains etc be kept clear of debris so that rainwater is carried away. If any cleaning is carried out in the vicinity of the FRP, it will be necessary to check that any solvents used will not cause damage. For some exposed structures, there may be a requirement to clean the surface to remove graffiti or other contaminants. Techniques such as water jetting or grit blasting, while appropriate for concrete surfaces, are not appropriate for FRP as they are likely to cause damage. Steam cleaning should not be used as it is likely to soften the adhesive.

REPAIR

When local areas of damaged composite are identified, they may be repaired by techniques such as resin injection (taking care not to further damage the material) or plate overlapping. When major damage is identified, it may be necessary to remove the defective material over a sufficiently large area such that material on the periphery is fully bonded. The concrete surface should then be prepared again and further FRP installed. It will be necessary to provide an adequate overlap between the new and old material at the periphery of the repaired area. Where these repair techniques are used, it is crucial to check the compatibility of the repair material with the materials already in place. In addition to compatibility, the repair material must have similar characteristics to the material in place. Such characteristics include fibre orientation, volume fraction, strength, stiffness and overall thickness.

SUMMARY

If correctly specified and installed, FRP strengthening systems should be very durable. The technique is highly dependent on the quality of workmanship and so testing must be carried out during installation as well as subsequently. The programme of inspection and monitoring must be agreed with the owner (or occupier) of the structure. It may be necessary to provide 'dummy' specimens for subsequent testing and to make special arrangements made for access to the structure.

REFERENCES
[1] Concrete Society. Design guidance for strengthening concrete structures using fibre composite materials, Technical Report 55, The Society, Crowthorne, 2000, 71 pp.

[2] Concrete Society. Diagnosis of deterioration in concrete structures, Technical Report 54, The Society, Crowthorne, 2000, 72 pp.

2.4 A new prepreg material used in conjunction with concrete to form a structural unit.

J Hulatt, L Hollaway and A Thorne
Department of Civil Engineering, University of Surrey, Guildford, UK

ABSTRACT
There is a rapidly growing interest in the use of advanced polymer composites (APC) in construction. This observation can be illustrated by the introduction of APCs into the fabrication of bridge decks, rebars for reinforcing concrete, externally retrofitting and upgrading reinforced concrete, cast iron and steel systems and the manufacture of all composite structures. One of the major reasons for this interest is the improved durability of composites over that of steel, a reduction of site labour costs and in construction time.

This paper will discuss the use of a new prepreg material (APC) developed and manufactured specifically for the construction industry. It will be used in a novel form of concrete/APC component which utilises the two materials to their best advantage. The illustrative unit is a Tee beam in which compression concrete is placed above the neutral axis and the tensile APC component is positioned below it. The paper will discuss two failure criteria, namely buckling of the webs and shear failure in the bond between the concrete and permanent APC shuttering. The preliminary results indicate that the failure criterion for this system is buckling of the webs followed by concrete crushing and buckling of the permanent GFRP shuttering.

INTRODUCTION
The interest in advanced polymer composites within the construction industry has grown rapidly in recent years. Applications for these types of materials include the manufacture of bridge deck systems [1], the rehabilitation of deteriorating RC, steel and cast iron structures [2], a replacement for steel in reinforced concrete structures [3] and the fabrication of all composite structures [4]. The lightweight APC materials allow a reduction in construction times and site labour costs. The inherent improvement in durability issues, compared with conventional steel, is also regarded as a significant advantage.

Prepreg (or pre-impregnated) materials have been used in many engineering disciplines with great success. However, standard aeronautical APC materials are regarded as being too refined. Consequently, a new heavyweight prepreg material with a larger ply thickness has been developed specifically for the civil engineering industry. This material has been combined with a traditional material, concrete, to form a novel structural unit. The unit, in the form of a T-beam, uses both material types to their maximum efficiency; the concrete is placed in compression and the APC is predominately used in tension. Research as part of an ongoing Brite-Euram* project is reported in this paper. Two different beam types with T cross-sections are introduced; one which is designed to fail by buckling of the webs and the other to fail the shear bond at the concrete and APC interface. Six beams have been manufactured and tested statically in 4-point bending: four of the web buckling design and two of the shear bond design.

The objectives of this paper are to determine the failure mechanisms of the two different beam types and to characterise the new heavyweight prepreg material. The manufacturing technique and testing regimes are discussed. The results obtained experimentally are then compared with a finite element (FE) analysis using PATRAN 2000 [5] as the pre-processor and ABAQUS v5.8 [6] as the analyser and post processor. These analyses also include a buckling prediction methodology incorporating eigenvalue extraction.

MATERIALS USED
The materials used in this study were prepregs of Unidirectional CFRP (xLTM65U/DWO164 /393 g/m^2/UD woven [90% warp/10% weft], 50% fibre volume fraction) and GFRP (xLTM65u/EBX1202 /1202 g/m^2/stitched biaxial, 50% fibre volume fraction) supplied by Advanced Composites Group (ACG). The resin system used in this project was a low temperature curing resin, suitable for civil engineering applications. The carbon and glass heavyweight fabrics developed by ACG and used in the project have a thickness of 0.43 and 0.54 mm respectively. Other prepreg materials used in the aeronautical, automotive etc. industries have typically thinner ply thicknesses (in the order of 0.14 and 0.3 mm for CFRP and GFRP respectively). This increase in the thickness of the ply layers is seen as a benefit in terms of reduced preparation and fabrication times for the beam manufacture. The GFRP used in this study consisted of off axis fibres at "plus" and "minus" 45° which were stitched to form a biaxial fabric. It was essential to produce a balanced lay-up for the +/-45° GFRP to avoid warping after curing. This was achieved by forming a "minus, plus, plus, minus" laminate (the minus or plus indicates the direction of the fibres in the top surface of the ply). This can be seen diagrammatically in Figure 1.

Material Testing
The two material types were characterised by longitudinal tensile testing in accordance with ASTM D3039M [7]. The coupon specimens, including basic dimensions and typical failure modes, are shown in Figure 2. A total of fourteen specimens of UD CFRP and ten specimens of +/-45° GFRP were tested in an Instron testing machine (model no. 1185). On a number of specimens the longitudinal stiffness was calculated by using electrical resistance strain gauges placed at the

* COMPCON project – European Community under the Ind. & Mat.Technologies Prog. (Brite-Euram III)

centre of the specimen. The loads and strains were recorded using a Schlumberger SI3531D data acquisition system.

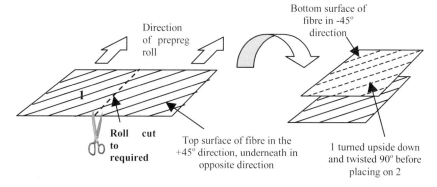

Figure 1: Diagrammatic representation of the lay-up of a +/-45° GFRP laminate.

The coupon test results are shown in Table 1. Two of the fourteen CFRP specimens failed outside the middle third and consequently the values of these results were removed from the average. The standard deviation of the longitudinal failure stress for the CFRP specimens is quite high; the maximum failure stress for one coupon was 912 MPa whilst the minimum recorded value was 696 MPa for another. This has been attributed to a flaw in the laminate that is often encountered in the development of a new prepreg material.

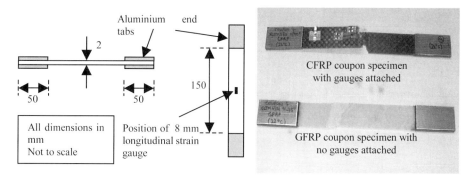

Figure 2: Diagrammatic representation of the tensile coupon test

Table 1: Showing results from material coupon tests

Material	Average Longitudinal Failure Stress (MPa)	Standard Deviation	Average Longitudinal Stiffness (GPa)	Standard Deviation
UD CFRP	809	71.4	86	3.9
+/-45° GFRP	125	4.2	13	0.12

SMALL SCALE BEAM MANUFACTURE

The six 1.5m span T-beams were manufactured using the vacuum bag technique. This method of manufacture has been described in Hulatt *et al* [8] on rectangular cross-sectional hybrid beams. A mould is first prepared from medium density fibreboard (MDF) to the correct dimensional requirements. A self adhesive PTFE sheet is used to cover the MDF tool to ease demoulding. The first layer of +/-45° GFRP (cut to the correct dimensions) is placed onto the mould before a "debulking" procedure is carried out. This is show diagrammatically in Figure 3.

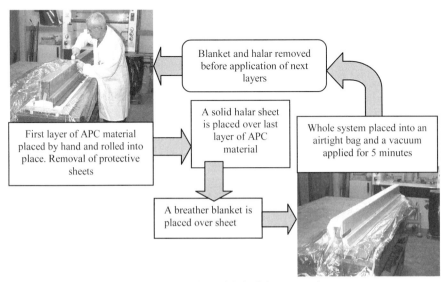

Figure 3: Diagrammatic representation of debulking procedure

Two layers of +/-45° GFRP (total thickness of 1.08 mm) and eight layers of UD CFRP (total thickness of 3.44 mm) were used in the web buckling beam design; the shear bond design had an additional two layers of +/-45° GFRP (total thickness of 2.16 mm). On completion of the beam lay-up, the whole system, still under vacuum, was placed in an oven and cured at a temperature of 65°C for 16 hours.

After demoulding, the beam was trimmed and the web stiffeners were placed in position. In the case of the web buckling beam design, two different stiffening methods were used to prevent buckling. The first method incorporated the utilisation of GFRP diaphragms (for beam numbers 1 and 2). These were made from two plies of GFRP and were bonded in place using a two-part 3M epoxy adhesive. They were placed at 75 mm centres with the exception of one location at which one stiffener was omitted; this position was chosen to induce buckling in this region (see Figure 4). At this location an attempt was made to measure, in three of the four beams tested, the extent and timing of the buckle, utilising strain gauges positioned at +/-45 degrees to the vertical.

For beam numbers 3 and 4, timber stiffeners replaced the GFRP diaphragms. The locations for the stiffeners was the same in all cases with both types of stiffening being bonded in position using the two-part 3M adhesive (see Figure 4). The second method of stiffening the webs was found to be more practical and less time consuming than utilising GFRP diaphragms.

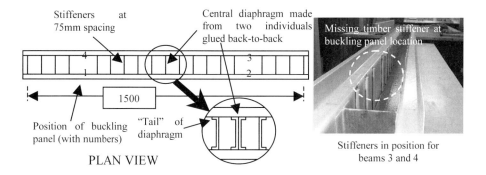

Figure 4: Diagrammatic representation showing position of diaphragms/stiffeners

Beam numbers 5 and 6 (the shear bond design) were stiffened using the original method of GFRP diaphragms. In addition, the missing diaphragms shown in the first four beams were included. This, together with the extra two layers of GFRP applied to the web as discussed in section 3.0, gave a much stiffer section.

A MDF plate was then placed on top of the diaphragms/stiffeners as a support for the concrete. An adhesive, Sikadur 31, was applied to the GFRP vertical and horizontal faces of the permanent shuttering before pouring the fresh concrete; the concrete had a target 28 day compressive strength of 40 N/mm^2.

SMALL SCALE BEAM TESTING
The six beams were tested statically in 4-point bending after the concrete had cured for 28 days. The loads were positioned to give a 300 mm length of pure bending at the centre of the beam with 600 mm shear spans at each end. The beams were instrumented with LVDT's and strain gauges to measure midspan deflection and strain on the concrete, GFRP webs and CFRP flange respectively. Additionally, for three of the four web buckling beams, strain gauges were added at the location of the buckling panel (described previously).

EXPERIMENTAL RESULTS
The results for the experimental beam tests are summarised in Table 2.

Table 2: Static beam test results

Beam No.	Web Buckling Design				Shear Bond	
	1	2	3	4	5	6
Max με (concrete)	-2470	-1950	-1074	-1193	-2514	-1882
Max με (flange)	2882	3336	2522	2273	3807	3150
Max deflection (mm)	12.6	13.2	8.86	8.78	9.86	9.85
Failure Load (kN)	34.9	40.2	28.3	30.3	44.4	42.1

The results of one buckling panel (number 2) for the experimental buckling analysis are shown in Figure 5.

Buckling Panel 2 (0-15 kN)

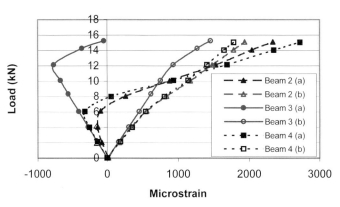

Where a refers to the strain gauge in the "plus" direction and b refers to the strain gauge in the "minus" direction

Figure 5: Plot showing results of gauges on buckling panel 2 for all beams

The results for panel 2 (see Figure 4) show that buckling appears to occur at a load of approximately 6 kN for beam numbers 2 and 4 but beam 3 shows a much higher value of 12 kN for this panel. However, the results from panel 4 for beam number 3 also indicated a buckling failure of 6 kN.

NUMERICAL ANALYSIS

To complement the experimental work, a finite element (FE) package was utilised to predict the behaviour of the beams. This was undertaken by using PATRAN 2000 [5] as a pre-processor, to set up the geometry and material properties, and ABAQUS v5.8 [6] as the analysis tool and post processor. Initially, a linear static analysis, with a serviceability load of 20 kN, was carried out on the three beam types and with the same loading arrangement as the experimental work. The material properties for the analysis were obtained from coupon specimens, as described previously, and from manufacturers data. The results of this analysis are shown in Table 3.

Table 3: Showing results of linear numerical analysis

| Description | Web Buckling Design | | Shear Bond design |
	GFRP	Timber	
Max με (concrete)	-720	-715	-600
Max με (flange)	1616	1472	1440
Max deflection (mm)	5.85	6.10	4.02

Additionally, an eigenvalue extraction method of buckling analysis was also carried out utilising the above FE packages. The results from this analysis of the three types of beams appeared to indicate buckling first occurred at values of 8.9 kN, 8 kN and 39.1 kN for both web buckling beam designs and the shear bond beam respectively.

DISCUSSION
The results of the two analyses for the hybrid beams are plotted in Figures 6, 7 and 8. Figure 6 is a plot of load against deflection and shows the difference in the stiffness for the two beam types. The reduction of the GFRP in the webs of the buckling beam designs combined with the missing diaphragm resulted in a greater deflection. However, with a serviceability limit of 20 kN, the deflections are approximately Span/250 and Span/375 for the Shear bond and Web buckling designs respectively. This is comparable to the maximum allowable serviceability deflection for an RC beam which is Span/250 [9]. The buckling resistance and ultimate failure loads of the hybrid beams are directly related to the thickness of the GFRP webs and the method of stiffening the sections. However, the results indicate that although the onset of buckling can be at relatively small loads, the overall stability and ultimate failure of the beam is at far greater loads.

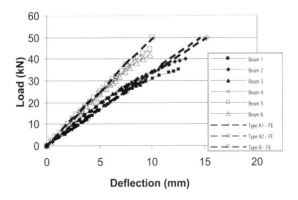

Figure 6: Plot showing load against deflection for the hybrid beams

Figure 7 shows the degree of non-linearity in the concrete. The results of the FE analysis showed a good agreement with experimental work in the linear portion of the plots. However, the results of the highly non-linear concrete indicate that a non-linear FE analysis must be undertaken; this currently is being investigated and the results will be reported later.

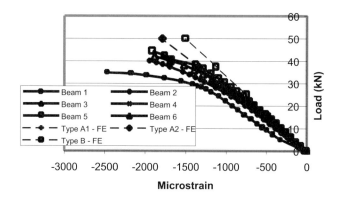

Figure 7: Plot showing load against microstrain on concrete

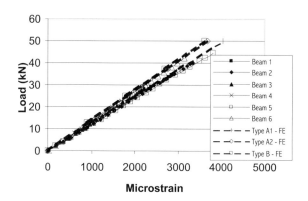

Figure 8: Plot showing load against microstrain on flange

Two different failure mechanisms were identified for each beam type. The web buckling beam design failed by delamination of the GFRP webs at the location of the buckling panel, followed by concrete failure. The shear bond beam design failed in the compressive concrete at the pure bending region with associated buckling of the permanent shuttering. Consequently, the use of the adhesive Sikadur 31 to bond the fresh concrete to the GFRP permanent shuttering was deemed successful in terms of creating the necessary shear transfer between the two materials.

CONCLUSIONS
The investigation concluded that the limiting design for these types of beams was that due to the web buckling; the failure was initiated by a buckling of the GFRP webs followed by concrete crushing. The shear bond design failed at a higher load than that of the former design, by a compressive failure of the concrete with an associated permanent shuttering buckling. Table 4 summarises the differences between experimental and FE analyses with a load of 20 kN. The manufacturing technique was deemed to be successful and the use of Sikadur 31 to bond the fresh concrete to the GFRP was adequate.

ACKNOWLEDGEMENTS
The authors gratefully acknowledge the financial support provided by the European Commission. Thanks are also due to the Partners of COMPCON, Dr H. Garden (Taylor Woodrow), Mr E. Shahidi & Dr T. Cordon (ACG), Ir F. Tolman (HBG), Ir T. Siemes & Ir G. van den Berg (TNO) and Mr F. Hue (Dragados).

Table 4: Comparisons between experimental and numerical analyses (20 kN load) – (for beams 1 to 6)

Description	WEB BUCKLING DESIGN						SHEAR BOND		
	1	2	FE	3	4	FE	5	6	FE
Midspan deflection (mm)	6.06	5.25	5.85	5.08	5.08	6.10	4.04	4.17	4.02
Microstrain on concrete	-922	-758	-720	-700	-719	-715	-639	-649	-600
Microstrain on flange	1676	1688	1616	1567	1483	1472	1720	1490	1440

REFERENCES
[1] Zhao, L., Karbhari, V. M., Seible, F. and Brostrom, M. (2001), Design and Evaluation of Modular Bridge Systems using FRP Composite Materials, Proc. FRPRCS-5, Ed C. Burgoyne, Thomas Telford, Cambridge, pp. 1143-1151.

[2] Garden, H. (2001), The Novel Use of Advanced Composite Materials in Civil Engineering Infrastructure, Proceedings of 2nd International Conference on Piping and Infrastructure, 10th-11th April '01, University of Newcastle.

[3] Alkhrdaji, T. and Nanni, A. (2001), Design, Construction and Field-Testing of an RC Box Culvert Bridge Reinforced with GFRP Bars, Proc. FRPRCS-5, Ed C. Burgoyne, Thomas Telford, Cambridge, pp. 1055-1064.

[4] Foster, D.C., Richards, D. and Bogner, B.R. (2000), Design and installation of a Fiber-reinforced Polymer Composite Bridge, Journal of Composites for Construction, Vol. 4, Part 1, pp 33-37.

[5] MSC/PATRAN (2000), Version 2000, MacNeil-Schwendler Co.

[6] ABAQUS (1998), Version 5.8, Hibbitt, Karlsson &Sorensen Inc, Pawtucket, RI.

[7]   ASTM D3039M (1995), Standard test method for tensile properties of polymer matrix composite materials, Vol. 15.03, pp 99-109, 1995a.

[8]   Hulatt, J., Hollaway, L and Thorne, A. (2000), Characteristics of Composite Concrete Beams, Proc. Bridge Management 4 – Inspection, maintenance, assessment and repair, Thomas Telford, London, pp. 483-491.

[9]   Mosley, W. H., Hulse, R. and Bungey, J. H. (1996), Reinforced Concrete Design to Eurocode 2, MacMillian Press Ltd.

2.5 Effect of flexural cracking on plate end shear stress in FRP-strengthened beams

H C Y Luk and C K Y Leung
Department of Civil Engineering, Hong Kong University of Science and Technology, Clear Water Bay, Hong Kong, China

ABSTRACT
For concrete beams strengthened with bonded FRP plates, failure often occurs as a result of local delamination at the end of the plate. Conventionally, such a failure mode has been ascribed to the presence of stress concentrations at the plate cut-off point, leading to failure of concrete in its vicinity. In the literature, various models have been developed to compute the elastic shear and normal stresses at the plate, and different failure criteria have been proposed. However, the predicted failure loads based on plate end shear and normal stresses are often in poor agreement with experimental data. This is due to the assumption of perfectly elastic behaviour up to the ultimate failure load. In reality, flexural cracking often occurs along the concrete beam, causing significant changes in the stress distribution at the FRP/concrete interface. In this work, experimental and analytical results will be presented to illustrate the change of interfacial stress distributions as loading is increased to induce flexural cracks beyond the end of the plate. Based on observations from the results, we found that the average shear stress over a given distance from the plate end could be a parameter governing delamination failure. An approximate method to calculate such an average stress was proposed. The computed values of critical shear stress were found to be similar for tested beams of various sizes.

INTRODUCTION
The bonding of fiber reinforced plastic (FRP) plates is an effective means for the strengthening of concrete beams. When the strengthened member is loaded in bending, final failure often occurs in a brittle manner through delamination of the plate from the concrete surface. A methodology to predict the delamination load is therefore of great practical significance.

Delamination is believed to be caused by high stresses along the FRP/concrete interface. Specifically, elastic stress analysis [1-3] indicates the presence of high shear and normal stresses at the cut-off point of the bonded plate. To deduce the failure load, some researchers [4, 5] proposed failure criteria in terms of the maximum interfacial shear and normal stresses at the plate end. Others compute the principal stresses in the surrounding concrete from the maximum interfacial stresses, as well as the maximum longitudinal stress at the bottom of the concrete beam. The applied load at delamination failure is then obtained with the use of a given failure envelope for concrete [6]. A review of the various approaches for failure prediction was recently conducted by El-Mihilmy and Tedesco [7]. The predicted failure loads were compared to experimental data, but the agreement was generally unsatisfactory. El-Mihilmy and Tesdesco [7] proposed the use of empirical factors and relations to modify the various stress components. While good agreement was achieved for the chosen set of experimental data, the applicability of the empirical factors to other test data is yet to be demonstrated.

Why is failure load not well correlated to the maximum shear and normal stresses at the concrete/FRP interface? A major reason is that the interfacial stresses are obtained from a purely elastic analysis, assuming no cracking in the concrete up to the failure load. In experiments, however, flexural or shear-flexural cracks can always be observed along the bottom of the beam. Since the first cracking moment of reinforced concrete members is only a fraction of its ultimate moment, flexural cracking may occur in a region very close to the support of the beam. The present investigation will focus on the effect of flexural cracking on the stress distribution along the FRP/concrete interface. Specifically, we will study the case with the plate terminated at a relatively large distance from the support so cracking occurs beyond the plate end before ultimate failure occurs. Through theoretical and experimental studies, it can be shown that the stress distribution predicted from an elastic analysis is very different to the one present at ultimate state. Based on observations from our results, a new delamination failure criterion, in terms of average interfacial shear stress, is proposed. Its applicability is then demonstrated with experimental results from beams of different sizes.

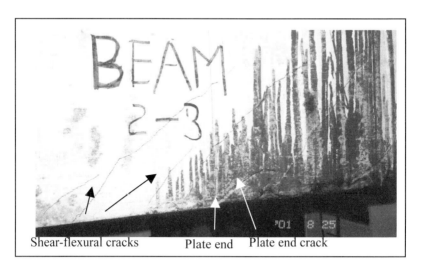

Figure 1 Photo of cracking around the bonded plate end in specimen MB2

EFFECT OF CRACKING ON PLATE END STRESSES
As the loading on a strengthened beam is increased, flexural and shear-flexural cracks can often be observed along the beam well before the ultimate loading is

reached.   The occurrence of cracking around the end of the bonded plate is illustrated in Figure 1. Specifically, due to the presence of stress concentration, a crack is usually found to form right at the plate end. As this crack is formed, the concrete right above the end region of the plate is unloaded, leading to significant changes of the stress distribution. To elaborate on this point, a finite element analysis is carried out.

Finite element simulation

Elastic finite element models are employed to calculate the plate end shear and normal stresses at the concrete-adhesive interface. Two cases, one without cracking and one with a discrete crack extending from the end of FRP plate into the concrete substrate, are analysed using the same mesh refinement.   The distributions of interfacial stresses for the two cases are shown in Figure 2 below.

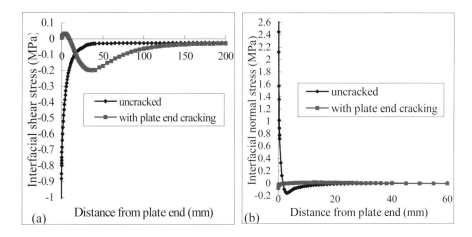

(a)

(b)

Figure 2 Comparison of plate end interfacial stresses between the cases with and without a plate end discrete crack at a specific load by elastic finite element models

Observation

From Figure 2, for the uncracked case, it can be observed that both the maximum shear and normal stresses are located at the end of the bonded plate and their magnitudes are very high. This result, with stress concentration at the cut-off point, is consistent with the linear elastic analytical models developed by Taljsten [3] and Malek *et al* [2] in which no cracking effect has been considered. However, with the consideration of plate end cracking, both the shear and normal stress distributions are significantly changed. The shear stress at the cut-off point is reduced while that within the plate is increased. The reduction in shear stress concentration is accompanied by an enlargement of the shear transfer zone (where there is significant shear stress). The normal stress concentration at the plate end also disappears. Indeed, after cracking, the normal stress along the plate is reduced to very small values. These small normal stresses are unlikely to affect ultimate failure and can be neglected in the delamination failure analysis.

Plate end interfacial shear stress distribution derived from experimental measurements

Table 1. Experimental beam dimensions

Specimen	Beam Length (mm)	Beam Width, B (mm)	Beam Depth, D (mm)	Tension Steel Area (mm^2)	Shear Span, a (mm)
LB1	7200	300	800	2268	2400
LB2	7200	300	800	2268	2400
MB1	3600	150	400	603	1200
MB2	3600	150	400	603	1200
SB1	1800	75	200	157	600
SB2	1800	75	200	157	600
SB3	1800	75	200	157	600

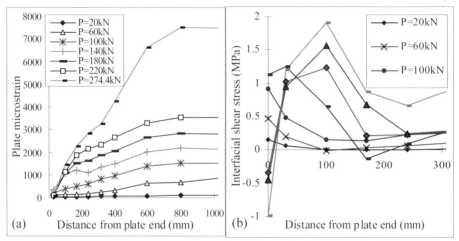

Figure 3 (a) Plate strain variation with applied load along the plate in specimen MB2 at failure end; (b) The corresponding predicted shear stress variation along the plate

We have recently conducted a set of tests with retrofitted beams of various sizes. The notations for the beams and their dimensions are summarized in Table 1. Each beam was strengthened with a carbon FRP plate (E=235 GPa, Tensile strength = 4.2 GPa) the thickness of which was 0.11% of the beam depth. The plate width was taken to be the same as that of the beam. The beams were tested under 4 point loading, and the plate was cut off at a distance of a/3 from the support, with 'a' being the length of the shear span. 19 strain gauges were placed along the FRP plate, with more gauges near the ends to monitor the strain concentration. Typical plots of strain vs distance from plate end are shown in Figure 3a and Figure 4a, for varying loading up to the ultimate failure load. Based on numerical differentiation of the FRP strain with the finite difference method, an approximate shear stress distribution along the plate can be obtained, and the results are shown in Figure 3b and Figure 4b.   While the shear stress values obtained in this manner are not expected to be very accurate, the general trend (increasing or decreasing with distance from the plate end) should be reliable.

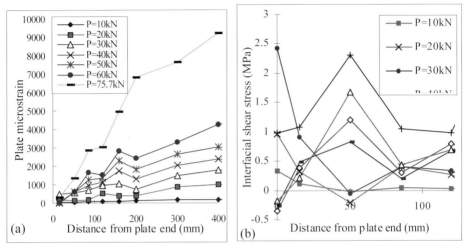

Figure 4   (a) Plate strain variation with applied load along the plate in specimen SB3 at failure end; (b) The corresponding predicted shear stress variation along the plate

Observation
From Figures 3b and 4b, it can be observed that at the initial stage where the loading magnitude is low and the plate end region is under elastic condition, the maximum interfacial shear stress is located at the plate cut-off point. With the increase in loading, cracking occurs at the plate end, causing the shifting of maximum shear stress to the interior of the plate. Also, as in the finite element analysis, the region of relatively high shear stress has grown in size.

DISCUSSION
During the elastic state, the concrete in the immediate vicinity of the FRP plate in a strengthened beam is under a state of combined shear and biaxial stresses including interfacial shear stress, interfacial normal stress and longitudinal tensile stress. To predict the initiation of cracking, the use of these stress components may be appropriate. However, delamination failure often occurs at a load much higher than that corresponding to initiation of the crack. From the results and observation discussed above, it can be concluded that with the formation of the plate end crack, the interfacial normal stress is significantly reduced. Also, cracking will lead to unloading of the longitudinal tensile stress at the bottom of the concrete beam. The only stress component that remains significantly high near the plate end is the interfacial shear stress. After cracking, the shear stress is no longer concentrated at

the cut-off point. Instead, the region of significant shear transfer is found to extend to a distance of roughly D/2 from the plate end, where D is the depth of the beam. (Note: the result for the 800mm beam is not shown, but the transfer zone is also found to be about 400mm in size). To predict delamination failure, it is hence inappropriate to consider the stress at a particular point. A new failure criterion, in terms of the average shear stress in the shear transfer region (which was found to be D/2 in size from our experiments) is therefore proposed. Once this average shear stress reaches a critical value, delamination failure is taken to occur. An approximate method to derive the average shear stress is discussed next.

AN APPROXIMATE METHOD TO CALCULATE THE AVERAGE PLATE END SHEAR STRESS

After cracking occurs at the plate end, the shear stress concentrations calculated from elastic models [1-3] are no longer valid. The exact determination of the post-cracking shear stress distribution requires finite element analysis, and is not appropriate for everyday engineering design. To find an approximate value for the post-cracking average shear stress near the plate end, we propose to follow the analytical approach of Malik and Saadatmanesh [2], which is based on the following governing equations.

$$\frac{d^2 f_p(x)}{dx^2} - \frac{G_a f_p(x)}{t_a t_p E_p} = -\frac{f_c(x) G_a}{t_a t_p E_c} \tag{1}$$

$$\frac{d^2 f_p(x)}{dx^2} - \frac{G_a f_p(x)}{t_a t_p E_p} = \frac{G_a M(x) y}{t_a t_p (EI)_{eff}} \tag{2}$$

Equation (2) is actually derived from equation (1). In the equations, $f_p(x)$ is the FRP stress as a function of distance from the plate end, while $f_c(x)$ is the longitudinal stress in the concrete right next to the interface. In equation (2), $M(x)$ is the moment at a particular section and y is the distance from the neutral axis to the bottom of the concrete beam. $(EI)_{eff}$ is the effective stiffness of the concrete beam. E_p and E_a are the Young's moduli for the FRP plate and the adhesive while t_p and t_a represent plate and adhesive thickness.

As ultimate failure is approached, significant flexural cracking should have occurred along the part of the concrete beam above the FRP plate. $(EI)_{eff}$ should therefore corresponds to the post-cracking value. To find $(EI)_{eff}$, a number of assumptions are made: (1) Plane cross sections before bending remain plane after bending; (2) small deformations; (3) the tensile strength of concrete is neglected once the concrete on the tensile side is cracked; (4) the strength contribution of the adhesive layer is ignored, but its geometrical presence is taken into account; and (5) complete composite action (no slip between composite plate and concrete beam) exists. Theoretically speaking, $(EI)_{eff}$ should be a function of the applied moment at a particular cross section. However, our numerical results show that for most common cases, $(EI)_{eff}$ drops abruptly from the uncracked value as the cracking moment is reached, but stay essentially constant with further increase in the applied moment. We therefore employ a constant value of $(EI)_{eff}$ for the whole cracked region of the beam. Assuming cracking to extend beyond the plate end, equation (2) can be solved directly with the known moment distribution along the beam and the value of $(EI)_{eff}$ for the cracked section. From the solution of equation (2), the stress at any point along the FRP plate can be obtained and the average plate end shear stress, τ_{ave} along a distance, L (=D/2) is given by Equation (3) below:

$$\tau_{ave} = \frac{t_p f_p(L)}{L} \tag{3}$$

Note that the shear stress distribution given by the above analysis will be of the same shape to that of the elastic analysis (which employ the uncracked value of $(EI)_{eff}$), but with higher magnitude everywhere along the plate. This is definitely not the same as the distributions shown in Figures 2, 3b and 4b. However, as shown in these figures, due to the increase in size of the shear transfer region, the average shear stress (within the transfer region) keeps increasing with applied load. The use of a smaller value of $(EI)_{eff}$ also effectively increases the average stress over the shear transfer region. The approximate average stress obtained as described above is not expected to be the same as the actual average shear stress, but if it reaches a consistent value as failure occurs in different specimens, it can be used as a parameter governing delamination failure. This aspect is further studied below.

PREDICTED AVERAGE SHEAR STRESSES AT FAILURE BY THE APPROXIMATE METHOD

Table 2. Average plate end shear stress (MPa) of different specimens over different length, L(mm) at failure by the approximate method

	L =50	L=100	L =150	L=200	L=250	L =300	L=350	L=400
LB1	5.54	2.97	2.09	1.65	1.39	1.22	1.09	1.00
LB2	5.62	3.02	2.13	1.68	1.41	1.24	1.11	1.01
MB1	3.07	1.71	1.26	1.03	0.90	0.81	0.74	0.82
MB2	3.05	1.70	1.25	1.03	0.89	0.80	0.74	0.79
SB1	1.89	1.14	0.89	0.81	1.06	1.26	1.41	1.51
SB2	1.89	1.14	0.89	0.81	1.06	1.25	1.41	1.50
SB3	1.82	1.09	0.86	1.09	1.37	1.58	1.74	1.83

With the use of the approximate method mentioned above and with the maximum failure load from the experiments, average plate end shear stress values over different lengths, L can be predicted in the various specimens. The results are shown in Table 2 above. From Table 1, the sizes of the LB, MB and SB specimens are respectively 800mm, 400mm and 200mm. The average stress should therefore be taken at L=400mm, 200mm and 50mm. Inspection of the values in Table 2 shows that the average shear stress at L =D/2 ranges from 1.03 MPa to 1.14 MPa for all the 7 tested beams. The consistency of this parameter for beams of various sizes show that the average shear stress can potentially be used as a parameter governing delamination failure in retrofitted beams. Further investigations should be carried out to check the applicability of this new delamination prediction approach to a wider set of experimental data.

CONCLUSION

In this paper, the effect of plate end cracking on stresses at the concrete/FRP interface is numerically and experimentally studied. After cracking occurs at the plate end, the stress distribution is found to change significantly from the elastic case. The stress concentration at the cut-off point predicted by elastic analysis no longer exists. The interfacial normal stress drops to very low values while the shear stress is reduced and 'spreads' over a shear transfer region with length about half the depth of the beam. The average shear stress over this region can hence be employed as a parameter governing delamination failure. Using an approximate method, the average shear stress at failure has been calculated for a number of retrofitted beams and the values are found to be very consistent. The results demonstrate the potential of a new approach for delamination prediction in FRP strengthened beams.

REFERENCES

[1] Roberts T.M. (1989). Approximate analysis of shear and normal stress concentrations in the adhesive layer of plated RC-beams. The Struct. Engr., 67 (12), 229-233.

[2] Malek A. M., Saadatmanesh H. and Ehsani M.R. (1998). Prediction of Failure Load of R/C Beams Strengthened with FRP Plate due to Stress Concentration at the Plate End. ACI Structures Journal, 95, 142-152.

[3] Taljsten B.(1997). Strengthening of Beams by Plate Bonding. ASCE J. of Materials in Civil Engineering, 9, 206-212.

[4] Ziraba Y.N., Baluch M.H., Basunbul I.A., Sharif A.M., Azad A.K. and Al-Sulaimani G.J.(1994) Guidelines toward the Design of Reinforced Concrete Beams with External Plates. ACI Structural Journal, 91 (6), 639-646.

[5] Quantrill R.J., Hollaway L.C. and Thorne A.M. (1996). Prediction of the Maximum Plate End Stresses of FRP Strengthened Beams: Part II. Magazine of Concrete Research, 48 (117), 343-351.

[6] Kupfer H.B. and Gerstle K.H.(1973). Behaviour of concrete under biaxial stresses. J. Engrg. Mech. Div., ASCE, 99 (4), 853-866.

[7] EI-Mihilmy M.T. and Tedesco J.W. (2001). Prediction of Anchorage Failure for Reinforced Concrete Beams Strengthened with Fiber-Reforced Polymer Plates. ACI Structural Journal, 98 (3), 301-314.

2.6   Torsional Strengthening of spandrel beams with CFRP laminates

P Salom[1], J Gergely[2] and D T Young
[1]*Zapata Engineering, Charlotte NC, USA*
[2]*UNC Charlotte, Charlotte NC, USA*

ABSTRACT
The present paper describes the experimental and analytical findings of a project focused on the structural strengthening of reinforced concrete (RC) spandrel beams using carbon fiber reinforced polymer (CFRP) composite laminates, and subjected to pure torsion.   Current torsional strengthening and repair methods are time and resource intensive, and quite often very intrusive. The proposed method however, uses composite laminates to increase the torsional capacity of concrete beams.

Six identical spandrel beams were built and tested.   Two of the beams were considered as baseline specimens, the remaining four were strengthened using three different composite laminates. In order to eliminate the flexural and shear forces on the beam, a close to pure twisting force was applied at the end of the beam specimen. In traditional cast-in-place construction the slab adjacent to the spandrel beam does not allow a complete wrap around the beam. To simulate this, the CFRP laminate was provided only on three sides of the beam, and special anchors were used to allow a continuous torsional shear flow around the specimens.   The experiments showed that this method increased the torsional strength of the spandrel beams up to 113%.   Excessive concrete cracking followed by composite delamination caused the specimen failures. An equation was also developed which, for most of the specimens, accurately predicted the composite contribution. This analytical procedure was based on the existing information on the torsional resistance provided by closed stirrups, and on the equations developed to design the composite shear retrofit of RC beams.

INTRODUCTION AND OBJECTIVES
Spandrel beams located at the perimeter of buildings carry loads from slabs, joists and beams from one side of the member only. This loading mechanism generates torsional forces that are transferred from the spandrel beams to the columns. Many existing reinforced concrete (RC) beams have been found to be deficient in torsional shear capacity and in need of strengthening.   In current practice, torsional strengthening of concrete members is achieved by one of the following methods: by increasing the member cross-sectional area combined with adding of transverse reinforcement; by using surface bonded steel plates and pressure grouting the gap

between plate and concrete element; or by applying an axial load to the member by post-tensioning. Although these methods will continue to be used in many more instances, carbon fiber reinforced polymer (CFRP) composites provide another option for strengthening. The present study verifies their effectiveness in increasing the torsional capacity of RC spandrel beams using FRP composites.

In order to investigate the torsional behavior of FRP strengthened RC spandrel beams, six beam specimens were designed, built and tested. The results from the two, unretrofitted baseline specimens (TB1 and TB6) were compared with the four composite retrofitted samples (TB2 through TB5), addressing the following factors: the amount, combination, and fiber orientation of the carbon FRP laminates, and the effects of composite anchors. An analytical procedure was developed to predict the increase in torsional strength provided by the composite laminates for a greater number of combinations of factors [1].

SPECIMEN AND TEST DETAILS
Six identical reinforced concrete spandrel beams were built. The beams were 2438 mm long with an L–shaped cross-section. Figure 1 shows the sample dimensions and the reinforcement details. The (102 mm x 203 mm) flange represented a section of the floor slab at the perimeter of a R/C diaphragm. The compressive strength of the concrete was 55 MPa, and the yield strength of the reinforcing steel was 414 MPa in all specimens. Out of the six specimens, two were not retrofitted with CFRP and served as baseline specimens. Four were retrofitted with a carbon fiber reinforced polymer (CFRP) composite laminate along the length of the beam and around the cross section, excluding the lateral face of the 102 mm flange. In order to simulate the presence of a slab in that location, the FRP was discontinued in this area.

The spandrel beam specimens were detailed following the current design guidelines outlined in the ACI 318-99 [2] code. Each specimen was mounted in the loading frame in such a way that a fixed end, and a torsional pinned end were simulated. A pure torque applied at a 1219 mm distance from the pinned end generated the torsional forces on the beam. The pinned end condition eliminated the bending moment and the shear force from the beam. In order to create a torsional deficiency, the specimens were built with insufficient transverse torsional reinforcement. The specimens were tested with a gradually increased half-cycle load (load in one direction only), with each load step applied three times. The only exception to this rule was the first baseline specimen (TB1), which was loaded with a monotonic force applied in one direction only.

STRENGTHENING DETAILS AND TEST RESULTS
Specimen TB2 was strengthened with a [0/90] laminate, where the [90] lamina was placed perpendicular to the longitudinal axis of the RC spandrel beam. No anchors were used for this specimen. The [0] lamina was provided to delay diagonal cracking, and functioned similarly to the conventional longitudinal torsional bars.

The [90] lamina provided additional torsional strength to the specimen, and performed similarly to the stirrups in a RC spandrel beam.

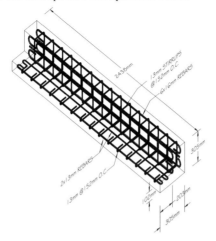

Figure 1. Specimen dimensions and reinforcement detail

Specimen TB3 was strengthened with a [±45] laminate and utilized a special laminate anchoring system. A 51 mm x 51 mm x 5 mm steel angle was used as the inside support for 6 mm diameter stainless steel anchor bolts along the longitudinal axis at the beam at 305 mm on center. A 51 mm x 5 mm steel plate was placed at the bottom of the beam, and connected to the angle by the bolts which passed through the beam. This arrangement provided a good anchoring for the laminate, and delayed a composite delamination initiated by tensile forces present in the inside corner of the spandrel beam. For simplicity, steel angles and plates were used in this project. However, in a real strengthening project stainless steel shapes would have been used to prevent galvanic corrosion in the steel. Specimen TB4 was strengthened in a way similar to specimen TB2, but TB4 used the anchoring system described for TB3. Specimen TB5 was strengthened only by one sheet of composite lamina placed perpendicular to the beam longitudinal axis (i.e. a [90] lamina), and anchored similarly to specimens TB3 and TB4.

Test results
Baseline specimen TB1 reached a maximum torque of 20.3 kN-m and a maximum rotation of 0.125 radians (or 7.2^0). As it was mentioned earlier, this was the only specimen in the experimental phase that was subjected to an increasing pushing force until failure. At the end of the monotonic load, the load was reversed, and a 6.8 kN-m torque was reached. Baseline specimen TB6 reached a maximum torque of 24.4 kN-m at an angle of 0.059 radians, or 3.4^0 (see Figure 2). Failure was determined as the point where the torque could not be sustained by the sample, and it was caused by forming extensive cracks at 45^0 along the longitudinal length of the

beam (see Figure 3).

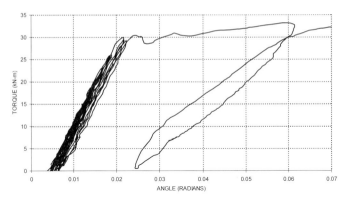

Figure 2. Baseline specimen TB6 results

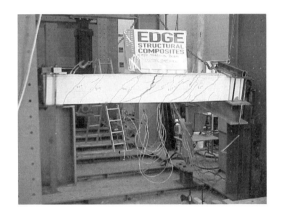

Figure 3. Specimen TB6 at failure

The testing procedure for the CFRP retrofitted beams was identical to the testing of baseline specimen TB6. In addition to the instruments on the steel reinforcement, strain gauges were also attached to the FRP laminate around the perimeter of the cross section, and were oriented in the fiber direction of the outside lamina. Specimen TB2 with the [0/90] composite laminate (no anchors) reached a maximum torque of 32.5 kN-m with a rotation of 0.16 radians. Failure occurred close to the fixed end of the sample due to an extensive torsional crack, which was followed by composite delamination. The peak composite tensile strain was 0.15%, well bellow the composite ultimate strain of 1.30%. It is important to note, that the stress values in the CFRP were negligible up to the load level corresponding to the formation of the first cracks for the baseline specimens. This result proves, that the retrofit material begins working only after sufficient cracking occurred in the concrete member.

Specimen TB3 with the [±45] anchored composite laminate reached a maximum torque of 43.4 kN-m at an angle of 0.11 radians (see Figure 4). The failure mode was very similar to the specimen TB2. The peak tensile strain in the FRP was 0.31%.

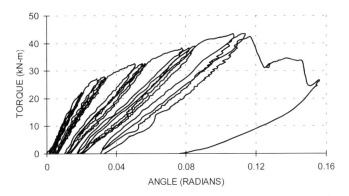

Figure 4. Specimen TB3 results

Table 1 shows the experimental results for all specimens. As it can be seen, the composite strengthened beams reached a significantly higher torsional resistance. As expected, the [±45] laminate was the most effective, followed by the [0/90] laminates with anchors, and finally, the [90] laminate had the least effect on the results. By using anchors, the torsional resistance of the beams was increased by 13% (TB2 vs. TB4), and less damage to the specimen was observed.

Table 1. Experimental Results

Specimen	Composite Laminate	Anchors Used	Maximum Torque (kN-m)
TB1	BASELINE	-	20.3
TB2	[0/90]	NO	32.5
TB3	[±45]	YES	43.4
TB4	[0/90]	YES	35.3
TB5	[90]	YES	33.9
TB6	BASELINE	-	24.4

ANALYTICAL STUDY

In this paper it was assumed that the composite laminates around the perimeter of the beam behave similarly to closed stirrups, where the anchors provided the closing leg on the fourth side. By combining the equation used to design the steel torsional reinforcement (from ACI 318-99) with the formula currently used to estimate the shear capacity of FRP jackets applied to concrete members [3], equation (1) was developed to obtain the torsional contribution from the CFRP laminate.

$$T_f = \frac{2A_o A_f f_{fe}}{s}(\cos\alpha + \sin\alpha) \tag{1}$$

where T_f = torsional strength provided by CFRP laminate (kN-m); $A_f = nw_f t_f$ area of composite laminate (m^2); n = number of composite plies; w_f = width of the composite laminate (m); t_f = thickness of one composite ply (m); A_o = area enclosed by CFRP transverse torsional reinforcement (m^2); $f_{fe} = \varepsilon_{fe}E_f$ effective composite stress (kPa); ε_{fe} = effective composite strain (m/m) (the absolute value was used); E_f = composite modulus of elasticity (kPa); α = angle between fiber orientation and beam longitudinal axis (degrees); s = spacing of composite laminates along the beam longitudinal axis (m).

Using the following CFRP material properties: t_f = 0.00058 m/layer, E_f = 103,862,000 kPa, ε_{fe} = 0.15%, and the specimen dimensions shown in Figure 1, the torsional resistance provided by the composite laminates for specimen TB2 (with a [0/90] anchored laminate) is 11.2 kN-m. This value compares well with the experimental result of 12.2 kN-m. Following the same procedure, the torsional resistance of the other retrofitted specimens can be calculated as well. In general, except for specimen TB5, the experimental data closely matched the analytical results.

CONCLUSIONS

From the analytical and experimental studies presented in the previous sections, the following conclusions can be made:

- The torsional strength of the retrofitted beams exceeded that of the baseline specimens by up to 113%. Although more experimental and analytical work is required in this area, these results proved that composites would increase the torsional strength of RC spandrel beams.

- The addition of anchors increased the composite contribution to the beam torsional resistance by an additional 27%. This increase was due to the shear flow in the anchors, which delayed the composite delamination.

- The 0° ply had a positive influence on the specimen overall behavior. Similar to the longitudinal torsional reinforcement in traditional RC beams, this ply influenced the crack propagation in the concrete, as well as improved its overall stiffness.

- As expected, the anchored [±45] laminate was the most efficient retrofit scheme for the spandrel beams. Although not always practical, the orientation of these laminates allowed the composite material to be stressed in the fiber direction.

- The analytical results were close to the experimental results. The developed equation combined the formulas for torsional steel reinforcement and the procedure used to estimate the shear strength provided by FRP laminates applied to concrete members.

ACKNOWLEDGEMENTS
The authors would like to acknowledge the financial support provided by Edge Structural Composites. The authors are also grateful to Metromont Prestress Company for donating the spandrel beam specimens.

REFERENCES
[1]   Salom, P.R. (2001). "Spandrel beams retrofitted with carbon fiber polymer and subjected to torsion." M.Sc. Thesis, Civil Engineering Department, UNC Charlotte.

[2]   ACI Committee 318 (1999). "Building Code Requirements for Reinforced Concrete (ACI 318-99) and Commentary (ACI 318R-99)." American Concrete Institute, Detroit, MI.

[3]   ACI Committee 440 (2001). "Guide for the Design and Construction of Externally Bonded FRP Systems for Strengthening Concrete Structures." American Concrete Institute, Detroit, MI.

2.7 Ductility and deformability of fibre composites strengthened reinforced concrete beams

D B Tann, R Delpak and E Andreou
University of Glamorgan, UK

ABSTRACT

The brittle failure mode seen in many Fibre Reinforced Polymer (FRP) composites strengthened concrete elements is due to a lack of system ductility, and is a prime concern to the structural engineers. This paper discusses the distinction between deformability and ductility of reinforced concrete (RC) beams strengthened by externally bonded FRP composites. High deformability index does not necessarily lead to good ductility, as very brittle failure modes of such beams have been observed in the experimental studies. An energy-based method was found to be more suitable for quantifying ductility levels of FRP strengthened RC members. It was found that acceptable degree of ductility for FRP strengthened concrete flexural elements could be achieved if the cross sectional area and the properties of the composites were designed at an optimised level.

INTRODUCTION

The word ductility initiates from Latin *ductili*s, which means the ability of metals and alloys to retain strength and freedom from cracks when their shape is altered due to applied stress. Today, this is being interpreted to describe the ability of any material to sustain *inelastic* deformation before fracture. Concrete itself is a brittle material, but it is generally accepted that conventionally reinforced concrete members can attain suitable ductile behaviour by proper design and detailing of steel reinforcement. The yield point of steel is thus treated as an important datum beyond which inelastic deformation of the RC member takes place, thus enabling the full stress and strain capacity of concrete to be developed before ultimate failure.

Since the early 1990s, the viability of using non-metallic FRP reinforcement instead of the conventional steel reinforcing bars has been widely researched [1]. The question of ductility for FRP reinforced concrete elements has been a topic of debate among many researchers [2-4]. Despite these attempts to address the issue, there is, to date, still a distinctive lack of general agreement as to how the ductility characteristics of such elements may be quantified and analysed. As for the ductility of FRP strengthened RC elements, there have been relatively fewer research activities focused on this important area [4, 5-7]. Yet, consideration of structural ductility is of predominant importance to any structural designer, as all appropriately designed structures should attain sufficiently ductile behaviour under ultimate load

ACIC 2002, Thomas Telford, London, 2002

conditions. This is to ensure (a) the redistribution of internal forces in a statically indeterminate structure when part of the structure reaches its ultimate capacity and (b) to provide sufficient warning so as to prevent the structure from sudden and brittle failure at ultimate limit state (ULS).

TRADITIONAL METHODS FOR DUCTILITY CALCULATION

From a structural designer's view point, ductility has been traditionally defined as the ability of a structure to sustain deformation before its failure under ultimate load. Kemp [8] described ductility as the "inelastic rotations through which critically stressed regions of a beam can deform in flexure before a loss of moment capacity occurs". It has been generally accepted that ductility can be measured by a dimensionless factor, often referred to as the ductility index, in several different forms within two broad categories.

Deformation Based Ductility

The ductility index, φ, of a structural element is determined as the ultimate deformation, Δ_u, divided by the corresponding values of deformation at the material yield point, Δ_y.

$$\varphi = \frac{\Delta_u}{\Delta_y} \qquad (1)$$

The yield point, in the case of RC elements, is the point when the steel reinforcement starts yielding for under reinforced sections. The broad term of deformation is a generic description of deflection, rotation, curvature or compressive strain.

Energy based ductility

In this approach, the total strain energy of a deformed element may be divided by the energy value at a reference point that is equivalent to the reinforcement yield stage. The quantity of energy may be obtained by integrating the area under the load deflection curves of the beam considered. This approach is in fact a variant of the deformation based method.

Both the above methods depend on the clear identification of the reinforcement yielding for the evaluation of ductility, and should lead to broadly similar values of ductility index for a given element.

DEFORMABILITY AND DUCTILITY

For RC members strengthened by externally bonded FRP composites, the structural ductility can no longer be expected to be reasonably assessed by any of the above traditional methods. This is due to difficulty in identifying the stage of internal steel reinforcement yield point, since the load-deflection curve of FRP strengthened flexural members may not exhibit apparent change of slope when the steel has yielded. The validity of the above is even more applicable for heavily FRP strengthened members. Shown in Figure 1 is a typical comparison of load-deflection responses of an under-reinforced beam with that of a similar beam strengthened by pultruded Carbon FRP (CFRP) plate.

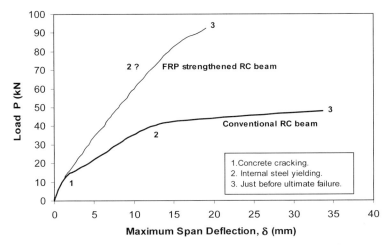

Figure 1: Load-Deflection of Conventional and FRP Strengthened RC Beams

As is illustrated, the conventional RC beam has a clear reduction of slope at the steel yield stage (point 2 on the curve). The FRP strengthened beam shows no such apparent behaviour even though a numerical calculation would indicate that the internal steel has indeed yielded. Such behaviour is due to the presence of FRP composites, which "take over" from the internal reinforcement and sustain the level of stiffness in the strengthened system after the yielding of steel.

Deformability of FRP strengthened elements
An alternative terminology should be used to define the deformation characteristics of FRP strengthened beam. The term "rotational capacity" can be used to distinguish deformation from ductility. However, this term could be misinterpreted as it refers only to the rotation of the beam and not deflection or strain deformation. A more appropriate description is therefore needed for loaded FRP strengthened elements undergoing large deformation but not necessarily displaying high ductility. The authors suggest that the term *deformability index* should be used.

The deformability index, φ_{df}, may be defined as the ratio of ultimate deformation to the deformation at a reference point as follows:

$$\varphi_{df} = \frac{\Delta_{95}}{\Delta_s} \tag{2}$$

In the above equation, Δ_{95} is the maximum generic deformation at 95% of the peak load, which may include deflection, rotation, curvature or compressive strains. The corresponding deformation at the reference point is expressed by Δ_s. The figure of 95% of pre-peak load is suggested here, since it has been observed during the laboratory tests that the ultimate deflection in the beams could be greatly influenced by the load rate and general configuration of the testing system. This may not accurately reflect the ultimate mechanical characteristics of the beam. The ductility index at 100% of ultimate load could be up to as high as 4-5 times that of the same beam at 95% of failure load. As for the reference point for deformation, after careful consideration and comparison with other alternatives, it was found that the deformation corresponding to the service load is best suited to be the denominator for the deformability index calculation. For the special case of equalised dead and live load, W_s is 67% of the design ultimate load.

Ductility for FRP strengthened elements
An energy based method for the determining ductility was first proposed by Naaman and Jeong (1995), expressed by the following equation, where E_{tot} and E_{el} are the total and elastic energy respectively.

$$\varphi_{du} = \frac{1}{2}\left(\frac{E_{tot}}{E_{el}} + 1\right)$$ (3)

For an ideal elastic-plastic material, Equations (3) yields to exactly the same value as Equation (1). For a material such as FRP composites with linear stress-strain responses up to failure, Δ_u equals to Δ_y, and the ductility index becomes unity, indicating that the element possesses virtually no ductility.

Grace *et al* [9] used an "energy ratio" method to quantify ductility of FRP reinforced beams. The energy ratio is defined as the ratio of inelastic energy to the total energy. If the energy ratio is 75% or greater, then the beam will exhibit a ductile failure. Fundamentally, there is no difference to the above procedure, since Grace's method is in effect a reversed form of the Naaman and Jeong approach [3]. If the energy ratio is 75%, than, E_{tot}/E_{el} would be 4 (1/{1-0.75}), and the ductility index as determined by Equation (5) would be 2.5, which naturally is an indication of acceptable ductility.

However, the above methods rely on the clear identification of the elastic potion of the energy, and in practice this is difficult for FRP strengthened members. Naaman and Jeong [3] suggested that the elastic energy could be estimated using an equivalent triangle area under the load-deflection curve. The two initial slops S_1 and S_2 (with the corresponding load of P_1 and P_2) of the load-deflection curve were weighted to define slope S for the equivalent unloading response.

$$S = \frac{P_1 S_1 + (P_2 - P_1)S_2}{P_2}$$ (4)

This empirical approach, although still somewhat arbitrary, leads to reasonable set of results. It is therefore recommended that the Naaman and Jeong's method should be used in determining structural ductility for FRP strengthened RC members.

EXPERIMENTAL VERIFICATION

In order to demonstrate the relationship before ductility and deformability, ten FRP strengthened RC beams, and a control beam, were tested to ultimate failure. All beams were 2.6 m long with an effective span of 2.4 m. The cross section was 100 mm by 200 mm. Two high-yield, 10 mm diameter bars were used as tension reinforcement (f_y = 600 N/mm^2), while 6 mm diameter bars at 100 centres were placed in the shear spans. A four-point load configuration was deployed, with the space between the two point loads being set at 1.25 m. All FRP composites lengths were set at 2.20 m, ie, 100 mm from each support. A summary of the test results is provided in Table 1.

Table 1: Summary of Test Beams and Results

Beam Ref.	Concrete f_{cu} (N/mm^2)	Externally Bonded FRP Composites		Failure at ULS		
		CFRP type[**]	A_p (mm^2)	Δ_u (mm)	Failure Load (kN)	Failure Mode [#]
Control	43.9	n/a	n/a	31.7	48.0	SY-CC
A1[*]	54.9	A	128	18.2	76.2	BTC
A2[*]	44.6	A	128	18.9	73.8	BTC
A3	48.9	A	128	20.9	89.9	BTC
A4[*]	49.6	A	128	19.8	74.4	BTC
A5[*]	49.6	A	128	19.4	82.0	BTC
B1	36.7	D	14.5	34.8	70.0	CC
B3	44.2	D	14.5	23.0	68.4	CC
B4	31.7	C	14.5	35.4	72.5	FR-CC
B5	33.1	D	14.5	32.9	76.2	CC
B6	30.0	C	14.5	34.9	62.5	FR-CC

[*]	These beams were preloaded (cracked) to service load before strengthening.
[#]	Failure mode: SY-CC Steel yielding followed by concrete crushing; CC-Concrete crushing; PTC-Premature tearing off of concrete cover from end of plate to centre of span; FR-CC-Fibre rupture.
[**]	CFRP Type: **A**-1.6 mm thick plate, UTS f_{pu} = 1500 Mpa, Elastic modulus E = 125 GPa; **C**- 0.14 mm thick fabric sheets, f_{pu} = 1750 MPa, E = 125 GPa; **D**-0.14 mm thick fabric sheets, f_{pu} = 4900 MPa, E = 230 GPa.

The maximum span deflections of all beams at 95% and 67% of the ultimate load were obtained from the load-deflection curves. Using equations (2), (3) and (4), the deformability and ductility indices of all B beams are evaluated and listed in Table 2.

It can be seen that for all beams strengthened by the prelaminated CFRP plates (A1-A5), the deformability index ranges from 1.57 to 1.70, with an average value of 1.65. This indicates that these beams can be expected to sustain further deformation of between 57% and 70% of the corresponding values at service load level before ultimate failure takes place. At an average deformability index of 1.65, beams A1 to

A5 failed in the brittle premature tearing-off mode. While group B beams have an average deformability index of 2.11, and these beams failed relatively more mildly even though concrete compression failure occurred in some of them. This indicates that in order to avoid the extremely brittle failure mode, a minimum deformability index of 2.5 is desirable.

Table 2: Comparison of Deformability (φ_{df}) and Ductility Indices (φ_{df}) of All Beams

Beam Ref.	$\delta_{.95}$ (mm)	δ_s (mm)	φ_{df}	E_{tot} (kNm)	E_{el} (kNm)	φ_{du}
Control	26.3	8.6	**3.10**	1.129	0.249	**2.77**
A1	16.8	10.1	**1.66**	0.792	0.552	**1.22**
A2	17.7	10.4	**1.70**	0.786	0.589	**1.17**
A3	18.8	11.2	**1.68**	1.115	0.699	**1.30**
A4	14.9	9.5	**1.57**	0.910	0.557	**1.32**
A5	16.9	10.3	**1.64**	0.921	0.664	**1.19**
B1	31.9	13.9	**2.29**	1.588	0.639	**1.74**
B3	23.0	11.5	**2.00**	0.954	0.581	**1.32**
B4	31.2	15.3	**2.04**	1.778	0.619	**1.93**
B5	27.5	13.4	**2.05**	1.616	0.628	**1.79**
B6	32.4	15.1	**2.15**	1.498	0.509	**1.97**

As for ductility, these results clearly show that the energy based ductility indices reflect the beam behaviour more realistically as these values take account of the unconsumed, elastic stored strain energy at the failure stage. By comparing, it is apparent that ductility index is generally of smaller value than the deformability index. Beams with large deformability index, such as that of beam B3, do not necessarily behave in a ductile manner, as the ductility index could be much lower.

Considering a pure elastic material such as FRP composites which behaves linearly until ultimate failure, the ductility index is unity since the elastic portion of the strain energy equals the total energy at failure. For all group A beams, the ductility index ranges from 1.17 to 1.32 with an average of 1.24. This relatively low figure suggests that brittle failure should be expected, and it was indeed the case that all group A beams failed in a extremely brittle manner.

The single CFRP fabric layer strengthened series B beams have an average ductility index of 1.75, which is 41% higher than the group A average. This confirms the laboratory observations that failure in this group of beams, as expected, had reduced brittleness than that of group A beams. It can be considered as near ductile behaviour and the authors suggest that a ductility index in the order of 2.0 can be accepted as the minimum value for practical strengthening design purposes. Shown in Table 3 is the general comparison of deformability and ductility for the two groups of beams.

Table 3: Comparison of Average Deformability and Ductility Indices

Beam Ref.	φ_{df}	φ_{du}	Failure Mode
Control Beam	3.10	2.77	Ductile
Group A beams	1.65	1.24	Extremely Brittle
Group B beams	2.11	1.75	Near ductile

It is clear that for FRP strengthened beams, whether plated or strengthened by fabric sheets, it is not possible to sustain the ductility level of the original unstrengthened elements. However, suitably designed element can possess acceptable ductile characteristics as demonstrated.

DISCUSSIONS AND CONCLUSIONS
It is important to distinguish between deformability and ductility index. High deformability is a prerequisite, rather than a sufficient condition, for good ductile characteristics. Ductile elements possess high deformability, while members with high deformability alone could still fail in a very brittle manner. Nevertheless, high deformability does provide some warning in that large elastic deformation may be observed before ultimate failure, and in order to achieve good ductility, deformability must first be ensured.

It must also be emphasised that the absolute value of the deformability or ductility index is less significant than the relative comparison with that of an unstrengthened control beam. One argument against using a ductility index for an FRP strengthened beam was presented by DiTommaso et al [10]. They pointed out that it should be solely the amount of inelastic energy that is used to quantify ductility, and the proportion of elastic energy to total energy is not important. In principle this is correct. However, the actual amount of inelastic energy varies in each element with different geometry, reinforcement and load configuration. It would be impractical to compare and determine which element is more ductile without the provision of the ductility indices for all members.

The present proposed concept of deformability index seems to be pertinent for application to FRP strengthened RC beams. The salient merit of this method is its design-based approach and ease of use. The deformability values also clearly indicate the possible further deformation scope after the service load and before the ultimate failure conditions.

Ductility and deformability of FRP strengthened members are influenced by a number of factors. Although confinement of concrete in compression does improve the ductility of FRP strengthened RC beams, the major influencing parameters are the material properties and cross sectional area of the strengthening FRP composite, as well as the properties of the elements to be strengthened. From what had been observed in the laboratory tests, the FRP under-strengthened sections possess better deformability and ductility behaviour than over strengthened beam, over-strengthening should, therefore, be avoided.

REFERENCES

[1]   Clarke, J. L. (Ed.)(1993), Alternative Materials for the Reinforcement and prestressing of Concrete, Blackie A & P, Glasgow, 204pp.

[2]   Saadatmanesh, H. and Malek, A. M. (1998), "Design guidelines for flexural strengthening of RC beams with FRP plates," ASCE J. Composite for Construction, V.2, No.4, pp158-164.

[3]   Naamam, A. E. and Jeong, S. M. (1995), "Structural Ductility of Concrete beams Prestressed with FRP Tendons," Proc. 2nd Int. RILEM Symp. (FRPRCS-2), L. Taerwe, Ed., RLEM, Cachan, France, pp379-384.

[4]   Grace, N. F., Abdel-Sayed, G., Soliman, A. K. and Saleh, K. R. (1999), "Strengthening Reinforced concrete beams using fibre reinforced polymer (FRP) laminates," ACI Structural J., V.96, No.5, pp865-874.

[5]   Razaqpur, G. A. and Ali, M. M. (1996), "Ductility and strength of concrete beams externally reinforced with CFRP sheets," Proc. Int. Conf. Advanced Composite Materials in Bridges and Structures, Ed., El-Badry, M. M., Canadian Society for Civil Engineers, Montreal, Canada, pp505-512.

[6]   Aridome, Y., Kanakubo, T., Furuta, T. and Matsui, M. (1998), Ductility of T-shaped beams strengthened by CFRP sheets," Japan Con. Inst, V20, pp117-124.

[7]   Pisanty, A. and Regan, P. E. (1998), "Ductility requirements for redistribution of moments in reinforced concrete elements and a possible size effect," Materials and Structures, V.31, Oct., pp530-535.

[8]   Kemp, A. R. (1998), "The achievement of ductility in reinforced concrete beams," Magazine of Concrete Research, V.50, No.2, June, pp123-132.

[9]   Grace, N. F., Soliman, A. K., Abdel-Sayed, G. and Saleh, K. R. (1998), "Behaviour and ductility of simple and continuous FRP reinforced beams," ASCE J. Composite for Construction, V.2, No.4, pp186-194.

[10] DiTommaso, A. Focacci, F. and Foraboschi, P. (1996), "Driven failure mechanisms in fibre-reinforced-plastic prestressed concrete beams for ductility requirements," Proc. Advanced Composite Materials in Bridges and Structures, Ed., El-Badry, M. M., Canadian Society for Civil Engineers, pp281-288.

2.8 Debonding in RC cantilever slabs strengthened with FRP strips

J Yao, J G Teng and L Lam
Department of Civil and Structural Engineering,
The Hong Kong Polytechnic University, Hong Kong, China

ABSTRACT

This paper presents the results of an experimental study on the debonding behaviour of reinforced concrete (RC) cantilever slabs strengthened with a fibre reinforced polymer (FRP) strip. The test program consisted of nine model cantilever slabs of different concrete strengths and different amounts of internal steel reinforcement, strengthened with a glass FRP (GFRP) or carbon FRP (CFRP) strip formed in a wet lay-up process, or a pultruded CFRP strip. All test slabs failed by complete debonding of the FRP strip following a process of debonding propagation, with those bonded with a wet lay-up strip showing a more ductile debonding process than those with a pultruded CFRP strip. The stresses in the FRP strips at final debonding obtained from the tests can be closely predicted by the Chen and Teng (2001) bond strength model developed on the basis of simple shear tests with a simple modification.

KEYWORDS

RC cantilever slabs, fibre reinforced polymer, FRP, strengthening, debonding

INTRODUCTION

A number of recent studies were conducted at The Hong Kong Polytechnic University on the strengthening of reinforced concrete (RC) cantilever slabs using fibre reinforced polymer (FRP) strips [1-3]. These studies showed that bonding glass FRP (GFRP) strips to the tension face of the slab, with proper anchorage of the strips at the fixed end of the slab, is very effective in enhancing the load carrying capacity. However, debonding of the FRP strips from the concrete is likely to occur, which starts from the fixed end as a result of the formation of the major flexural crack there and propagates towards the free end. This debonding mode generally limits the strength of the slab unless additional anchorage measures are implemented, and has been the focus of a previous study where GFRP-strengthened RC cantilever slabs were examined in detail [2]. This paper presents the results of further tests in which RC cantilever slabs were bonded with strips of three different FRPs, with the emphasis being on the stress level in the FRP strip at final debonding and the ability of the recent bond strength model of Chen and Teng [4] to predict this stress level. The tests covered different FRP strip sizes, different concrete strengths and different amounts of longitudinal steel reinforcement.

EXPERIMENTAL DETAILS
Test specimens
Nine model cantilever slabs (Figure 1 and Table 1) in three series were prepared and tested in this study. To avoid difficulty associated with the uneven debonding of FRP strips [2], the slab was bonded with only one FRP strip at the mid-width of the slab, with the strip anchored in a precast slot in the supporting wall using epoxy cement mortar (Figure 1). The GS series consisted of two slabs of two different concrete strengths, each bonded with a GFRP strip formed in a wet lay-up process. The CS series also consisted of two slabs of two different concrete strengths, each bonded with a wet lay-up carbon FRP (CFRP) strip. The CP series consisted of five slabs, each bonded with a pultruded CFRP strip. Two parameters were varied in the CP series: the concrete strength and the steel reinforcement ratio. Slabs CP1 and CP2 were reinforced with four longitudinal mild steel bars of 10 mm in diameter, corresponding to a nominal steel ratio of 0.70%, while all other slabs had half this ratio. Within each series, the same FRP strip width and thickness were used, but across the three series different widths and thicknesses were used. The two sizes for the wet lay-up GFRP and CFRP strips respectively were chosen to give almost identical total tensile capacity of the strip. The size of the pultruded CFRP strips was as supplied. Further details of the test specimens are given in Table 1 and Yao *et al* [5].

The properties of concrete and FRP are given in Table 2. The compressive strength f_{cu} and elastic modulus E_c of concrete were determined on the day of testing the slab using cubes and cylinders. The tensile strength f_{frp} and elastic modulus E_{frp} of the wet lay-up FRPs were from tensile tests of flat coupons following ASTM D3039/D3039M-95 [6], and were calculated on the basis of the nominal thickness given in Table 1. The tensile properties of pultruded CFRP strips were supplied by the manufacturer. The ultimate strain of FRP ε_{frp} was obtained by dividing f_{frp} by E_{frp}. The tensile strength and elastic modulus of the longitudinal steel bars from tensile tests were 343 MPa and 208 GPa, respectively.

Loading and instrumentation
The load was applied by a hydraulic jack through a steel rod of 50 mm in diameter at 1000 mm from the wall surface. Eleven strain gauges (gauge length = 10 mm) were installed on the FRP strip along the centre line at 15, 50, 100, 150, 250, 350, 450, 550, 650, 750 and 850 mm from the wall surface respectively. In addition, three strain gauges (gauge length = 60 mm) were installed on the compressive face of concrete, two at 50 mm and one at 350 mm from the wall surface. Deflections at the loading position and at 500 mm from the wall surface were measured by two displacement transducers. Three additional displacement transducers were installed on the supporting wall to record the very small movements of the wall. Only the net deflections at the loading position, found by removing the effects of the small wall movements, are discussed in this paper. An overall view of a test slab under loading is shown in Figure 2.

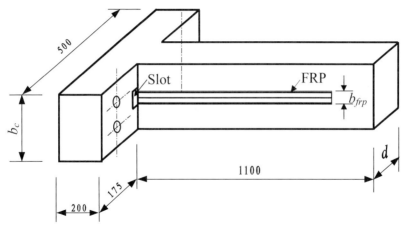

Figure 1 Dimensions of test slabs

Table 1. Details of test slabs

Slab	Dimensions		Longitudinal steel bars			FRP strip		
	Width, b_c (mm)	Depth, d (mm)	No.	ϕ (mm)	Net concrete cover (mm)	Type	Width, b_{frp} (mm)	Nominal thickness t_{frp} (mm)
GS1	302.0	151.2	2	10	28.3	Wet lay-up GFRP	89.7	1.27
GS2	302.7	151.3			33.5			
CS1	303.0	150.8	2	10	30.5	Wet lay-up CFRP	50	0.165
CS2	301.8	150.5			36.5			
CP1	301.5	150.5	4	10	28.1	Pultruded CFRP	50	1.2
CP2	303.6	151.9			35.6			
CP3	302.7	150.0			36.8			
CP4	304.5	150.3	2	10	25.1			
CP5	304.4	149.0			26.5			

Figure 2 Overall view of a test slab during loading

Table 2. Material properties of test slabs

Slab	Concrete		FRP		
	f_{cu} (MPa)	E_c (MPa)	f_{frp} (MPa)	E_{frp} (GPa)	ε_{frp} (%)
GS1	28.2	25.5	269	20.5	1.31
GS2	67.7	34.0			
CS1	26.8	28.2	3720	271	1.37
CS2	76.0	35.5			
CP1	33.8	27.7	2800	165	1.70
CP2	47.1	28.8			
CP3	15.8	20.3			
CP4	57.7	32.9			
CP5	32.0	27.9			

Table 3. Summary of test results

Slab	Peak load and associated maximum strain of the FRP strip		Maximum strain of the FRP strip at final debonding (10^{-6})	Type of debonding
	Load (kN)	FRP strain (10^{-6})		
GS1	10.0	9574	9574	I
GS2	8.95	10667	10667	II
CS1	8.51	11055	11055	I
CS2	8.79	11152	11152	II
CP1	19.95	5420	5420	I
CP2	17.58	4298	4298	I
CP3	13.31	5240	5277	I
CP4	13.54	5425	5425	II
CP5	10.00	3761	3761	I

TEST RESULTS

All slabs experienced FRP debonding that initiated near the fixed end and propagated towards the free end, and eventually failed by the complete debonding of the FRP strip. The peak loads and associated maximum strains of the FRP strip as well as the maximum strains of the FRP strip at final debonding are given in Table 3 for each slab. The load-deflections curves of the slabs are shown in Figure 3.

For the two series GS slabs, the propagation of debonding of the FRP strip from the concrete was a gradual process. As a result, the slabs showed ductile load-deflection responses (Figure 3a). The initiation and propagation of debonding could be easily observed, as the GFRP strip changed from yellow to white when debonding occurred, as can be seen in Figure 4. Similar to the series GS slabs, the two series CS slabs also showed gradual propagation of debonding of the FRP strip from the concrete and ductile load-deflection responses (Figure 3a). Separation of the FRP strip from the concrete along the strip edges provided signs of the initiation and propagation of debonding during the tests, although the CFRP did not show any colour change due to debonding.

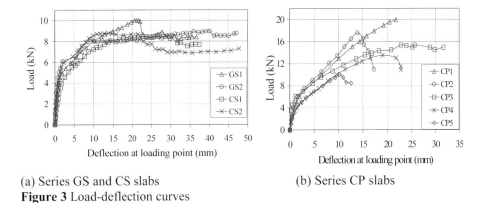

(a) Series GS and CS slabs   (b) Series CP slabs

Figure 3 Load-deflection curves

The CP series included five test slabs. Slabs CP1 and CP2 were reinforced with four longitudinal steel bars of 10 mm in diameter, and slabs CP3, CP4 and CP5 were reinforced with two longitudinal steel bars of the same size. All the five slabs were strengthened with a pultruded CFRP strip of the same size. Unlike the GS and CS slabs, debonding of the FRP strip was a rapid process with little warning. Consequently, they showed brittle load-deflection responses (Figure 3b). The initiation of debonding was difficult to identify as there was no visible change of the test specimen in the initial stage of debonding. The variation of the strain distributions along the CFRP strip of slab CP1 is shown in Figure 5, which provides useful information of the debonding process.

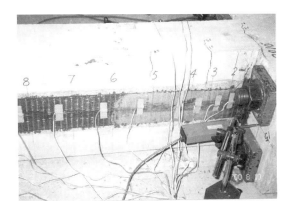

Figure 4 Debonding of GFRP strip

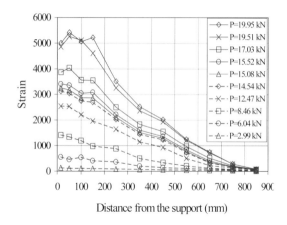

Distance from the support (mm)

Figure 5 FRP strain distribution along the strip on slab CP 1

It should be mentioned that although all the slabs experienced complete debonding of the FRP strip, two types of debonding were observed: (a) debonding in the concrete at 1-5 mm from the adhesive-to-concrete interface (Type I debonding) (Figure 6a) and (b) debonding basically along this interface with little concrete attached to the debonded FRP strip (Type II debonding). Type II debonding occurred in slabs with higher concrete strengths (Table 3). For two otherwise identical slabs, Type II debonding due to a weak interface is expected to occur at a smaller stress level in the FRP. This influence is confirmed in the comparisons given in the next section.

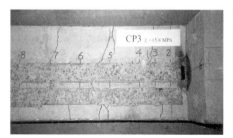
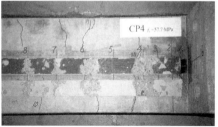

(a) Debonding in the concrete   (b) Debonding along the interface
Figure 6 Two types of debonding failure

COMPARISON WITH CHEN AND TENG'S MODEL

Chen and Teng [3] proposed an equation to predict the stress in the FRP at bond failure based on simple shear tests of FRP-to-concrete bonded joints. The stress state in the FRP strip in the present cantilever slabs is expected to be similar to that in the simple shear tests on which the Chen and Teng equation is based. This equation is given by

$$\frac{\sigma_{frp}}{f_{frp}} = \frac{0.427\beta_{frp}\beta_L}{\varepsilon_{frp}}\sqrt{\frac{\sqrt{f_c'}}{E_{frp}t_{frp}}} \tag{1}$$

where ε_{frp} = rupture strain of FRP, σ_{frp} = stress in FRP at bond failure, β_L = bond length coefficient, β_{frp} = FRP width coefficient, E_{frp} = elastic modulus of FRP, f_{frp} = tensile strength of FRP, f_c' = cylinder compressive strength of concrete, and t_{frp} = thickness of FRP, respectively. β_{frp} is given by

$$\beta_{frp} = \sqrt{\frac{2 - b_{frp}/b_c}{1 + b_{frp}/b_c}} \tag{2}$$

where b_{frp} = width of FRP and b_c = width of concrete slab. β_L is given by

$$\beta_L = \begin{cases} 1 & \\ \sin\dfrac{\pi L}{2L_e} & \end{cases} \quad\text{if}\quad \begin{array}{l} L \geq L_e \\ \\ L < L_e \end{array} \tag{3}$$

where L = bond length of FRP, and L_e = effective bond length of FRP given by

$$L_e = \sqrt{\frac{E_{frp}t_{frp}}{\sqrt{f_c'}}} \tag{4}$$

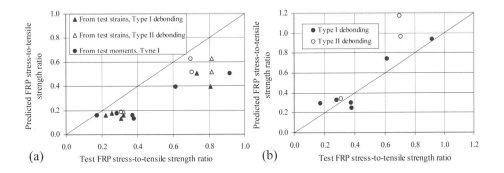

(a)   (b)

Figure 7 Test results versus predictions of Chen and Teng's model

Figure 7 shows a comparison between the predictions of Chen and Teng's model and the test results. The test results were obtained in two different ways: directly based on the strain readings and back-calculated from the ultimate moments of the cantilever slabs through section analysis adopting the plane section assumption. This section analysis is based on BS8110 [7] using the equations given in Teng *et al* [8]. The test results are also differentiated according to the types of debonding (Figure 7). In calculations using Chen and Teng's model, the cylinder compressive strength of concrete f_c' was taken as 0.8 times the cube compressive strength.

It can be seen that the predictions of Chen and Teng's model follow the trend of the test results and are conservative (Figure 7a). In general, the predictions of Chen and Teng's model are less conservative for Type II debonding results, indicating the detrimental effect of an adhesive-to-concrete interface which is weaker than the substrate concrete. By modifying the coefficient in Chen and Teng's model from 0.427 to 0.796 which is the best-fit value for the Type I debonding test results back-calculated from the ultimate test moments, much closer predictions of the Type I debonding test results can be obtained (Figure 7b). The best-fit value of the coefficient for all test results back-calculated from the ultimate test moments is 0.727.

CONCLUSIONS

This paper reported briefly the results of an experimental study on the debonding behaviour of FRP-strengthened RC cantilever slabs. The following conclusions may be drawn based on the results presented here. All the test slabs failed by complete debonding of the FRP strip from the concrete. The slabs strengthened with wet lay-up GFRP and CFRP strips showed a more gradual debonding process and a ductile load-deflection response. The slabs strengthened with pultruded CFRP plates showed a rapid debonding process and a brittle load-deflection response. The maximum stresses in the FRP strips at final debonding from the tests compared reasonably well with Chen and Teng's [4] bond strength model and can be more closely predicted by this model with a modified coefficient. A more detailed description with a more in-depth interpretation of the test results is given in Yao *et al* [5].

ACKNOWLEDGEMENTS

The authors wish to thank The Hong Kong Polytechnic University for its financial support provided through a PhD studentship awarded to the first author and a postdoctoral fellowship awarded to the third author. They also wish to thank Dr S.T. Smith for his value assistance.

REFERENCE

[1]   Teng J.G., Lam L., Chan W. and Wang J. (2000). Retrofitting of Deficient RC Cantilever Slabs Using GFRP Strips. Journal of Composites for Construction, ASCE 4:2, 75-84.

[2]   Teng J.G., Cao S.Y. and Lam L. (2001). Behaviour of RC Cantilever Slabs Bonded with GFRP Strips. Construction and Building Materials 15, 339-349.

[3]   Lam L. and Teng J.G. (2001). Strength of Cantilever RC Slabs Bonded with GFRP Strips. Journal of Composites for Construction, ASCE, 5:4, 221-22.

[4]   Chen, J.F. and Teng, J.G. (2001). Anchorage Strength Models for FRP and Steel Plates Bonded to Concrete. Journal of Structural Engineering, ASCE, 127:7, 784-791.

[5]   Yao, J., Teng, J.G. and Lam, L. (2002). Debonding in FRP-Strengthened RC Cantilever Slabs. To be published.

[6]   ASTM D3039/D3039M-95a (1995) Standard Test Method for Tensile Properties of Polymer Matrix Composite Materials, American Society for Testing and Materials (ASTM), Philadelphia, U.S.A.

[7]   BS 8110 (1997). Structural Use of Concrete, Part 1. Code of Practice for Design and Construction, British Standards Institution, London, U.K.

[8]   Teng, J.G., Chen, J.F., Smith, S.T. and Lam, L. (2002) FRP-Strengthened RC Structures, John Wiley and Sons Ltd, UK, November, 238 pp.

2.9   Anchorage length of near-surface mounted FRP rods in concrete

L De Lorenzis, A Rizzo and A La Tegola
University of Lecce, Italy

ABSTRACT
The use of near-surface mounted (NSM) Fiber Reinforced Polymer (FRP) rods is a promising technology for increasing flexural and shear strength of deficient concrete, masonry and timber members. In order for this strengthening technique to perform effectively, bond between the NSM reinforcement and the substrate material is a critical issue. Aim of this project was to investigate the mechanics of bond between NSM FRP rods and concrete, and to analyze the influence of the most critical parameters on the bond performance. Following up to previous investigations, a different type of specimen was designed in order to obtain a test procedure as efficient and reliable as possible. Among the investigated variables were: type of FRP rod (material and surface pattern), groove-filling material, bonded length, and groove size. Results of the first phase of the project are presented and discussed in this paper.

INTRODUCTION
A new strengthening technique based on fiber-reinforced polymer (FRP) composites has been recently emerging as a valid alternative to externally bonded FRP laminates. It will be referred to as near-surface mounted (NSM) FRP rods. Embedment of the rods is achieved by grooving the surface of the member to be strengthened along the desired direction and to the desired depth and width. The groove is filled half-way with epoxy or cementitious paste, the FRP rod is then placed in the groove and lightly pressed, so forcing the paste to flow around the bar and fill completely the space between the bar and the sides of the groove. The groove is then filled with more paste and the surface is leveled.

The use of NSM FRP rods is an attractive method for increasing flexural and shear strength of deficient reinforced concrete members and masonry walls and, in certain cases, can be more convenient than using FRP laminates. Application of NSM FRP rods does not require surface preparation work (other than grooving) and implies minimal installation time compared to FRP laminates. The use of customized grooving tools can allow technicians to cut the appropriate grooves in one pass, whereas the choice of high-viscosity epoxies as groove-filling material allows to easily gun the material in the groove even when strengthening members for positive moments. Another advantage is the feasibility of anchoring these rods into members adjacent to the one to be strengthened. Furthermore, this technique becomes particularly attractive for strengthening in the negative moment regions of slabs and decks, where external reinforcement would be subjected to mechanical and environmental damage and would require protective cover which could interfere with the presence of floor finishes.

The literature currently available on NSM FRP rods is relatively limited. Beside field applications and experimental field projects, laboratory projects on NSM FRP rods for structural strengthening of concrete members have been carried out during the past three years [1-6].

PREVIOUS BOND TESTS ON NSM FRP RODS IN CONCRETE
Bond of NSM rods in concrete has been investigated in different test series [2-4, 6].

Yan *et al* [3] performed bond tests of NSM sandblasted Carbon FRP (CFRP) rods using a specimen consisting of one 152x152x203-mm concrete block with grooves cut on two opposite faces. Load was applied to the NSM rods while a steel frame restrained the concrete block. Two different failure modes were recorded: the specimens with the two shorter bonded lengths failed by rupture of the concrete at the edge of the block, those with the longest bonded length failed at the rod-epoxy interface (pull-out). The problem of this setup is that a small eccentricity between the grooves (or the placement of the rods) on the two sides of the block can easily induce flexural effects and thus alter the bond behavior and the ultimate load of the specimen. Furthermore, in the pictures shown by the authors it appears that, due to the closeness of the bonded length to the edge of the concrete block, the behavior in the case of shorter bonded lengths resembles that of a fastening anchorage, which is not appropriate when studying bond of reinforcement. This suggests that the distance between the edge of the concrete block and the beginning of the bonded length must be properly chosen in order for "proper" bond failure to occur.

Another series of bond tests was performed by Warren [2]. Concrete blocks (150 by 150 by 300 mm) were cast and slots were saw cut on opposite faces of the blocks. Two blocks in each test setup were aligned in a wood frame and CFRP sandblasted rods were encapsulated in the slots. A hydraulic ram was placed between the blocks to apply a force on a strip midway between the rods so that the four embedded lengths reacted against each other. Two problems are envisioned in this setup. The first one is again eccentricity, both between the grooves on the two blocks at the same side, and between the two sides of the same block. Another problem seems to arise from the specimen size. From the picture shown by the author, it appears that, when failure occurs in the concrete surrounding the groove, the propagation of the cracks is influenced by the limited cross-sectional dimensions of the specimen.

De Lorenzis and Nanni [5] performed beam pull-out tests on NSM CFRP and Glass FRP (GFRP) ribbed and sandblasted rods, varying bonded length and groove size. The specimens were unreinforced concrete beams with an inverted T-shaped cross-section. Each beam had an NSM FRP rod applied to the tension face and oriented along the longitudinal axis of the beam. One side of the beam was the test region, with the NSM FRP rod having a limited bonded length. The rod was fully bonded on the other side of the beam. Such test method allows to avoid all problems of eccentricity arising in the direct pull-out tests. The appropriate choice of the position of the bonded length allows to avoid fastening-type failures and flexural cracking within the bonded length. Nevertheless, it presents some practical limitations, such as that the specimen configuration does not allow to monitor the loaded-end slip; the specimen dimensions, although optimized for a bending-type bond test, are not as small as those of a direct pull-out test; it is not possible to conduct the test in slip-control mode; finally, unless the specimen is tested upside down, it is not possible to visually inspect the behavior of the bonded joint during loading.

OBJECTIVE
In order for the NSM-rods strengthening technique to perform effectively, bond between the NSM reinforcement and the substrate material is a critical issue. Aim of this work was to investigate the mechanics of bond between NSM FRP rods and concrete, and to analyze the influence of the most critical parameters on the bond performance. In particular, the objective of the bond test series was twofold:
- develop a new specimen in which all the limitations encountered in the specimens of the previous series would be reduced or, when possible, eliminated;
- investigate the effect on the bond performance of variables neglected in the previous test series, such as, for example, the type of groove-filling material. Furthermore, after the ribbed and sandblasted rods used for the previous series, another type of CFRP reinforcement was studied, sanded and spirally wound with a carbon fiber ribbon. With this addition, the three most widespread surface configurations of FRP rebars are covered.

EXPERIMENTAL TESTS
The specimen used for this investigation is shown in Figure 1. It consists of a C-shaped concrete block with a square groove in the middle for embedment of the NSM rod. The dimensions of the block were designed, upon a preliminary phase of testing, to ensure that no transverse cracking would occur in the specimen before bond failure of the NSM rod and that, in case of bond failure by concrete cracking, the crack pattern would not be influenced by the specimen width. Dimension s_p in Figure 1 depends on the groove size and is such that the FRP rod, once applied, is situated in the middle of the specimen cross-section. The distance u is such that no fastening-type failure occurs in the concrete above the bonded length. The applied load is reacted by means of four steel threaded rods at the four corners of the block inserted into a stiffened steel plate.

This specimen offers the advantages of the direct pull-out type of test (manageable specimen size, possibility to conduct the test in slip-control mode and to measure both loaded-end and free-end slip, visual access to the test zone during loading). At the same time, it does not need to use two or even four bonded lengths of FRP rod for reaction and thus minimizes the problem of eccentricity, the preparation time and the use of material.

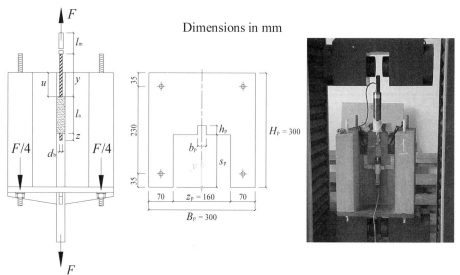

Figure 1. Test Specimen

The test variables were: groove-filling material (epoxy paste and cementitious mortar), bonded length, groove size, and surface configuration of the rod (spirally wound with addition of sand and ribbed). A total of 36 specimens were tested, as shown in Table 1. Of the ribbed type, both CFRP and GFRP rods were tested, whereas the spirally wound rods were all made of CFRP. Apart from the different material properties, CFRP and GFRP ribbed rebars differed for the height of the surface deformations (ribs), more pronounced for the GFRP rods than for the CFRP ones. For the purpose of measuring the bond strength, the actual dimensions of the rod rather than the nominal ones were significant, therefore, the diameter values reported in Table 1 are not the nominal ones. Spirally wound and ribbed rods had a nominal diameter of 7.5 mm and 9.5 mm, respectively. However, the first ones had an actual diameter of approximately 8 mm. For the ribbed rods, a conventional diameter was computed as the average between the maximum and minimum diameters obtained by including and excluding the rib height on both sides of the core. Also reported in the table are the values of the ratio groove-size-to-bar-diameter, which has been termed k. Such parameter (related to the ratio cover-thickness-to-bar-diameter) is expected to have similar significance to the ratio cover-thickness-to-bar-diameter in the context of bond of internal reinforcement in concrete [7].

Prior to starting the bond test series, a detailed material characterization was carried out. The concrete had an average compressive strength of 22 MPa and a tensile strength (calculated as 85% of the experimental splitting strength) of 2 MPa, the epoxy paste had a direct tensile strength of 28 MPa and the cement mortar had a flexural tensile strength of 6.3 MPa. The ribbed GFRP rods had 873 MPa tensile strength and 37.17 GPa Young's modulus, the ribbed CFRP rods had 2014 MPa tensile strength and 109.27 GPa Young's modulus, and the spirally wound CFRP rods had 2214 MPa tensile strength and 174.71 GPa Young's modulus.

Table 1. Test Program

Specimen Code	*Rod Type*	*Filling Material*		b_p (mm)	d_b (mm)	k	l_a (n° of d_b)
SW/k1,25/l04	Sp. Wound	Epoxy	Cem. Mortar	10	8.0	1.25	4
SW/k1.50/l04	Sp. Wound	Epoxy	Cem. Mortar	12	8.0	1.50	4
SW/k2.00/l04	Sp. Wound	Epoxy	Cem. Mortar	16	8.0	2.00	4
SW/k2.50/l04	Sp. Wound	Epoxy	Cem. Mortar	20	8.0	2.50	4
SW/k1.25/l12	Sp. Wound	Epoxy	Cem. Mortar	10	8.0	1.25	12
SW/k1.50/l12	Sp. Wound	Epoxy	Cem. Mortar	12	8.0	1.50	12
SW/k2.00/l12	Sp. Wound	Epoxy	Cem. Mortar	16	8.0	2.00	12
SW/k2.50/l12	Sp. Wound	Epoxy	Cem. Mortar	20	8.0	2.50	12
SW/k1.25/l24	Sp. Wound	Epoxy	Cem. Mortar	10	8.0	1.25	24
SW/k1.50/l24	Sp. Wound	Epoxy	Cem. Mortar	12	8.0	1.50	24
SW/k2.00/l24	Sp. Wound	Epoxy	Cem. Mortar	16	8.0	2.00	24
SW/k2.50/l24	Sp. Wound	Epoxy	Cem. Mortar	20	8.0	2.50	24
RC/k1.24/l04	Ribbed/CFRP	Epoxy	Cem. Mortar	14	11.3	1.24	4
RC/k1.59/l04	Ribbed/CFRP	Epoxy	Cem. Mortar	18	11.3	1.59	4
RC/k2.12/l04	Ribbed/CFRP	Epoxy	Cem. Mortar	24	11.3	2.12	4
RG/k1.27/l04	Ribbed/GFRP	Epoxy	Cem. Mortar	14	11.0	1.27	4
RG/k1.64/l04	Ribbed/GFRP	Epoxy	Cem. Mortar	18	11.0	1.64	4
RG/k2.18/l04	Ribbed/GFRP	Epoxy	Cem. Mortar	24	11.0	2.18	4

For this phase of testing, the grooves had square shape and were pre-formed rather than saw-cut after hardening, therefore, their surface was smooth. In a second phase currently underway, other specimens are being tested in which the grooves have been roughened so that the surface layer of mortar is removed and the aggregate is visible. It could be noted that, when repairing/strengthening a concrete member, it is often required that the geometry of the member be reconstructed with the addition of new concrete, therefore, the grooves might be pre-formed rather than cut even in a real situation. If the bond failure mode was by splitting of the cover and/or concrete cracking or pull-out at the interface between rod and filling material, there would be no difference between the two ways of grooving.

The specimen was instrumented with two LVDTs, to monitor slip of the NSM bar with respect to the concrete at both the loaded end and the free end of the bonded length. Testing was conducted in displacement-control mode on a 200-kN universal testing machine.

EXPERIMENTAL RESULTS

Test results in term of ultimate load and failure mode are reported in Table 2 (where τ_{av1u} = average bond stress at the epoxy – concrete interface at failure; τ_{av2u} = average bond stress at the rod – epoxy interface at failure). Due to lack of space, only results for specimens with epoxy resin have been reported herein. Different failure modes were encountered depending on the test variables: pull-out at the interface between concrete and groove-filling material, splitting of the cover with no concrete cracking, pull-out at the rod-mortar interface, and cracking of the concrete surrounding the groove accompanied by formation of splitting cracks in the epoxy cover (Figure 3).

For specimens with epoxy resin as groove-filling material and spirally wound or ribbed CFRP rods, failure at the epoxy – concrete interface was the critical mechanism in all cases. This was due to the smooth surface of the pre-formed grooves and possibly to the negative effect of the form oil used for the specimens. When this type of failure occurred, the typical bond stress vs. slip curves are shown in Figure 3-a (τ_{av1} = average bond stress at the epoxy – concrete interface; τ_{av2} = average bond stress at the rod – epoxy interface; the loaded-end and free-end slip are both reported but practically coincide in the example shown due to the limited bonded length). They are characterized by a first steep ascending branch followed by a descending branch approaching a constant value of bond stress. The peak corresponds to loss of adhesion between epoxy and concrete, whereas the asymptotic value of bond stress (corresponding to the asymptotic load P_∞ in Table 2) is due to the residual friction between the two materials. The average bond stress at the epoxy-concrete interface corresponding to both ultimate and asymptotic load (that is, *approximately* to adhesion and friction) decreases as bonded length and groove size increase, due to the non-uniform distribution of bond stresses.

Table 2. Test Results for Specimens with Epoxy Paste

Spec. Code	P_{max} [kN]	P_∞ [kN]	τ_{av1u} [MPa]	τ_{av2u} [MPa]	*Failure Mode*
SW/k1.25/l04	9.462	4.790	9.529	13.386	Epoxy-Concr. Interf.
SW/k1.50/l04	11.794	4.360	10.712	16.685	Epoxy-Concr. Interf.
SW/k2.00/l04	12.074	5.500	8.189	17.081	Epoxy-Concr. Interf.
SW/k2.50/l04	13.393	6.600	7.343	18.947	Epoxy-Concr. Interf.
SW/k1.25/l12	25.714	9.700	9.291	12.126	Epoxy-Concr. Interf.
SW/k1.50/l12	26.340	10.200	7.804	12.421	Epoxy-Concr. Interf.
SW/k2.00/l12	23.195	7.730	5.270	10.938	Epoxy-Concr. Interf.
SW/k2.50/l12	32.157	11.800	5.881	15.164	Epoxy-Concr. Interf.
SW/k1.25/l24	36.022	15.700	6.313	8.493	Epoxy-Concr. Interf.
SW/k1.50/l24	49.703	12.600	7.483	11.719	Epoxy-Concr. Interf.
SW/k2.00/l24	49.677	14.701	5.575	11.713	Epoxy-Concr. Interf.
SW/k2.50/l24	58.615	17.700	5.360	13.821	Ep.-Co. Int.+splitting
CR/k1.24/l04	12.460	6.500	7.529	10.958	Epoxy-Concr. Interf.
CR/k1.59/l04	12.834	5.930	6.124	11.287	Epoxy-Concr. Interf.
CR/k2.12/l04	16.625	8.700	6.072	14.621	Epoxy-Concr. Interf.
GR/k1.27/l04	11.221	-	6.726	9.868	Splitting
GR/k1.64/l04	11.414	-	5.340	10.038	Spl. + Concr. Crack.
GR/k2.18/l04	13.067	7.900	4.672	11.492	Epoxy-Concr. Interf.

Specimens with ribbed GFRP rods, having more pronounced ribs with respect to the ribbed CFRP rods, experienced a gradual change in failure mode as the depth of the groove increased, passing from splitting of the epoxy cover to cracking of the concrete surrounding the groove (accompanied by the formation of a longitudinal splitting crack in the epoxy close to the loaded end), to failure at the epoxy-concrete interface. In the latter case, the cover depth offered enough resistance to splitting and the controlling failure mechanism shifted to the epoxy-concrete interface. In the case of splitting failure, the bond stress vs. slip curve has a steep ascending branch and then fails abruptly as the cover splits (Figure 3-b). For specimens with cement mortar, splitting of the cover was more frequent than for epoxy, due to the lower tensile strength of the material. Also failure at the mortar-rod interface was experienced in some cases for spirally wound rebars. However, the ultimate load of these specimens was in all cases lower than that of epoxy-filled specimens.

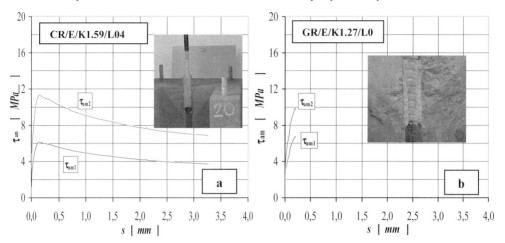

Figure 3. Typical Average Bond Stress vs. Slip Curves for Specimens Failed at the Epoxy – Concrete Interface (a) or by Splitting and Concrete Cracking (b).

CONCLUSIONS

A new specimen was developed for the study of bond between NSM FRP reinforcement and concrete. This specimen offers the advantages of the direct pull-out type of test while minimizing the problem of eccentricity, the preparation time and the use of material. A first series of 36 specimens was tested to investigate the effect on the bond performance of groove-filling material, bonded length, groove size, and surface configuration of the rod. For specimens with epoxy resin and spirally wound or ribbed CFRP rods, failure at the epoxy–concrete interface was the critical mechanism in all cases, due to the smooth surface of the grooves. The average bond stress at the epoxy-concrete interface corresponding to both ultimate and asymptotic load (that is, *approximately* to adhesion and friction) decreased as bonded length and groove size increased, due to the non-uniform distribution of bond stresses. Specimens with ribbed GFRP rods experienced a gradual change in failure mode as the depth of the groove increased, passing from splitting of the epoxy cover to cracking of the concrete surrounding the groove to failure at the epoxy-concrete interface. For specimens with cement mortar, splitting of the cover was more frequent than for epoxy, due to the lower tensile strength of the material. However, the ultimate load of these specimens was in all cases lower than that of epoxy-filled specimens. Further tests with roughened grooves and analytical modeling are currently underway.

ACKNOWLEDGEMENTS
The authors wish to acknowledge MAC Italia S.p.A. for supplying the materials used in this experimental program and FICES S.p.A for casting the concrete specimens free of charge. This investigation was supported by the Italian Ministry of Research (MURST) under program COFIN2000.

REFERENCES
[1]   Warren, G.E. (1998), Waterfront Repair and Upgrade, Advanced Technology Demonstration Site No. 2: Pier 12, NAVSTA San Diego, Site Specific Report SSR-2419-SHR, NFESC, Port Hueneme, CA.

[2]   Warren, G.E. (2000), Waterfront Repair and Upgrade, Advanced Technology Demonstration Site No. 3: NAVSTA Bravo 25, Pearl Harbour, Site Specific Report SSR-2567-SHR, NFESC, Port Hueneme, CA.

[3]   Yan, X.; Miller, B.; Nanni, A.; and Bakis, C.E. (1999), "Characterization of CFRP Bars Used as Near-Surface Mounted Reinforcement", Proc. 8th International Structural Faults and Repair Conference, M.C. Forde, Ed., Engineering Technics Press, Edinburgh, Scotland, 1999, 10 pp., CD-ROM version.

[4]   De Lorenzis, L., and Nanni, A. (2001), "Shear Strengthening of RC Beams with Near Surface Mounted FRP Rods," ACI Structural Journal, Vol. 98, No. 1, January 2001.

[5]   De Lorenzis, L., and Nanni, A. (2001), "Characterization of FRP Rods as Near-Surface Mounted Reinforcement", ASCE Journal of Composites for Construction, Vol. 5, No. 2, May 2001, pp. 114-121.

[6]   De Lorenzis, L., Nanni, A. (2001), "Bond of Near-Surface Mounted FRP Rods to Concrete", ACI Structural Journal, in press.

[7]   Tepfers, R. (1973), A Theory of Bond Applied to Overlapped Tensile Reinforcement Splices for Deformed Bars, Publication 73:2, Division of Concrete Structures, Chalmers University of Technology, Gothenburg, Sweden, 328 pp.

2.10 Behaviour of FRP strengthened flexural RC elements subjected to degradation influences

R Delpak, D B Tann, E Andreou, M Abuwarda and D K Pugh
University of Glamorgan, UK

ABSTRACT
A series of experiments have been conducted to evaluate the performance of flexural RC elements strengthened by FRPs and the impact of certain degradation issues on the performance of such members. The study examines; (a) imperfections within the FRP/concrete bondline at the strengthening stage, (b) the influence of cyclic loading on FRP strengthened members and, (c) the strengthening of degraded beams due to environmental attacks.

The experimental results showed that; (i) the introduction of air-voids between the interface of the composite and concrete bond-line, results in reductions in the ultimate capacities of the retrofitted beams when compared to perfectly bonded FRP beams, (ii) cyclic loading is seen to mitigate the advantages of the external FRP strengthening and the improvements gained in ductility performance of the beams, since the deformations are not as high as in the case of statically loaded elements, (iii) the beam element exposed to an aggressive acidic environment appeared to exhibit an increase in ultimate capacity (79%), when compared to the control element.

In conclusion, these experimental studies show that the ductility of RC members was significantly enhanced by the use of FRPs. It was also shown that even in the case of structures subjected to an acidic environment, the FRP strengthening technique is effective and that strengthened members can reach high ultimate capacities.

KEYWORDS
Fibre Reinforced Polymers; Degraded RC members; Acid and mechanical degradation; Flexure strength; Strengthening; Cyclic loading.

INTRODUCTION
For over a decade, Fibre Reinforced Polymers (FRPs) have been applied to structural strengthening and have attracted considerable attention. The most significant advantage of an FRP strengthened system is that it consists of a non-corrosive combination of materials, so that it provides a more durable solution than strengthening or upgrading by steel. Although the advanced composites are relatively costly, in many situations FRP lamination is the most cost-effective

solution to a strengthening problem since labour and whole life costs can be dramatically reduced.

This paper, which has a practical focus, is the result of controlled experimental investigations designed to address the issues regarding the use of FRPs for strengthening flexural members [1]. In addition, there are some practical aspects and design considerations for which only a limited or no study has been reported. These performance aspects relate to the following impairments:

- Imperfections within the FRP/concrete bondline due to air-voids (blisters).
- The degrading influence of cyclic loading on FRP strengthened members.
- Strengthening of sections that have been exposed to aggressive environment, such as acidic attack due to the burning of fossil fuels.

EXPERIMENTAL PROGRAMME

Experiments consisted of load testing twelve beams to failure, all of which were cast at the University of Glamorgan Structural Laboratories. The specimens were separated into four groups according to the intended function of each specimen.

The control group was used as "datum" beams. Four beams were tested for this purpose; two in pristine condition, which were cured and tested in normal conditions and the remainder, were identically strengthened with two layers of Aramid (AFRP) fabric laminates.

The first group was used to assess the influence of bond imperfections (Air-Voids/"Blisters") on strengthened beam performance. Of the two beams tested, both were strengthened with two layers of AFRP. Blisters consisting of 2.04% and 5.29% (of bonded area) were formed between the concrete tension faces and the FRP in order to represent the imperfection in the bond line. These blisters were created using 1mm deep rings, cut from plastic pipes. This was regarded as a satisfactory simulation, since the diameter of trapped air could be controlled accurately.

The second group was used to assess the influence of cyclic loading on strengthened beams. Four beams were tested under repeated loading, two of which were identically strengthened with two layers of AFRP and the remaining were used as control. A sinusoidal cyclic loading regime of 3Hz frequency was applied for 15 minutes at selected levels of loading. This frequency was chosen in accordance with BS 5400, C.P. for bridge design. Cyclic loading was applied at 2kN, 6kN, 10kN and every 10kN interval until failure. The loading ranged between ±10% of the average jack force level. At higher loads the P-to-P load was smoothly increased.

The third group was tested to assess the strengthening of degraded beams due to environmental attacks. Two degraded beams were tested, which had been submerged in a sulphuric acid (of pH 2) for 19 months and then ambient cured for 15 months. One of the beams was tested as control, without external strengthening and the other strengthened with two layers of AFRP.

(a) Beam dimensions - The element length and cross-section were specified to be sufficiently large to avoid scale effects and to be of "handleable" dimensions in order to avoid logistic difficulties and wastage of materials. It was decided that 100mm width × 200mm depth and overall length of 2600mm were of appropriate proportions. Specimens Dimension is shown in Figure 1 (a).

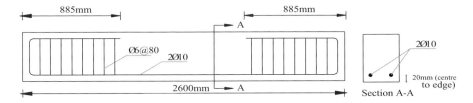

Figure 1 (a): Steel reinforcement and beam sectional detail.

(b) Concrete - The concrete, from which the twelve beams were made, contained Ordinary Portland Cement with the mix proportions of water/cement ratio (by weight) /coarse aggregate/fine aggregate = 0.5:1:2:4. The concrete mix proportions were chosen from the industry-guided mix after the mix design exercises had been carried out in the early stage of the project. This design was chosen since it was required to model the concrete strength as available in 1950s and 60s. No admixtures were used and slump was aimed to be between 30-50mm. Cubes and cylinders were available for testing at the time of each beam load-test, see Table 1.

(c) Internal Steel Reinforcement - Two T10 mm bars were used as the main tension reinforcement and 18R6 mm links at 80 mm c/c were provided for shear strengthening. The steel area ratio (ρ) for tension reinforcement was 0.79%, which is regarded as under-reinforced, see Figure 1 (a).

(d) Kevlar® Aramid fibre - The Fibre Reinforced Polymer (FRP) fabric used for external strengthening of the RC elements, was Kevlar® AK 60-10. The main characteristics of Kevlar® 49, as given in Table 2. Two layers of Kevlar® fabric, 100mm wide and 2.2 m long, were required to be impregnated, so that they could be "over lap" bonded. These fabrics were applied to the concrete tension surface and the bonding pressure was applied by small rollers to achieve satisfactory bonding between AFRP and the concrete substrate. The required number of FRP laminates used were based on the previous FRP-layer/optimisation study; see Andreou et al [2-4].

(e) Polymer Resins - Two types of polymers resins were used, which were Primer and Adhesive (saturant) resins of two-part cold epoxy variety. These had good, (a) wetting and lamination characteristics, (b) physical, (c) adhesive and, (d) thermal and chemical resistance properties. The technical properties of both resins are given in Andreou et al [2].

Table 1: Mechanical properties of concrete

Category		f_{cu} (28) N/mm^2	E_{comp} (28) N/mm^2	f_{cu} (test) N/mm^2	E_{comp} (test) N/mm^2	Age at testing (weeks)
Control Beams	Control (pristine)	44	36000	NA	NA	16
	AFRP (optimum)	51.5	39500	NA	NA	13
Air voids	AFRP (2.04%)	48.5	35000	NA	NA	NA
	AFRP (5.29%)			NA	NA	NA
Cyclic Load	Control (cyclic)	47.5	37500	36.75	31500	85
	AFRP (cyclic)	NA	NA	38	36000	85
Acid Exposed	Control (degraded)	NA	NA	41.5	58000	145
	AFRP (degraded)	NA	NA			145

Table 2:   Characteristics of Kevlar® Laminates and design values

Characteristics	Value
Density	1.44 (g/cm^3)
Typical weight	415 (g/m^2)
Design tensile strength	2100 (N/mm^2)
Design tensile modulus	124 (kN/mm^2)
Elongation at brake	2.4 %
Design thickness	0.29 (mm)
Design load capacity /ply	60 (ton f/m)
Tested load capacity / ply	75 (ton f/m)
Design stiffness / ply	36 (kN/mm)

(f) Instrumentation and Load Tests - All beams were tested under a four-point loading configuration up to failure. The mechanical span of the beams was 2.4m and the point loads were applied centrally on the beam spaced at 1.26m.

Six rows and three columns of Demec discs were bonded on each vertical face of the beam in the pure bending zone. Also Linear Variable Displacement Transducers (LVDTs) were placed at ten locations along the beam soffit and the deflection readings under each load increment were electronically recorded, Figure 1 (b). The instrumentation enabled the determination of moment-curvature (MC) relationships both from strain and displacement variations, Andreou et al [2].

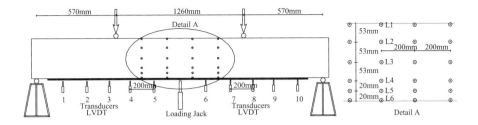

Figure 1 (b) Location of deflection transducer and spacing of Demec gauges

RESULTS AND DISCUSSIONS
The outcome from the tests on FRP strengthened beams, all with optimum composite content, is given in Table 3.

Table 3: Summary of beam tests for performance optimisation-mean values

Category		Beam No.	P_{ult} kN	δ_{ult} mm	P_{serv} kN	M_{ult} (ultimate moment) kNm	C_{ult} (ultimate curvature) 1/mm
Control Beams	Control (pristine)	106, 107	53.5	27	26	15.4	61.1×10^{-6}
	AFRP (optimum)	108, 109	86.5	39	35	24.9	70.7×10^{-6}
(i)-Air voids	AFRP (2.04%)	152	81.5	40	40	23.4	60.2×10^{-6}
	AFRP (5.29%)	153	80	35	40	23	55.0×10^{-6}
(ii)-Cyclic Load	Control (cyclic)	145, 143	53	22	28	15.2	22.0×10^{-6}
	AFRP (cyclic)	142, 144	76	25	33	21.8	23.0×10^{-6}
(iii)-Acid Exposed	Control (degraded)	154	52.5	34	34	15.1	48.0×10^{-6}
	AFRP (degraded)	155	93.5	40	40	26.9	96.0×10^{-6}

(i) Air-Voids (Blisters) group - Table 3 data show a drop in ultimate capacity of blistered beams of about 7% compared to the perfectly bonded FRP beams 108 and 109. It also shows a 1.5kN difference between the failure loads of the two-blistered beams (2.04% and 5.29%). Figure 2 (a), (b) and (c) display the load-deflection graphs for both blistered beams. Figure 2 (d) clearly shows the reductions in moment and curvature due to the concrete/FRP bond imperfections.

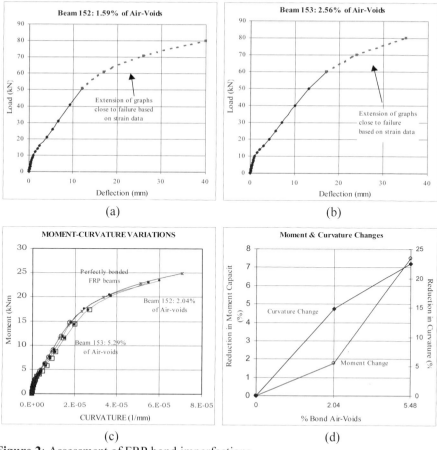

Figure 2: Assessment of FRP bond imperfections

(ii) Cyclic Loading - Table 3 shows the satisfactory performance of control beams with monotonic static loading. In comparison elements subjected to cyclic loading are adversely influenced deformation-wise, without any losses in strength. Aramid strengthened beams, in both cases, have an improved performance over the control, but the beams loaded cyclically have a decrease in strength of about 12% when compared to the statically loaded Aramid members. All significant loss of performance is in deformations and curvatures.

The MC curves of Figure 3, confirm the conclusions drawn from examining Table 3. The drop in ductility is also obvious in the case of strengthened beams, while the improvement in performance is not as satisfactory as in the case of statically loaded beams.

(iii) Beams subjected to acidic environment - The results in Table 1 indicate that the concrete was of an increased stiffness where the concrete cube strength is about

48.5N/mm^2. The results in Table 3 showed an increase in load carrying capacity of the strengthened elements, while the FRP strengthened pristine and degraded beams had strength of about 62 % and 79% respectively, both higher than control beam.

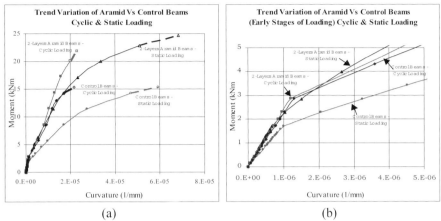

(a)   (b)

Figure 3: MCs of cyclically and static loaded beams

Figure 4 presents the MC graph for this condition. It shows an improvement in the elastic behaviour for both strengthened and control degraded beams compared to the air cured elements. This suggests that the concrete stiffness has developed in the case of three-and-a-half-year old beams. The results so far are intriguing. If acid soaking embrittles the concrete fabric in short-term, then it may well be an explanation of the loaded performance. Verification of the influence of the acidic environment on beam and steel reinforcement physical condition were carried out after testing which showed no corrosion of reinforcing steel.

CONCLUSION
The present study has been based on testing twelve reinforced concrete beams to failure. The research carried out has been focused on identifying the basic issues regarding the use of FRPs for strengthening flexural members. The study has investigated (a) imperfections within the FRP/concrete bondline at the strengthening stage, (b) the influence of cyclic loading on FRP strengthened members and, (c) the strengthening of degraded beams due to environmental attack.

The conclusions from the recorded result, analyses and observations are as follow:
o   *Air-Voids/"Blisters".* The introduction of air-voids between the interface of FRP and concrete bond-line results in reductions of up to 8% of the ultimate capacities of the retrofitted beams, when compared to perfectly bonded FRP beams. There appears to be a variation linking the extent of blistering to reduction in beam performance.

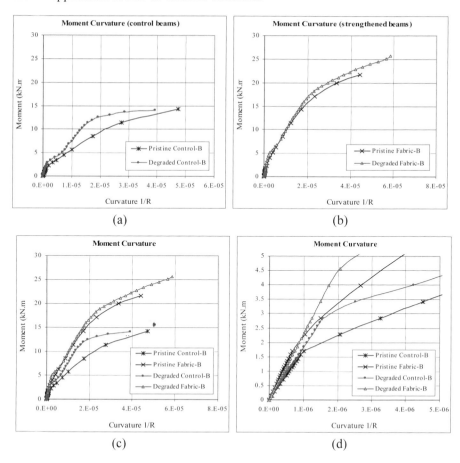

Figure 4: Moment Curvature diagrams

o   *Cyclic Loading.* Strengthened beams have increased ultimate capacities and delayed crack formation, but they also exhibit reduced deflections at failure. The cyclic loading is seen to mitigate the advantages of the external FRP strengthening and the improvements gained in ductility performance of the beams, since these are not as high as in the case of statically loaded elements.

o   *Beam subject to acidic environment.* Strengthening with Kevlar® aramid appears to result in increased stiffness of RC flexural members. Even in cases where the structure has been exposed to aggressive environments, the FRP strengthening method seems to be an effective way of increasing the structural performance. This manner of simulating the acidic attack does not appear to show any concrete degradation.

The results of this report have shown that even in the case of partially degraded structures, the FRP strengthening technique is effective and that strengthened members can reach high operating ultimate capacities.

ACKNOWLEDGEMENTS

The authors wish to acknowledge the following:

o   Du Pont SA (Switzerland) for providing the main funding and resources.
o   MBT (UK) for help, advice and very generous contributions in kind.
o   Mr S. Chevalier in allowing citation of some of his project data.
o   Finally, University of Glamorgan by securing up-to-date logging, monitoring and data retrieval facilities, without which this work would have been unfulfilled.

REFERENCES

[1]   S Chevalier, (2001). Influence of Air Bubbles on Composite Strengthening, BEng (Hons., Civil Eng.) Project, Division of Civil Engineering, School of Technology, May 2001, University of Glamorgan, Pontypridd, Wales, UK

[2]   E Andreou, R Delpak, et.al. (2000a). Performance Evaluation of RC Structural Elements, Strengthened by Advanced Composites. Transfer Report, University of Glamorgan, Pontypridd.

[3]   E Andreou, R Delpak, et.al. (2000b). Report on: Full size laboratory Tests on RC Beams Strengthened With Kevlar and analytic Modelling of the Full Loaded Characteristics – Paper I: Experimentation and Calibration, The Institution of Structural Engineers, University of Cardiff, Sept 2000.

[4]   E Andreou, R Delpak, et.al. (2000c). The Application of Composite based on Kevlar for the Strengthening of RC Beams, Concrete Communication Conference 2000: The 10th BCA Annual Conference on Higher Education and the Concrete Industry, Birmingham, BCA.

2.11 Interfacial stresses in plated RC Beams under arbitrary symmetric loads: a high-order closed-form solution

J Yang[1], J F Chen[2], and J G Teng[3]
[1]School of the Civil Engineering, University of Leeds, UK
[2]School of the Built Environment, Nottingham University, Uk
[3]Department of Civil and Structural Engineering,
The Hong Kong Polytechnic University, China.

ABSTRACT

A new popular method for retrofitting reinforced concrete (RC) beams is to bond a fibre-reinforced polymer (FRP) plate to their soffit. An important failure mode of such strengthened members is the debonding of the FRP plate from the concrete due to high interfacial stresses near the plate ends. This paper presents a closed-form solution for interfacial stresses in simply supported beams bonded with a thin plate and subjected to arbitrary symmetric loads. The salient features of this new analysis include the consideration of non-uniform stress distributions in and the satisfaction of the stress boundary conditions at the ends of the adhesive layer.

INTRODUCTION

The technique of enhancing the flexural capacity of reinforced concrete (RC) beams by bonding a fibre reinforced polymer (FRP) plate to their soffit has become popular over the last few years. Debonding failure due to high interfacial stresses between the FRP plate and the concrete near the ends of the plate is one of the important failure modes in such strengthened RC beams [1]. A good understanding of these interfacial stresses is thus important for the development of suitable design methods.

A number of approximate analytical solutions exist for these interfacial stresses in simply supported FRP plated RC beams as reviewed and compared in Smith and Teng [2] who also presented a more general solution. This type of solutions is based on the assumption that the shear and normal stresses are uniform across the thickness of the adhesive layer. Although this assumption reduces the complexity of the problem, leading to relatively simple closed-form solutions, it violates the free surface condition at the ends of the adhesive layer.

More recently, Rabinovich and Frostig [3] developed a high-order analysis in which the adhesive layer is treated as an elastic medium with negligible longitudinal stiffness. This leads to shear stresses which are uniform and transverse normal stresses which vary linearly across the thickness of the adhesive layer. While the free surface condition at the ends of the adhesive layer is precisely satisfied in this analysis, it does not produce explicit expressions for the interfacial stresses and its correctness is also in doubt [4]. They proposed an alternative analytical approach which led to closed-form expressions. This analysis is limited to only uniformly distributed loads and equal end moments. Since many other loading conditions exist in practice, there is a need to extend Shen *et al*'s approach to other loading conditions. In this paper, an extension of Shen *et al*'s approach to the case of arbitrary symmetric loads for simply supported RC beams is presented. Singular loads (e.g. point loads) and loads with discontinuities (e.g. patch loads) are treated as equivalent distributed loads using the distribution theory [5, 6]. The extension of Shen *et al*'s approach to arbitrary anti-symmetric loads is presented in Yang *et al* [7]. Because all loads can be decomposed into a symmetric and an anti-symmetric component in linear elastic analysis, these two solutions together cover all loading conditions.

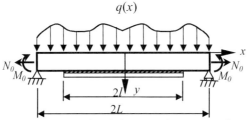

Figure 1 A plated beam under symmetric

METHOD OF SOLUTION
Geometry and Loading
Consider a simply supported RC beam with a length of $2L$. The bonded plate has a length of $2l$. The beam is subjected to an axial force N_0, a pair of end moments M_0 and a symmetrically distributed arbitrary transverse load $q(x)$ (Figure 1). A global Cartesian co-ordinate system x-y is used with the origin located at the middle of the upper surface. For ease of reference, superscripts [1], [2] and [3] are used to denote the plate, the adhesive layer and the RC beam in the strengthened beam (e.g. their thicknesses are denoted by $h^{[1]}$, $h^{[2]}$ and $h^{[3]}$ respectively) (Figure 2a). Similarly, superscripts (0), (1), (2) and (3) are used to denote the surfaces and interfaces (e.g. their widths are $b^{(0)}$, $b^{(1)}$, $b^{(2)}$ and $b^{(3)}$ respectively) (Figure 2b). The interface widths are so denoted for convenience of presentation, while in reality, $b^{(0)} = b^{(1)} = b^{(2)}$. A local coordinate system $x^{[i]}$-$y^{[i]}$ is adopted for each of the three layers, with the origin located at the geometrical centre of each layer (Figure 2a).

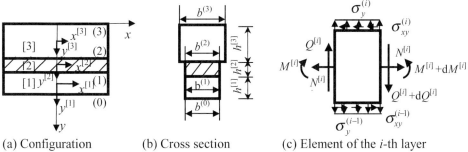

(a) Configuration   (b) Cross section   (c) Element of the i-th layer

Figure 2 Geometry and notation of a strengthened beam.

Equilibrium Equations and Boundary Conditions
For the i-th layer (Figure 2c), equilibrium considerations lead to the following relations between the interfacial stresses and the stress resultants acting on the layer

$$\frac{dN^{[i]}(x)}{dx} = b^{(i)}\sigma_{xy}^{(i)}(x) - b^{(i-1)}\sigma_{xy}^{(i-1)}(x) \tag{1a}$$

$$\frac{dQ^{[i]}(x)}{dx} = b^{(i)}\sigma_{y}^{(i)}(x) - b^{(i-1)}\sigma_{y}^{(i-1)}(x) - \begin{cases} 0 & (i \neq 3) \\ q(x) & (i = 3) \end{cases} \tag{1b}$$

$$\frac{dM^{[i]}(x)}{dx} = Q^{[i]}(x) - \frac{h^{[i]}}{2}\left[b^{(i-1)}\sigma_{xy}^{(i-1)}(x) + b^{(i)}\sigma_{xy}^{(i)}(x)\right] \tag{1c}$$

where $N^{[i]}(x)$, $Q^{[i]}(x)$ and $M^{[i]}(x)$ are the axial force, shear force and bending moment respectively in the i-th layer and $\sigma_{xy}^{(i)}(x)$ and $\sigma_{y}^{(i)}(x)$ are the shear and transverse normal stresses respectively at the i-th interface. In Equations 1a-c and the rest of this paper, the superscript in $x^{[i]}$ is omitted because the global and the three local co-ordinate systems share the same horizontal axis.

The boundary conditions at the ends of the plate are

$$N^{[i]}(\pm l) = \begin{cases} 0 & (i \neq 3) \\ N_0 & (i = 3) \end{cases}; \quad Q^{[i]}(\pm l) = \begin{cases} 0 & (i \neq 3) \\ \mp \int_0^l q(x)dx & (i = 3) \end{cases};$$

$$M^{[i]}(\pm l) = \begin{cases} 0 & (i \neq 3) \\ M_0 + L\int_0^L q(x)dx - l\int_0^l q(x)dx - \int_l^L xq(x)dx & (i = 3) \end{cases} \tag{2a-c}$$

Solution Procedure
The interfacial normal and shear stresses may be expressed as Fourier series

$$\sigma_{y}^{(i)}(x) = \sum_m a_m^{(i)} \cos\frac{m\pi x}{l}; \quad \sigma_{xy}^{(i)}(x) = \sum_m b_m^{(i)} \sin\frac{m\pi x}{l} \quad (i = 1, 2) \tag{3a,b}$$

where $a_m^{(i)}$ and $b_m^{(i)}$ are unknown Fourier coefficients, and m=1, 2, ...∞. Because equilibrium requires the integration of the interfacial normal stress to be zero, the constant term in Equation 3a has been set to zero.

Substituting Equations 2 and 3 into 1 yields

$$N^{[i]}(x) = \sum_m \left(\frac{l}{m\pi}\right)\left[b^{(i)}b_m^{(i)} - b^{(i-1)}b_m^{(i-1)}\right]\left[(-1)^m - \cos\frac{m\pi x}{l}\right] + \begin{cases} 0 & i \neq 3 \\ N_0 & i = 3 \end{cases} \tag{4a}$$

$$Q^{[i]}(x) = \sum_m \left(\frac{l}{m\pi}\right)\left[b^{(i)}a_m^{(i)} - b^{(i-1)}b_m^{(i-1)}\right]\sin\frac{m\pi x}{l} - \begin{cases} 0 & i \neq 3 \\ \int_0^x q(x)dx & i = 3 \end{cases} \tag{4b}$$

$$M^{[i]}(x) = \sum_m \left(\frac{l}{m\pi}\right)\left\{\frac{l}{m\pi}\left[b^{(i)}a_m^{(i)} - b^{(i-1)}a_m^{(i-1)}\right] - \frac{h^i}{2}\left[b^{(i)}b_m^{(i)} + b^{(i-1)}b_m^{(i-1)}\right]\right\}\left[(-1)^m - \cos\frac{m\pi x}{l}\right]$$

$$+ \begin{cases} 0 & i \neq 3 \\ M_0 + L\int_0^L q(x)dx - x\int_0^l q(x)dx - \int_l^L xq(x)dx - \int_x^l \int_x^L q(x)dxdx & i = 3 \end{cases} \tag{4c}$$

Note that $a_m^{(0)} = b_m^{(0)} = a_m^{(3)} = b_m^{(3)} = 0$ here because no interfacial stresses exist on the top and bottom surfaces of the strengthened beam.

To facilitate further calculations, the following parameters are introduced

$$\chi^{(i)} = \frac{b^{(i)}}{b^{(1)}}; \quad \xi^{(i)} = \left(\frac{m\pi}{l}\right)\frac{\chi^{(i)}b_m^{(i)}}{a_m^{(1)}}; \quad \eta^{(i)} = \frac{\chi^{(i)}a_m^{(i)}}{a_m^{(1)}}; \quad \theta^{[i]} = \frac{\chi^{(i)}E_x^{[i]}}{E_x^{[1]}} \tag{5a-d}$$

where $E_x^{[i]}$ ($i = 1, 2, 3$) is the Young's modulus in the x-direction of the i-th layer. Because $b^{(0)} = b^{(1)} = b^{(2)}$ and $a_m^{(0)} = b_m^{(0)} = a_m^{(3)} = b_m^{(3)} = 0$, $\chi^{(0)} = \chi^{(1)} = \chi^{(2)} = \eta^{(1)} = 1$ and $\xi^{(0)} = \eta^{(0)} = \xi^{(3)} = \eta^{(3)} = 0$. The compatibility of longitudinal strain and curvature at the interfaces between the layers together with the equilibrium equation for each layer leads to the following relations

$$\xi^{(1)} = \frac{3[2A_1 - A_2(h^{[1]} - 2h_0)]}{h^{[1]}[3A_1 - A_2(h^{[1]} - 3h_0)]}; \quad \xi^{(2)} = \left\{1 + \frac{\theta^{[2]}h^{[2]}[2A_1 - A_2(2h^{[1]} + h^{[2]} - 2h_0)]}{h^{[1]}[2A_1 - A_2(h^{[2]} - 2h_0)]}\right\}\xi^{(1)} \tag{6a,b}$$

$$\eta^{(2)} = 1 + \frac{h^{[2]}}{2}(\xi^{(1)} + \xi^{(2)}) + \frac{\theta^{[2]}(h^{[2]})^3 A_2}{6h^{[1]}[2A_1 - A_2(h^{[2]} - 2h_0)]}\xi^{(1)} \tag{6c}$$

where

$$A_1 = \begin{cases} 0 & N_0 = 0 \\ \left(b^{(1)}\sum_{i=1}^3 \theta^{[i]}h^{[i]}\right)^{-1} & N_0 \neq 0 \end{cases}; \quad A_2 = 6\frac{2\sum_{i=1}^3 h^{[i]} - h^{[3]} - 2h_0}{b^{(1)}\left[\sum_{i=1}^3 \theta^{[i]}(h^{[i]})^3 + 6\sum_{i=1}^3 \theta^{[i]}h^{[i]}\left(2\sum_{j=1}^i h^{[j]} - h^{[i]} - 2h_0\right)^2\right]}$$

$$h_0 = b^{(1)}A_1\sum_{i=1}^3 \theta^{[i]}h^{[i]}\left(\sum_{j=1}^i h^{[j]} - \frac{h^{[i]}}{2}\right) \tag{6d-f}$$

The stress distribution in the beam and the bonded plate can be found using the classical Euler-Bernoulli beam theory. Since the adhesive layer is treated as an elastic continuum without any body forces, the following equilibrium equations also have to be satisfied within the adhesive layer

$$\frac{\partial \sigma_x^{[2]}}{\partial x} + \frac{\partial \sigma_{xy}^{[2]}}{\partial y^{[2]}} = 0 \,, \qquad\qquad \frac{\partial \sigma_{xy}^{[2]}}{\partial x} + \frac{\partial \sigma_y^{[2]}}{\partial y^{[2]}} = 0 \qquad\qquad \text{(7a,b)}$$

From Equations 4-7, the stresses in the beam can be obtained as

$$\sigma_x^{[i]} = \frac{1}{\chi^{(i)}} \left\{ \frac{1}{h^{[i]}} \left(\xi^{(i)} - \xi^{(i-1)} \right) - \frac{12 y^{[i]}}{\left(h^{[i]}\right)^3} \left[\frac{h^{[i]}}{2} \left(\xi^{(i)} + \xi^{(i-1)} \right) - \left(\eta^{(i)} - \eta^{(i-1)} \right) \right] \right\}$$

$$\sum_m \left(\frac{l}{m\pi} \right)^2 a_m^{(1)} \left[(-1)^m - \cos\frac{m\pi x}{l} \right] + \begin{cases} 0 & (i = 1, 2) \\ \sigma_x^a & (i = 3) \end{cases} \qquad\qquad \text{(8)}$$

$$\sigma_{xy}^{[i]} = \frac{1}{\chi^{(i)}} \left\{ -\frac{1}{4} \left(\xi^{(i)} + \xi^{(i-1)} \right) + \frac{3}{2h^{[i]}} \left(\eta^{(i)} - \eta^{(i-1)} \right) - \frac{y^{[i]}}{h^{[i]}} \left(\xi^{(i)} - \xi^{(i-1)} \right) \right.$$

$$\left. + \frac{6\left(y^{[i]}\right)^2}{\left(h^{[i]}\right)^3} \left[\frac{h^{[i]}}{2} \left(\xi^{(i)} + \xi^{(i-1)} \right) - \left(\eta^{(i)} - \eta^{(i-1)} \right) \right] \right\} \sum_m \left(\frac{l}{m\pi} \right) a_m^{(1)} \sin\frac{m\pi x}{l} + \begin{cases} 0 & (i = 1, 2) \\ \sigma_{xy}^a & (i = 3) \end{cases} \qquad \text{(9)}$$

$$\sigma_y^{[i]} = \begin{cases} 0 & (i = 1, 3) \\ \zeta(y^{[i]}) \sum_m a_m^{(1)} \cos\frac{m\pi x}{l} & (i = 2) \end{cases} \qquad\qquad \text{(10)}$$

where

$$\zeta(y^{[i]}) = \frac{1}{\chi^{(i)}} \left\{ -\frac{h^{[i]}}{8} \left(\xi^{(i-1)} - \xi^{(i)} \right) + \frac{1}{2} \left(\eta^{(i-1)} - \eta^{(i)} \right) + \frac{y^{[i]}}{4h^{[i]}} \left[\left(\xi^{(i-1)} + \xi^{(i)} \right) h^{[i]} - 6 \left(\eta^{(i-1)} - \eta^{(i)} \right) \right] \right.$$

$$\left. + \frac{\left(y^{[i]}\right)^2}{2h^{[i]}} \left(\xi^{(i-1)} - \xi^{(i)} \right) + \frac{\left(y^{[i]}\right)^3}{\left(h^{[i]}\right)^3} \left[-h^{[i]} \left(\xi^{(i-1)} + \xi^{(i)} \right) + 2 \left(\eta^{(i-1)} - \eta^{(i)} \right) \right] \right\}, \quad (i = 2) \qquad \text{(11)}$$

and $\sigma_x^{[i]}$, $\sigma_y^{[i]}$ and $\sigma_{xy}^{[i]}$ ($i = 1, 2, 3$) are the normal stresses in the $x^{[i]}$ and $y^{[i]}$ directions and the shear stress in the $x^{[i]}$-$y^{[i]}$ plane respectively in the i-th layer, σ_x^a and σ_{xy}^a are stresses in the un-strengthened beam caused by the same applied loads and given by

$$\sigma_x^a = \frac{N_0}{b^{(3)} h^{[3]}} + \frac{12 y^{[3]}}{b^{(3)} \left(h^{[3]}\right)^3} \left[M_0 + L \int_0^L q(x) dx - x \int_0^l q(x) dx - \int_l^L x q(x) dx - \int_x^l \int_x^l q(x) dx dx \right] \qquad \text{(12)}$$

$$\sigma_{xy}^a = -\frac{24 y^{[3]}}{b^{(3)} \left(h^{[3]}\right)^3} \left[\left(h^{[3]}\right)^2 - 4\left(y^{[3]}\right)^2 \right] \int_0^x q(x) dx \qquad\qquad \text{(13)}$$

Equations 5b and 5c show that all coefficients in Equations 3a and b can be related to $a_m^{(1)}$. These unknown coefficients may now be determined by minimising the total complementary energy of the composite beam in the strengthened range. Due to symmetry, only one half of the beam is considered. The total complementary energy in half of the strengthened portion is

$$U = \frac{b^{(1)}}{2} \sum_{i=1}^{3} \chi^{(i)} \int_{-h^{[i]}/2}^{+h^{[i]}/2} \int_{0}^{l} \left[\frac{1}{E_x^{[i]}} \left(\sigma_x^{[i]} \right)^2 + \frac{1}{E_y^{[i]}} \left(\sigma_y^{[i]} \right)^2 - \frac{2v_{xy}^{[i]}}{E_x^{[i]}} \sigma_x^{[i]} \sigma_y^{[i]} + \frac{1}{G_{xy}^{[i]}} \left(\sigma_{xy}^{[i]} \right)^2 \right] dx dy^{[i]} \quad (14)$$

where $E_y^{[i]}$, $G_{xy}^{[i]}$ and $v_{xy}^{[i]}$ are, respectively, the Young's modulus in the y-direction, the shear modulus and Poisson's ratio in the $x-y$ plane of the i-th layer. Minimising U in terms of $a_m^{(i)}$ by setting $\partial U/\partial a_m^{(i)} = 0$ leads to

$$S_1 \left(\frac{l}{m\pi} \right)^2 (-1)^m \sum_{n=1,2,\cdots} \left(\frac{l}{n\pi} \right)^2 a_n^{(1)} (-1)^n + \left[S_2 \left(\frac{l}{m\pi} \right)^4 + S_3 \left(\frac{l}{m\pi} \right)^2 + S_4 \right] a_m^{(1)} = S(m) \quad (15)$$

where

$$S_1 = \sum_{i=1}^{3} \frac{b^{(1)}}{E_x^{[i]}} \left\{ \frac{\left[\xi^{(i)} - \xi^{(i-1)} \right]^2}{h^{[i]}} + \frac{12}{\left(h^{[i]} \right)^3} \left[\frac{h^{[i]}}{2} (\xi^{(i)} + \xi^{(i-1)}) + \eta^{(i-1)} - \eta^{(i)} \right]^2 \right\} \quad (16)$$

$$S_2 = S_1/2 \quad (17)$$

$$S_3 = \sum_{i=1}^{3} \frac{b^{(1)} h^{[i]}}{G_{xy}^{[i]}} \left\{ \left(\xi^{(i)} \right)^2 \left[\frac{1}{2} \left(\frac{\chi^{(i)}}{\chi^{(i-1)}} - 1 \right)^2 + \frac{1}{15} \right] + \frac{1}{h^{[i]}} \xi^{(i-1)} \left(\eta^{(i)} - \eta^{(i-1)} \right) \left[\frac{\chi^{(i)}}{\chi^{(i-1)}} - \frac{11}{10} \right] \right.$$
$$+ \frac{1}{15} \xi^{(i)} \left(\xi^{(i)} - \xi^{(i-1)} \right) + \frac{1}{10 h^{[i]}} \xi^{(i)} \left(\eta^{(i)} - \eta^{(i-1)} \right) + \frac{3}{5 \left(h^{[i]} \right)^2} \left(\eta^{(i)} - \eta^{(i-1)} \right)^2 \right\}$$
$$- \frac{v_{xy}^{[2]} b^{(1)} h^{[2]}}{15 E_x^{[2]}} \left[2 \left(\left(\xi^{(1)} \right)^2 + \left(\xi^{(2)} \right)^2 \right) + \xi^{(1)} \xi^{(2)} \left(14 - 15 \chi^{(2)} \right) + \frac{18}{h^{[2]}} \left(\xi^{(1)} + \xi^{(2)} \right) \left(1 - \eta^{(2)} \right) \right.$$
$$+ \frac{15 \chi^{(2)}}{h^{[2]}} \left(\xi^{(1)} \eta^{(2)} - \xi^{(2)} \right) + \frac{18}{\left(h^{[2]} \right)^2} \left(1 - \eta^{(2)} \right)^2 \right] \quad (18)$$

$$S_4 = \frac{b^{(1)} h^{[2]}}{420 E_y^{[2]}} \left\{ \left(h^{[2]} \right)^2 \left[70 \left(\chi^{(2)} - 1 \right)^2 \left(\xi^{(1)} \right)^2 + 2 \left(\xi^{(2)} + \xi^{(1)} \right) \left(\xi^{(2)} - 6 \xi^{(1)} \right) + 14 \chi^{(1)} \left(\xi^{(1)} \right)^2 \right. \right.$$
$$+ 7 \left(4 - 3 \chi^{(2)} \right) \xi^{(1)} \xi^{(2)} \right] + h^{[2]} \left[2 \left(1 - \eta^{(2)} \right) \left(67 \xi^{(1)} + 11 \xi^{(2)} \right) + 7 \chi^{(2)} \xi^{(1)} \left(30 \chi^{(2)} + 2 l \eta^{(2)} - 46 \right) \right.$$
$$\left. - 35 \chi^{(2)} \xi^{(2)} \right] + \left[210 \chi^{(2)} \left(\chi^{(2)} + \eta^{(2)} - 1 \right) + 78 \left(\eta^{(2)} - 1 \right)^2 \right] \right\} \quad (19)$$

$$S(m) = \left(\frac{l}{m\pi}\right)^2 \left\{\frac{12}{\left(E_x^{[3]}h^{[3]}\right)^3}\left(\frac{h^3}{2}\xi^{(2)}+\eta^{(2)}\right)\left[(-1)^m\left(M_0+L\int_0^L q(x)dx-\int_l^L xq(x)dx\right.\right.\right.$$

$$\left.\left.-\frac{1}{2l}\int_0^l\left(l^2+x^2\right)q(x)dx\right)-\frac{l}{(m\pi)^2}\int_0^l q(x)\left(2-(-1)^m+\cos\frac{m\pi x}{l}\right)dx\right]+\frac{(-1)^m N_0\xi^{(2)}}{E_x^{[3]}h^{[3]}}$$

$$\left.+\frac{1}{G_{xy}^{[3]}l}\left(\frac{6\eta^{(2)}}{5h^{[3]}}+\frac{11\xi^{(2)}}{10}-\frac{\chi^{(3)}\xi^{(2)}}{\chi^{(2)}}\right)\left[\int_0^l q(x)\left((-1)^m-\cos\frac{m\pi x}{l}\right)dx\right]\right\} \qquad (20)$$

Let $\displaystyle S_0 = \sum_{n=1,2,\cdots}\left(\frac{l}{n\pi}\right)^2 a_n^{(1)}(-1)^n$; $\displaystyle 2\beta = \left(\frac{l}{\pi}\right)^2\frac{S_3}{S_4}$; $\displaystyle \alpha^2 = \left(\frac{l}{\pi}\right)^4\frac{S_2}{S_4}$ $\qquad (21)$

Equation 15 can be re-written as

$$\frac{S_4}{m^4}\left[\frac{l^2 m^2 S_0 S_1}{\pi^2 S_4}(-1)^m + a_m^{(1)}\left(m^4+2\beta m^2+\alpha^2\right)\right] = S(m) \qquad (22)$$

It may be noted that $\beta > \alpha$ for all practical cases.

NUMERICAL EXAMPLES
Steel Plated RC Beam under Uniformly Distributed Load
The first example problem is an RC beam bonded with a steel plate subjected to a uniformly distributed load (UDL) $q=15$N/mm given in Roberts and Haji-Kazemi [8]. For this problem, the present solution reduces to that of Shen et al [4]. The emphasis here is on the performance of the present approach by comparing its predictions with those from the solution of Smith and Teng [2] and the finite element (FE) results of Teng et al [8]. The geometrical and material properties of this beam are [9]: $L=1200$mm, $l=900$mm, $h^{[1]}=h^{[2]}=4$mm; $h^{[3]}=150$mm; $b^{(1)} = b^{(2)} = b^{(3)} = 100$mm, $E_x^{[1]} = E_y^{[1]}=200$GPa, $E_x^{[2]} = E_y^{[2]}=2$GPa, $E_x^{[3]} = E_y^{[3]}=20$GPa, and $v_{xy}^{[2]}=0.25$. In their FE analysis, Teng et al (2002b) used the Poisson's ratios of $v_{xy}^{[3]}=0.17$ and $v_{xy}^{[1]}=0.3$ respectively for concrete and steel which were not specified by Roberts and Haji-Kazemi [9].

Figure 3 shows the comparison of interfacial stresses predicted by the three different approaches. The interfacial shear stresses predicted by the present solution are almost identical for the plate-to-adhesive (PA) interface and the adhesive-to-concrete (AC) interface (Figure 3a). They increase towards the free edge and then reduce rapidly to zero at the free edge, complying with the free surface condition.

The present predictions agree well with the FE predictions [8] in general, but significant differences are seen within the close vicinities of both plate ends. The FE predictions for the AC interface, the mid-adhesive (MA) section and PA interface are significantly different from each other. This seems to suggest that the current assumption of parabolic shear stress variation across the thickness of the adhesive layer (see Equation 7a) is insufficient to accurately model the local stress fields near the plate ends, although the peak interfacial shear stress along each interface from the present solution is only slightly smaller than the corresponding FE prediction. It may be noted that the FE predictions do not reduce to zero at the AC and the PA interfaces at the plate ends, which is attributable to stress singularity here [8]. Smith and Teng [2] predicts a monotonically increasing interfacial shear stress from the mid-span of the beam towards the ends of the plate, with a maximum value at the ends of the plate, which does not satisfy the free surface condition there. In general, the two solutions are in close agreement, although the peak interfacial shear stress predicted by Smith and Teng's solution [2] is about 1/3 larger than the present predictions.

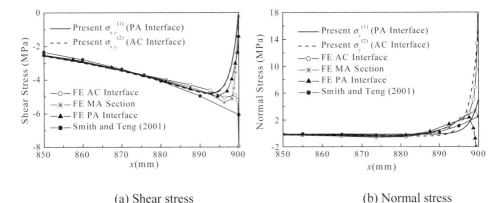

(a) Shear stress   (b) Normal stress

Figure 3. Interfacial stresses in a beam bonded with a steel plate and subjected to a UDL

For the interfacial normal stress, the predictions of all three methods are very small apart from the very small zones close to the plate ends. There is no difference between the AC and PA interfaces in Smith and Teng's prediction because this is so assumed. However, both the present and the FE results show that this stress is clearly non-uniform across the thickness of the adhesive layer close to the plate ends. For the AC interface, the present results are in close agreement with the FE results but are much higher than those from Smith and Teng's solution. The interfacial normal stress is much lower at the PA interface than at the AC interface according to the present solution. The FE predictions go further. The interfacial normal stress becomes negative (compressive) at the PA interface near the plate end. The consistently higher tensile normal stress at the AC interface, compared to that at the PA interface, may be the reason why debonding at the PA interface was rarely observed in tests as noted by Rabinovich and Frostig [3].

CFRP plated RC Beam under Patch Loads
A CFRP plated RC beam under two patch loads is considered here to demonstrate the application of the new closed-form solution. The strengthened beam has the

following properties: L=1500mm, l=1200mm, $b^{(1)}$=$b^{(2)}$=$b^{(3)}$=200mm, $h^{[1]}$=4mm, $h^{[2]}$=2mm, $h^{[3]}$=300mm, $E_x^{[1]}=100$GPa, $E_x^{[2]}=E_y^{[2]}=2$GPa, $E_x^{[3]}=$ $E_y^{[3]}$=30GPa, $G_{xy}^{[1]}$=5GPa, $v_{xy}^{[2]}$=0.35 and $v_{xy}^{[3]}$=0.17. All these parameters, except the Poisson's ratios for the adhesive layer and the shear modulus for the FRP plate for which common values are assumed here, were given by Smith and Teng.

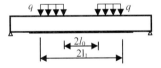

Figure 4 A strengthened beam under patch loads

The strengthened beam is under two patch loads symmetrically positioned on the beam (Figure 4). The patch loads start at l_0 and end at l_1 from the mid-span of the beam. The locations of the patch loads are thus defined by l_0 and l_1 whilst the total load on the beam = $2q(\tilde{l}_1 l_0)$. Only l_0=0 (i.e. a single patch load centred around the mid-span of the beam) is considered here. Three patch sizes (l_1 = 1, 600 and 1200 mm) are considered to investigate the effect of the patch size on interfacial stresses. For ease of comparison, the interfacial shear and normal stresses are normalised against a nominal stress defined by $\sigma^* = q(\tilde{l}_1 l_0)/lb^{(3)}$ which is equal to the maximum shear force in the beam $q(\tilde{l}_1 l_0)$ divided by $lb^{(3)}$.

The dimensionless interfacial shear stress $\tilde{\sigma}_{xy}^{(i)} = \sigma_{xy}^{(i)}/\sigma^*$ increases from the mid-span of the beam towards the ends of the patch load and this increase is linear except for the case of a concentrated load with x/l =1/1200 (Figure 5a) following the linear variation of shear force in this zone. This shear stress then remains constant within the constant shear force zone. Close to the plate ends, it increases very rapidly before reducing back to zero. The interfacial shear stress along the AC interface is indistinguishable from that along the PA interface in Figure 5. The interfacial normal stress along the AC interface is significantly higher than that along the PA interface. The distributions of the normalised interfacial stresses close to the plate ends are almost identical numerically and are seen in Figure 5 to be independent of the patch size. As the shear force and the bending moment at the plate ends are the same for the three patch sizes for the same total load on the beam, this observation suggests that both the interfacial shear and normal stress distributions near the plate ends are completely controlled by the shear force and moment at the plate ends in practical terms.

CONCLUSIONS
A new high-order interfacial stress analysis has been presented for simply supported reinforced concrete (RC) beams bonded with a soffit plate and subjected to general symmetric loads. The salient features of the new analysis include the consideration of non-uniform stress distributions in and the satisfaction of the free surface

condition at the ends of the adhesive layer. The solution methodology is general in nature and may be applied to the analysis of other types of composite structures.

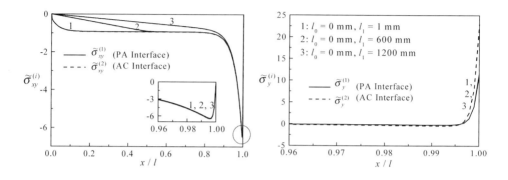

(a) Shear Stress   (b) Normal Stress

Figure 5 Effect of patch size on interfacial stresses in a strengthened beam

Two numerical examples have been presented to demonstrate the validity and the application of the present solution. The numerical results for a beam subject to a symmetrically positioned patch load of three different patch sizes acting within the strengthened part of the beam showed that both the interfacial shear and normal stress distributions near the plate ends are completely controlled by the shear force and bending moment at the plate ends.

REFERENCES

[1]   Teng, J.G., Chen, J.F., Smith, S.T. and Lam, L. (2002a), FRP strengthened RC Structures, Wiley, Chichester, U.K.

[2]   Smith, S.T. and Teng, J.G. (2001) "Interfacial stresses in plated beams", Engineering Structures, 23, 857-871.

[3]   Rabinvich, O. and Frostig, Y. (2000) "Closed-form high-order analysis of RC beams strengthened with FRP strips", Journal of Composites for Construction, ASCE, 4, 65-74.

[4]   Shen, H.S., Teng, J.G. and Yang, J. (2001) "Interfacial stresses in beams and slabs bonded with thin plate", Journal of Engineering Mechanics, ASCE, 127, 399-406.

[5]   Schwarz, L. (1966), Thorie des Distributions, Hermann, Paris.

[6]   Yavari, A., Sarkani, S. and Moyer, E.T. (2000) "On applications of generalized functions to beam bending problems", International Journal of Solids and Structures, 37, 5675-5705.

[7]   Yang, J., Teng, J.G. and Chen, J.F. (2002) "Interfacial stresses in FRP-plated beams under arbitrary loads", in preparation.

[8]   Teng, J.G., Zhang, J.W. and Smith, S.T. (2002b) "Interfacial stress in RC beams bonded with a soffit plate: a finite element study", Construction and Building Materials, in press.

[9]   Roberts, T.M. and Haji-Kazemi, H. (1989) "Theoretical study of the behavior of reinforced concrete beams strengthened by externally bonded steel plates", Proceedings of the Institution of Civil Engineers, Part 2, 87, 39-55.

2.12 Premature failure of RC continuous beams strengthened with CFRP laminates

S A El-Refaie, A F Ashour, and S W Garrity
Department of Civil and Environmental Engineering,
University of Bradford, UK

ABSTRACT
This paper investigates the premature failure of carbon fibre reinforced polymer (CFRP) laminates attached to the tension face of continuous reinforced concrete beams. Sixteen reinforced concrete continuous beams with different arrangements of internal and external reinforcement were tested to failure. All test specimens had the same geometrical dimensions and were classified into three groups according to the internal steel reinforcement. Each group had one unstrengthened control beam designed to fail in flexure. The length, thickness, form and position of the CFRP laminates were the main parameters investigated.

Three premature failure modes of beams with external CFRP laminates were observed, namely laminate rupture and laminate separation and peeling failure of the concrete cover adjacent to the laminate. Comparisons between failure loads obtained from experiments and the results of theoretical analysis are presented. The ductility of all strengthened beams was reduced compared with that of the respective unstrengthened control beam.

INTRODUCTION
Using CFRP laminates has proved to be an effective means of upgrading and strengthening reinforced concrete beams. However, premature failures such as laminate rupture and debonding can significantly limit the capacity enhancement and prevent the full ultimate flexural capacity of the retrofitted beams from being attained. Several factors have been found to affect the resistance against peeling failure such as the tensile strength of concrete, plate thickness, plate length and the relative stiffnesses of the plate, adhesive and concrete [1, 2].

Although many in-situ reinforced concrete beams are of continuous construction, there has been very little research into the behaviour of such beams with external reinforcement [3-7]. In this paper, test results at failure of 16 RC continuous beams strengthened with different arrangements of CFRP laminates are presented. The flexural capacity of each of the beams tested is estimated and compared with that obtained from experiments.

ACIC 2002, Thomas Telford, London, 2002

TEST PROGRAMME

The test specimens consisted of sixteen reinforced concrete continuous beams classified into three groups according to the arrangement of the internal steel reinforcement as given in Table 1. Beam geometry and reinforcement as well as the loading and support arrangement are illustrated in Figure 1.

Table 1. Details and properties of materials used

Group no.	Beam no.	Main longitudinal steel*		Top CFRP over the central support		Bottom CFRP at mid-span		f_{cu} (N/mm^2)
		Top	Bottom	No.	L_1 (m)	No.	L_2 (m)	
H	H1	2T8	2T20	-	-	-	-	24.0
	H2			2	2.0	-	-	43.5
	H3			6	2.0	-	-	33.0
	H4			10	2.0	-	-	33.2
	H5			6	1.0	-	-	46.0
	H6			2	3.0	2	1.0	44.0
S	S1	2T20	2T8	-	-	-	-	26.0
	S2			-	-	2	2.0	42.9
	S3			-	-	6	2.0	33.3
	S4			-	-	6	3.5	42.8
	S5			-	-	10	3.5	24.4
E	E1	2T16	2T16	-	-	-	-	24.0
	E2			1	2.5	-	-	43.6
	E3			-	-	1	3.5	47.8
	E4			1	2.5	1	3.5	46.1
	E5			6	**2.5**	-	-	44.7

Note:

* Yield strengths of *8, 16* and *20 mm* steel bars are *505, 520 and 510 N/mm^2* respectively and their modulus of elasticity is *200 kN/mm^2*.

All beams had the same geometrical dimensions: 150mm wide × 250mm deep × 8500mm long. Vertical links of 6mm bar diameter at 100mm centres were provided throughout each beam length in order to prevent shear failure. The position, length, thickness and form (sheets or plates) of the CFRP laminates were the main parameters investigated as summarised in Table 1. Beams H1, S1 and E1 had no external reinforcement and were used as control specimens. Beams E2, E3 and E4 were strengthened with CFRP plates of 1.2mm thickness, 100mm width and 2500 N/mm² tensile strength; whereas the rest of the beams were strengthened with CFRP sheets, each layer of 0.117mm thickness, 110mm width and 3900 N/mm² tensile strength. The CFRP laminates applied to the top face of the beams were placed symmetrically about the central support and those applied to the bottom face of the beam were positioned symmetrically about the centres of both spans. The compressive strength f_{cu} of the concrete obtained from testing three 100 mm cubes is given in Table 1.

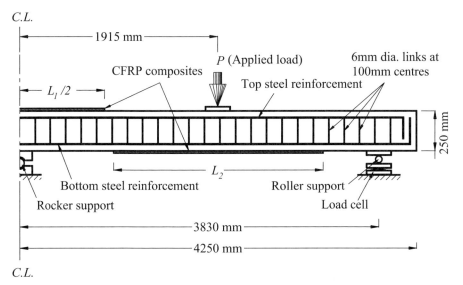

Figure 1. Test rig and reinforced concrete beam details

TEST RESULTS

Each test beam, which comprised two equal spans of 3830mm, was loaded as shown in Figure 1. Test results are summarised in Table 2 and four different observed failure modes are described below.

Failure Mode 1: Conventional ductile flexural failure due to yielding of the internal tensile steel reinforcement followed by concrete crushing at both the central support and mid-span sections (control beams H1, S1 and E1).

Failure Mode 2: Beams H2 (see Figure 2) and H6 exhibited tensile rupture of the CFRP sheets over the central support at 80% and 70% of each beam failure load, respectively. Sudden tensile brittle rupture of the CFRP sheets was followed by flexural failure in beam H2 and by peeling failure of the concrete cover attached to the soffit sheets in beam H6.

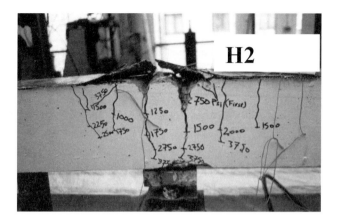

Figure 2. Tensile rupture of the CFRP sheets (beam H2)

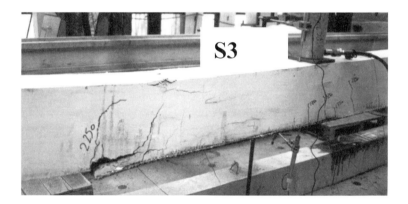

Figure 3. Peeling failure of the cover concrete (beam S3)

Failure Mode 3: Peeling failure of the concrete cover along the steel reinforcement level adjacent to the external CFRP laminates (see Figure 3) occurred for a large number of strengthened beams in the test programme as given in Table 2. It was brittle, explosive and was followed immediately by beam failure. Extending the CFRP sheet to cover the entire hogging zone such as in beams H3, E2, E4 and E5 or the entire sagging zone such as in beams S4 and E3 did not prevent peeling failure of the cover concrete.

Failure Mode 4: CFRP sheet separation without concrete attached was noticed in beams S2, S5 and E4 (bottom plates). CFRP laminate separation occurred along the adhesive/concrete and CFRP laminate/adhesive interfaces. This may be as a result, in part, of workmanship problems. It was accompanied by a little noise along the adhesive level before beam failure indicating impending failure of the adhesive.

ENHANCEMENT OF LOAD AND MOMENT CAPACITIES

The failure load and moment results for all test specimens are summarised in Table 2. The hogging and sagging bending moments given in Table 2 were calculated on the basis of satisfying equilibrium using the measured end support reaction and the mid-span applied load. Table 2 also gives the ultimate load enhancement ratio, ξ (the ratio between the ultimate load of a strengthened beam and that of the corresponding unstrengthened control beam) and the ultimate moment enhancement ratio, η (the ratio between the ultimate moment of a strengthened section in a strengthened beam and that of the corresponding section in the respective unstrengthened control beam). It should be noted that the enhancement ratio of a beam with a compressive strength higher than that of the control beam is likely to be slightly larger than would be the case if the same compressive strength of concrete was used for both beams.

Table 2 shows that using CFRP composites for the strengthening of continuous beams is an effective technique; the load and moment capacities can be increased by factors of up to 2.04 and 3.34, respectively, as was the case for beam S4. For a specified length of CFRP sheets, there was an optimum number of CFRP layers above which the beam load capacity was either decreased (beams H3 and H4) or not improved (beams S2 and S3), as indicated in Table 2. Increasing the length of the CFRP sheets was found to increase the load capacity of the strengthened beam as in the case of beams H3, H5, S3 and S4. However, there was also an optimum length of the CFRP sheets above which increasing the CFRP sheet length was not effective. An example of this is seen with beams H2 and H6; in both cases tensile rupture of the CFRP sheets occurred at an applied load of 121.5 kN. By comparing the ultimate load enhancement ratio of a strengthened beam and the moment enhancement ratio of a strengthened section in the same beam, it can be concluded that the latter was significantly higher than the former. Such a conclusion is not valid for simply supported beams strengthened with external reinforcement where the moment and load enhancement ratios are the same.

BEAM DUCTILITY

The deflection ductility index, μ_Δ, used for simply supported beams is adopted to measure the ductility of the continuous beams tested by the authors as given in Table 2. The deflection ductility index, μ_Δ, is defined as:

$$\mu_\Delta = \frac{\Delta_u}{\Delta_y} \qquad (1)$$

where Δ_u is the mid-span deflection at beam ultimate load and Δ_y is the mid-span deflection at the lower yielding load of the tensile reinforcement over the central support or the beam mid-span. As can be seen from Table 2, all strengthened beams exhibited less ductility than the corresponding unstrengthened control beams. Only beam H2 showed nearly similar ductility at failure as that of the control beam H1 because both beams failed in flexure at the ultimate load level. However, if the ductility was considered at tensile rupture of the CFRP sheets, beam H2 would have less ductility than that of the control beam H1. Increasing the thickness and length of CFRP laminates was found to decrease the beam ductility.

Table 2. Experimental results at failure of test specimens

Beam no.	Failure mode	P_u (kN)	ξ	M_h (kNm)	M_s (kNm)	η	μ_Δ
H1	1	138.0	1.00	21.21	56.78	1.00	4.10
H2	2 and 1	152.3 (121.5*)	1.10 (0.88*)	31.60	61.00	1.42 (h)	4.34 (1.83*)
H3	3	172.9	1.25	46.48	59.56	2.19 (h)	1.35
H4	3	162.6	1.18	53.07	51.32	2.50 (h)	1.74
H5	3	162.6	1.18	35.00	64.27	1.65 (h)	2.64
H6	2 and 3	172.9 (121.5*)	1.25 (0.88*)	28.26*	70.24	1.33 (h) 1.24 (s)	3.46 (1.71*)
S1	1	83.6	1.00	57.77	11.13	1.00	12.85
S2	4	121.8	1.46	71.28	22.67	2.04 (s)	6.21
S3	3	121.8	1.46	66.90	24.72	2.22 (s)	2.93
S4	3	170.5	2.04	88.97	37.15	3.34 (s)	2.52
S5	4	111.7	1.34	50.18	28.36	2.55 (s)	1.00#
E1	1	149.7	1.00	54.49	**44.41**	**1.00**	6.12
E2	3	178.6	1.19	79.78	45.64	1.46 (h)	3.58
E3	3	207.0	1.38	53.56	72.35	1.63 (h)	3.21
E4	3 and 4	231.4	1.55	77.00	72.29	1.41 (h) 1.63 (s)	2.10
E5	3	174.6	1.17	77.42	44.87	1.42 (h)	2.60

P_u = ultimate load; M_h and M_s = ultimate sagging and hogging moments; ξ and η = ultimate load and moment enhancement ratios, respectively; (h) = central support section and (s) = mid-span section.
*Tensile rupture of the CFRP sheets occurred over the central support.
#There was no ductility.

PREDICTIONS OF FLEXURAL CAPACITY OF TEST SPECIMENS
Several analytical methods have been developed for predicting peeling failure of simply supported beams strengthened with external plates. These methods, however, are cumbersome and inconsistent with experimental results [8, 9]. In addition, they are not directly applicable to continuous beams. In the following, the flexural load capacity of each of the beams tested by the authors is calculated and compared with

that obtained from experiments to calculate the reserve capacity if pre-mature failure was prevented. The prediction of the ultimate load capacity resisted by the test specimens was based on satisfying equilibrium conditions and assuming that both the mid-span and central support sections reached their flexural capacities M_{us} and M_{uh}, respectively as shown in Figure 4a. Accordingly, from equilibrium considerations, the applied point load, P_t, at the beam mid-span is calculated from:

$$P_t = \frac{2}{L}(M_{uh} + 2M_{us})$$
(2)

When Equation (2) is applied to beams H4, H5, S2 and S3, it is found that the bending moment at the unstrengthened section next to the sheet end (the end near the end support for soffit sheets) is higher than the moment capacity of that section. In such cases, the point load, P_t, for beams H4 and H5 was calculated from:

$$P_t = \frac{2}{L}\left(\frac{M_{us}L_1 + M_{he}L}{L - L_1} + 2M_{us}\right)$$
(3)

and that for beams S2 and S3 was determined from:

$$P_t = \frac{2}{L}\left(M_{uh} + \frac{2L}{L - L_2}M_{se}\right)$$
(4)

where M_{he} and M_{se} are the moment capacities of the hogging and sagging unstrengthened sections at the CFRP sheet end (see Figures 4b and 4c, respectively) and L_1 and L_2 are the lengths of the CFRP sheets over the central support and at the beam soffit, respectively.

The calculated flexural capacities of different critical sections as required by Equations 2, 3 and 4 are given in Table 3. These calculations were based on the analytical method presented elsewhere [6, 10]. In this method, a linear strain distribution at failure for the strengthened section is assumed and using the appropriate constitutive relationships for different materials, the forces acting on the cross section can be established. When equilibrium of forces is satisfied, the flexural capacity of the section can be calculated.

Table 3 presents the theoretical flexural capacity, P_{th} (= $2P_t$), obtained from Equations 2, 3 and 4 for the test beams and the ratio, λ_l (= P_u / P_{th}), between experimental and theoretical load capacities. The comparison shows that, apart from beam S5, the strengthened beams tested were close to developing their full flexural capacity.

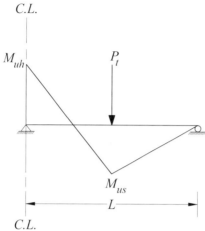

(a) Critical sections: over support hogging and mid-span sagging

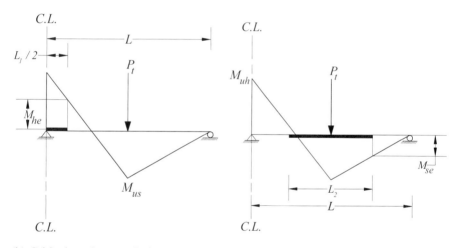

(b) Critical sections: end plate
hogging and mid-span sagging

(c) Critical sections: over support
hogging and end-plate sagging

Figure 4. Ultimate hogging and sagging moments at the critical sections of the continuous beams tested

CONCLUSIONS

Based on the work described in this paper, the following conclusions are drawn:

1. All strengthened beams exhibited a higher beam load capacity but lower ductility compared with their respective unstrengthened control beams.

Table 3. Theoretical flexural moment and load capacities of test specimens

Beam no.	M_{uh} (kNm)	M_{us} (kNm)	P_{th} (kN)	λ_l
H1	12.52[•]	50.58	118.73	1.16
H2	33.83	60.33	161.35	0.94
H3	69.96	57.83	193.86	0.89
H4	87.08 (11.20[*])	57.83	210.19	0.77
H5	75.30 (11.67[*])	60.51	165.22	0.98
H6	33.96	70.24	182.18	0.95
S1	49.78	10.68	74.30	1.13
S2	60.23	33.81 (11.61[#])	113.66	1.07
S3	57.98	70.21 (11.22[#])	109.60	1.11
S4	60.23	74.63	218.77	0.78
S5	54.25	81.06	225.97	0.49
E1	41.20	41.20	129.09	1.16
E2	82.97	42.44	175.30	1.02
E3	42.64	85.24	222.58	0.93
E4	84.17	84.17	263.72	0.88
E5	83.35	42.48	175.78	0.99

[•] Moment capacity calculated based on ultimate strength of steel bars.
[*] Moment capacity of hogging unstrengthened section at the central support laminate end.
[#] Moment capacity of sagging unstrengthened section at the soffit laminate end.

2. Sudden, brittle peeling failure of the concrete cover adjacent to the CFRP sheets was the dominant failure mode of the strengthened beams tested. However, most beams were close to achieving their flexural capacity.

3. Increasing the CFRP sheet length to cover the entire hogging or sagging zones did not prevent peeling failure of the CFRP laminate.

4. Increasing the length of the CFRP laminates was found to be ineffective when tensile rupture of the CFRP sheets was the failure mode.

5. Unlike simply supported beams, the enhancement of the bending moment capacity of a continuous beam due to external strengthening was found to be higher than that in the load capacity of the continuous beam.

ACKNOWLEDGEMENTS
The first author acknowledges the grant provided from the Egyptian government to fulfil this research. The experimental work described in this paper was conducted in the Heavy Structures Laboratory in the University of Bradford; the assistance of the

laboratory staff is acknowledged. The authors are grateful to Weber and Broutin (UK) Ltd. for providing the CFRP reinforcement and the associated priming and bonding materials for the research.

REFERENCES

[1] Malek, A. M., Saadatmanesh, H. and Ehsani, M. R. (1998), "Prediction of failure load of R/C beams strengthened with FRP plate due to stress concentration at the plate end", ACI Structural Journal, pp. 142-152.

[2] Buyukozturk, O. and Hearing, B. (1998), "Failure behaviour of precracked concrete beams retrofitted with FRP", Journal of Composites for Construction, ASCE, August, pp. 138-144.

[3] El-Refaie, S. A., Ashour, A. F. and Garrity, S. W. (2000), "Tests of reinforced concrete continuous beams strengthened with carbon fibre sheets", The 10th BCA Annual Conference on Higher Education and the Concrete Industry, 29-30 June, Birmingham, UK, pp. 187-198.

[4] El-Refaie, S A., Ashour, A F. and Garrity, S W. (2001a), "Strengthening of reinforced concrete continuous beams with CFRP composites", The International Conference on Structural Engineering, Mechanics and Computation, Cape Town, South Africa, 2-4 April, pp. 1591-1598.

[5] El-Refaie, S. A., Ashour, A. F. and Garrity, S. W. (2001b), "Sagging strengthening of continuous reinforced concrete beams using carbon fibre sheets", The 11th BCA Annual Conference on Higher Education and the Concrete Industry, Manchester, UK, 3-4 July, pp. 281-292.

[6] El-Refaie, S. A. (2001), "Repair and strengthening of continuous reinforced concrete beams", PhD thesis, University of Bradford, UK, 207p.

[7] Arduini, M., Nanni, A., Tommaso, A. D. and Focacci, F. (1997), "Shear response of continuous RC beams strengthened with carbon FRP sheets", Non-Metallic (FRP) Reinforcement for Concrete Structures, Proceedings of the Third International Symposium (FRPRCS-3), Sapporo, Japan, October, Vol. 1, pp. 459-466.

[8] Roberts, T. M. (1989), "Approximate analysis of shear and normal stress concentrations in the adhesive layer of plated RC beams", The Structural Engineer, The Institution of Structural Engineers, London, June, Vol. 67, No. 12/20, pp. 229-233.

[9] El-Mihilmy M. T. and Tedesco, J. W. (2001), "Prediction of anchorage failure for reinforced concrete beams strengthened with fiber-reinforced polymer plates" ACI Structural Journal, May, pp. 301-314.

[10] Ashour, A. F. (2001), "Flexural strength of reinforced concrete beams strengthened with FRP laminates", The 11th BCA Annual Conference on Higher Education and the Concrete Industry, Manchester, UK, 3-4 July, pp. 267-280.

2.13 Behaviour of concrete beams strengthened with near surface mounted reinforcement, NSMR

A Carolin and B Täljsten
Luleå University of Technology and Skanska, Sweden

INTRODUCTION

In the last decades it has become more and more important to repair and strengthen existing concrete structures. The reasons for this are numerous and vary often for each and every case. A great amount of research is also going on all over the world in this field. Many different methods can be suitable for repair and strengthening, shotcrete with steel fibres, additional reinforcement with concrete overlays, and post-tension just to mention a few. One repair and strengthening method that has been growing rapidly for at least the last 10 years is plate bonding with fibre reinforced polymers, FRPs, [1, 2]. It has shown to be a competitive method both regarding structural performance and economical aspects. FRP, also referred to as composites, is a material with high stiffness and strength and it serves as reinforcement when bonded on to a structures surface. The material is lightweight, do not corrode and can come in almost any length and dimension. The method has been found to work both for shear and flexure strengthening. NSMR, near surface mounted reinforcement is, when possible to use, a refined method of laminate plate bonding where the reinforcement is placed in cut grooves in the concrete cover. Further description of NSMR can be found in Täljsten and Carolin [3], Nanni [4], and Rizkalla and Hassan [5]. The method has also been found to be suitable for post-tensioning, see Nordin *et al* [6]. Advantages and drawbacks of NSMR can vary between the objects and must always be considered, Carolin *et al* [7].

One very important question from the owners of structures is if it possible to keep the structure in service during the strengthening process. Full-scale applications have shown that this is possible, but there is lack of understanding how cyclic loads during strengthening, for example traffic loads, affects the final strengthening effect [8]. If the structure can be completely unloaded including the self weight of the structure it will enhance better utilization of the FRP's contribution before the structure reaches critical stages, yielding of reinforcement, concrete failure and so on, see Täljsten [9].

If epoxy is handled in the wrong way it can be harmful to man. Therefore, it is of interest to investigate if the epoxy can be substituted with concrete mortar. Previous tests have shown that it is possible to strengthen with NSMR and use a concrete mortar as bonding agent, Carolin *et al* [7].

Furthermore, structures strengthened in reality are subjected too much more harsh conditions then laboratory beams. In Sweden, it is of certain interest to understand how the strengthening system is affected by cold climate.

This paper presents laboratory tests that investigate the effect of live loads during strengthening process with NSMR, the effect of cold climate and use of cement mortar as bonding agent.

EXPERIMENTAL PROGRAM
The tests have been undertaken on rectangular reinforced 4 m long concrete beams presented in Figure 1. A total of 10 beams were tested, 3 beams were strengthening with dynamic loads during the strengthening process, 3 beams where tested in cold climate and 2 beams were non-strengthened reference beams. The beams are presented in Table 1. The beams that were tested at low temperature were placed in a climate chamber at –28 °C for 48 hours prior testing and then tested to failure at the same temperature. The other beams were tested at + 22 °C.

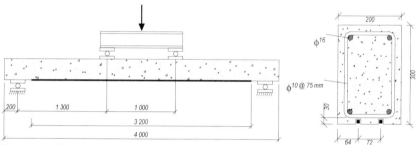

Figure 1: Test specimens

Properties of the used materials can be found in Table 2. The properties of the composites and the adhesive are given by the supplier and are not tested. The composites are made up of carbon fibres and vinyl ester matrix. The concrete quality is tested on 150 mm cubes. Empirical relation to the tested split strength obtains the tensile strength of the concrete. The quadratic rods bonded with cement mortar were covered with a thin layer of quartz sand to improve the anchorage with the bonding agent (cement mortar). Therefore, the slots in these beams were wider 20x20 mm^2, instead of 16x16 mm^2. All beams had an age of approximately 240 days. The beams were equipped with strain gauges on the concrete, the internal steel bars and on the fibres. The beams were subjected to four-point bending both during the strengthening period and when they were taken into failure. In addition to the strain gauges midpoint deflection and support settlement were registered with LVDTs.

Strengthening
Five beams were strengthened while unloaded. The beams had their grooves cut, cleaned and half filled with adhesive before the rods were placed, so that the groove were completely filled and the rod had adhesive on three sides. The length of the

NSMR-rods, 3200 mm, was chosen to be critical for anchorage since it is easy to believe that the anchorage can be affected of the cyclic loads. This length gives rods that end between the supports and will therefore not be significantly affected of the bearings.

Table 1: Test specimens

	Curing condition	Concrete strength [MPa]	Slot size [mmxmm]	Strengthening reinforcement	Adhesive
Reference beam	-	61	-		-
Reference beam cold	-	58	-		-
				BPE®	BPE®
Bsstat	static	61	16x16	NSMR 101 S	465 epoxy
Bscold	static	53	16x16	NSMR 101 S	465 epoxy
BScem-stat	static	58	20x20	NSMR 101 QS	cement
BScem-cold	static	70 / 83*	20x20	NSMR 101 QS	cement
Bsdyn	dynamic	68	16x16	NSMR 101 S	epoxy
BScem-dyn	dynamic	66	20x20	NSMR 101 QS	cement
Bmstat	static	68	16x16	NSMR 101 M	465 epoxy
Bmdyn	dynamic	67	16x16	NSMR 101 M	465 epoxy

*) Tested in cold climate

Table 2: Material properties

	Compressive strength [MPa]	Tensile strength [MPa]	Young's module [GPa]
Concrete	63[1]	3.8[1]	42
Steel	515[2]	515[2]	210
Adhesive[3]			
BPE® 567 epoxy	93[3]	46[3]	7[3]
BPE® 465 epoxy	103[3]	31[3]	7[3]
BPE® cement	45[3]	9[3]	26.5[3]
Composite[3]			
BPE® Quality M		2000[3]	250[3]
BPE® Quality S		2800[3]	160[3]

[1] Average, normal temperature
[2] Yielding
[3] Suppliers data

The beam strengthened with cement mortar adhesive, had the slots saturated with water before strengthening to get the best performance of the cement mortar. After strengthening the mortar was kept moist for 21 days.   Three beams were strengthened in the same manner but with live loads acting during curing of the adhesive. Due to the size of the beams they were mounted in the load set-up prior to strengthening. The beams were strengthened upside down to facilitate the monitoring and visual inspection. In order to keep the test set-up steady a load of 5 kN was chosen as the lower limit. The load program for the live loads started 20 minutes after the mixing of the two adhesive compounds had commenced. This time allowed the strengthening system to be mounted in the same way for all the beams. Every 108 seconds one "sinus shaped" load cycle with a maximum of 40 kN was applied and then was the beam unloaded to 5 kN. The load level was chosen as 60 % of the load for yielding of internal steel bars for the control beam. It is only of interest to simulate trucks since the load and deformation from private cars can be neglected. The "sinus shape" of the load-time curve is to simulate the global behaviour of a bridge rather than isolating each axel on the vehicle. The beams were subjected to the live load for 72 hours and then unloaded and stored until they were tested to failure. The beams strengthened with epoxy adhesive showed no signs of damage immediately after live loading. The beam strengthened with cementitious adhesive had cracks in the adhesive all along the length of the rods. All strengthened beams except the beam with cementitious adhesive were allowed to cure one week after the strengthening commenced until tested to failure. The cementitious bonding agent was allowed to cure for 28 days.

RESULTS AND EVALUATION
Strengthening
The effect of the strengthening system during curing of the adhesive is shown in Figure 2. The figure shows on the left vertical axis the strains over the cross-section versus time plotted for the peak load for every load cycle, beam *BS-dyn*. The mid-point deflection as function of time is also plotted in the same figure with numbers on the right vertical axis.

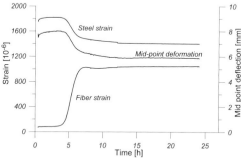

Figure 2: Typical cross-section strains at peak load and mid point deformations for every load cycle versus time from mixing of epoxy.

It can be found from Figure 2 that initially the strains are slightly increasing due to crack distribution in the concrete. Then the strains are quite stable until initiation of polymer setting, which seems to occur approximately at 4 hours. As the epoxy cures the composite rods starts to be stressed. Meanwhile the steel bars are unloaded to an equivalent degree. After setting of the polymer, another 4 hours, the beam has reached its strengthen capacity and the peak strains are unaffected during the remaining load cycles. The decrease of mid-point deflection and steel strains for all beams during hardening of the adhesive can be found in Table 3. It is obvious that the beams, except for *BScem-dyn*, had gain stiffness from the applied FRP. The mid-point deflections had decreased which is the best indicator of the strengthening effect. Also the steel strains have decreased but the measured values are affected by the distances to the closest crack on each side of the gauge, which will be different for all beams.

Table 3: Level of mid-point deflections and steel strains for all beams during hardening of adhesive

	Mid-point deflection			*Steel strain*		
	Before hardening [mm]	*After hardening* [mm]	*Decrease* [%]	*Before hardening* [10^{-6}]	*After hardening* [10^{-6}]	*Decrease* [%]
BMdyn	8.4	5.2	38	1820	1270	30
BSdyn	8	5.9	26	1820	1400	23
BScem-dyn	8.7	8.7	0	1880	1880	0

Failure tests
The load deflection plots for the beams strengthened by NSMR and epoxy are shown in Figure 3.

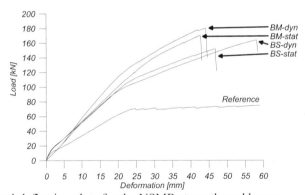

Figure 3: Load-deflection plots for the NSMR strengthened beams

Figure 3 shows in general that strengthening with Near Surface Mounted Reinforcement is an effective strengthening method. It also shows that the dynamic loads during the curing of the epoxy do not affect the strengthening. Beam *BMdyn* failed by fibre rupture in one rod and anchorage failure in the concrete for the other rod. Beam *BMstat* failed by severe concrete damage where the part of the concrete beneath internal steel was detached. The other beams failed by normal anchorage failure in the concrete. In Figure 4 it is shown that a cementitious mortar also works as bonding agent when the beam is strengthened while unloaded. In fact, it is possible to achieve the same capacity with cement as with epoxy. It is also obvious that the same mortar is not useful if dynamic loads during the strengthening are prevailing.

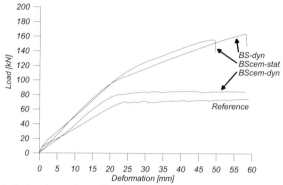

Figure 4: Load-deflection plots for the NSMR strengthened beams with cement adhesive

Beam *BScem-dyn* was acting like the Reference beam. Even if the rods were almost loose in the groove the roughness was able to transfer forces to the rods that actually became stressed. After the steel started to yield the strain in the rods remained at the same level as the deformation of the beam increased. Beam *BScem-stat* failed by fibre rupture in one of the rods. Figure 5 shows the load deflection behaviour in cold climate for tests up to failure.

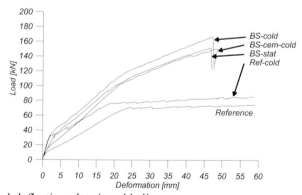

Figure 5: Load-deflection plots in cold climate

It is clear that the fracture energy of the concrete increases in lower temperature. The crack levels of the beams are increased about 3 times, both for strengthened and reference beams. The ultimate capacity is not significantly affected, at least not decreased, by the low temperature, as it is possible to be fear. *Beam BS-cold* failed by bondslip and *BS-cem-cold* by fiber rupture in one rod and anchorage failure in the other. The reference beam did not increase its ultimate load after initiation of yielding and test was aborted when concrete crushing occurred. Ultimate loads and failure descriptions are summarized for all beams in Table 4.

Table 4: Summarized results of test to failure

	Ultimate load [kN]	Deflection at ultimate load [kN]	Failure description
Reference beam	72*	24*	Yielding and large deformation
Reference beam cold	76*	20*	Yielding and large deformation
BSstat	153	47	Bond slip
BScold	166	47	Bond slip
BScem-stat	157	48	Fibre rupture in one rod. Anchor failure in the other
BScem-cold	150	49	Fibre rupture in one rod. Anchor failure in the other
BSdyn	164	58	Anchor failure in the concrete
BScem-dyn	80*	22*	Bond slip. Yielding and large deformation
BMstat	169	39	Anchor failure in the concrete
BMdyn	180	44	Fibre rupture in one rod. Anchor failure in the other

*) yield load

The next step in the evaluation is to compare the test results with theoretical derivations, when it comes to anchorage for the NSMR-rods.

CONCLUSIONS
For the beams strengthened NSMR with epoxy, the cyclic loads are not significantly affecting the strengthening effect. The small differences noticed in the tests can be due to normal scatter of the failure mode. The use of cementitious mortar together with NSMR is not suitable when cyclic loads are prevailing during hardening of the adhesive. Cementitious bonding agent together with NSMR does work when it cures under static conditions. Lower temperatures, neither for epoxy nor cementitious bonding agent, do not affect the strengthening effect.

ACKNOWLEDGEMENTS
Lars-Erik Lundbergs Foundation, The Swedish Building and Development Fund (SBUF) and Skanska AB have provided financial support. For help with the laboratory tests special attention should be given to Håkan Johansson and Georg Danielsson at TESTLAB, Luleå University of Technology. The assistance of Professor Lennart Elfgren is also greatly appreciated. Last but not least the students Arvid Hejll, Otto Norling and Mattias Clarin should be thanked for their never-ending energy working with this project.

REFERENCES
[1]   Meier, U *et al* (1993): *"CFRP bonded sheets"*. Editor Nanni, A., Amsterdam: Elsevier. ISBN 0-444-89689-9. pp 423-434.

[2]   Neale, KW. and Labossiére, P., 1997, *"State-of-the-art report on retrofitting and strengthening by continuous fibre in Canada"*, Non-Metallic (FRP) Reinforcement for Concrete Structures, Japan Concrete Institute, 1997, pp 25-39.

[3]   Täljsten, B. and Carolin, A. (2001): "CFRP - Strengthening. Concrete Beams Strengthened with Near Surface Mounted CFRP Laminates" *Fibre reinforced plastics for reinforced concrete structures, FRPRCS-5*, Cambridge (Edited by Chris Burgoyne), pp 107-116.

[4]   Nanni, A. (2001): "North American design guidelines for concrete reinforcement and strengthening using FRP: Principles, applications and unresolved issues" *International Conference on FRP Composites in Civil Engineering*, Volume 1, J.-G. Teng (Ed). ISBN: 0-08-043945-4 pp 61-72.

[5]   Rizkalla, S. and Hassan T. (2001): "Various FRP strengthening techniques for retrofitting concrete structures" *International Conference on FRP Composites in Civil Engineering*, Volume 2, J.-G. Teng (Ed). ISBN: 0-08-043945-4 pp 1033-1040.

[6]   Nordin, H., Täljsten, B. and Carolin, A. (2001): "Concrete beams strengthened with prestressed near surface mounted reinforcement (NSMR)" *International Conference on FRP Composites in Civil Engineering* Volume 2, J.-G. Teng (Ed). ISBN: 0-08-043945-4 pp 1067-1075.

[7]   Carolin, A., Nordin, H. and Täljsten, B. (2001): "Concrete beams strengthened with near surface mounted reinforcement of CFRP" *International Conference on FRP Composites in Civil Engineering*, Volume 2, J.-G. Teng (Ed). ISBN: 0-08-043945-4 pp 1059-1066.

[8]   Carolin, A (2001): "Strengthening of concrete structures with carbon fibre reinforced polymers –Shear strengthening and full-scale applications" Licentiate thesis. Luleå University of Technology, Division of structural engineering. 118 pp

[9]   Täljsten, B. (2002): *"Strengthening of existing concrete structures with externally bonded Fibre Reinforced Polymers – design and execution"*. Technical report. Luleå University of Technology, Division of structural engineering. Under printing.

Session 3 : **Application of FRP to metallic structures**

3.1 Early age curing under cyclic loading - an investigation into stiffness development in carbon fibre reinforced steel beams

S S J Moy
University of Southampton, Southampton, UK

INTRODUCTION

When a carbon fibre composite (CFRP) plate is bonded to a metallic substrate there is a curing period during which the adhesive develops its full strength. For commonly used epoxy adhesives (Sikadur PBA31 for example) at normal temperatures this period will be about 24 hours.

If a railway underbridge is strengthened in this manner it is not usually economic to close the bridge during the curing period. Thus the adhesive will be subjected to cyclic loading each time a train passes over the bridge. It was found that trains crossed a London Underground bridge at Acton in West London every two to three minutes during the curing period. This paper reports research [1,2] which was undertaken to investigate the development of stiffness in CFRP reinforced steel beams subjected to cyclic loading during early age cure of the adhesive. The research has shown that the final strength of the adhesive is reduced and under certain conditions bond will not develop between the CFRP and the steel.

GENERAL DESCRIPTION OF THE SPECIMENS

Two test programmes were carried out. In the first five steel beams, each 1.2m long, were reinforced with a single carbon fibre composite plate, 0.98m long, attached to the tension flange. The steel beams were 127×76UB13 sections, with a flange thickness of 7.6mm. The reinforcing plates were fabricated using K13710 ultra-high modulus carbon fibres and were 7.6mm thick and 76mm wide. The ends of the plates were not tapered because calculations had indicated that peel would not be a problem. Four specimens were cured under load; the fifth was prepared in the same manner but was used as a control specimen and was not subjected to load during adhesive cure.

In the second test programme a total of ten specimens were available. They were identical in terms of sizes and dimensions to the first set and were prepared in five pairs. One specimen of the pair was cured under cyclic load while the other was retained as a control.

The preparation of all the specimens was identical. The surface of one flange of the steel beam was grit-blasted to the SA2½ standard (the common industrial standard for surface preparation). Within one hour the carbon fibre plate was attached to the steel flange using two-part epoxy adhesive (Sikadur BPA31). After mixing a layer of adhesive was applied to the steel and carbon plate; the adhesive was shaped as shown in Figure 1. When the carbon was placed on the steel and pressed into position the shaping allowed air to be pushed out resulting in an even layer of adhesive about 1.5mm thick over the whole contact area. Six small G-clamps were used to hold the carbon in place although they were only done up finger-tight. Specimens to be cured under load were immediately transferred to the loading rig, control specimens were stored and left to cure.

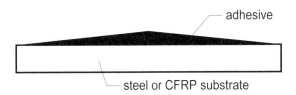

Figure 1. Shaped adhesive on substrate

TEST SET-UP
The tests were carried out in a servo-hydraulic Instron 8032 test machine. A heavy steel beam was placed on the lower platen to support the test specimen via two rollers placed 1m apart during the first set of tests, and 1.1m apart during the second set. A sinusoidal load was applied to the specimen at a frequency of 0.25Hz, and this was continued for up to 160 hours. At intervals the load cycling was stopped and a static test was carried out to measure the load-deflection response of the specimen.

In the first test four point bending was used with a minimum load of about 1kN and a maximum of 50kN. During an unattended period overnight the spreader beam moved such that load was applied only to the outstand parts of the loaded flange, resulting in localised damage to the flange. It was decided that all subsequent tests would use three point bending. In the subsequent tests the minimum load was kept as 1kN but a range of values were used for the maximum load as shown in Table 1.

During the static tests mid-span deflections of the specimens were measured using two displacement potentiometers and also by the platen displacement. The latter included the deflection of the test machine (small) and the spreader beam and was only used as a check. The potentiometers were mounted so that they gave specimen displacements relative to the roller supports. Plate 1 shows the four point bending test and the instrumentation. The maximum load during the static tests was kept to the maximum cyclic load except in the final test after full adhesive cure when the load was taken up to at least 50kN.

Table 1. Load details for test specimens

Specimen	Minimum Load (kN)	Maximum Load (kN)
First set of tests		
50-4point	1	50
50/1	1	50
50/2	1	50
70/1	1	70
Second set of tests		
25/1	1	25
34/1	1	34
42/1	1	42
50/3	1	50
62/1	1	62

The control beams were allowed to cure undisturbed and were then subjected to a static test identical to and at the same time as the final test on the corresponding cyclically loaded specimen.

LAP-SHEAR TESTS
Lap-shear specimens were prepared from beams 50/1, 50/2 and the control beam 50/C1 [3]. Preparation involved cutting off the strengthened flange and machining the cut edge smooth. Three specimens were cut from each end and the centre of the flange giving a total of nine specimens per beam. Finally slots were machined through the steel on one side and the carbon plate on the other, giving a shear area of approximately 25mm by 20mm on each specimen. The ends of the specimen were clamped in the jaws of a test machine and load was increased at a constant rate until a failure occurred within the shear area.

TEST RESULTS
Deflections were recorded using a data logger and were loaded into Excel files for analysis and presentation. Figure 2 shows the load-deflection graphs from specimen 50/1 plotted from tests at different ages. Figure 3 shows the graphs for specimen 70/1. Figure 4 shows stiffness versus time graphs for specimens 34/1and 34/C and figure 5 shows the corresponding graphs for specimens 42/1 and 42/C. Table 2 summarises the stiffness results from the beam tests. Table 3 summarises the results of the lap-shear tests.

DISCUSSION OF RESULTS
There were experimental problems which had an influence on the test results. The displacement potentiometers were mounted beside the specimen on steel bars resting on the support rollers. There was a small amount of movement of the bars which would have contributed to the variability of the results.

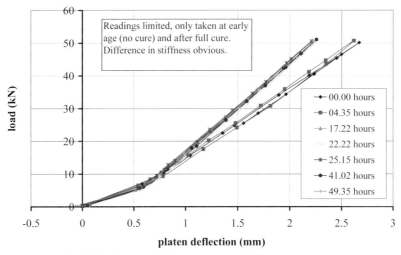

Figure 2. Load - deflection graphs for specimen 50/1

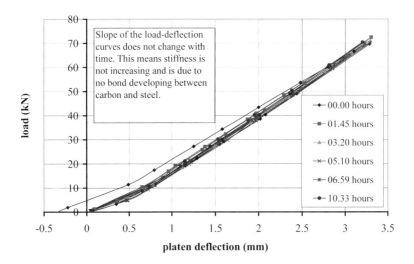

Figure 3. Load deflection graphs for specimen 70/1

Figure 2 shows typical load- deflection graphs obtained during the static tests. In this case results are shown before and after the adhesive had cured and the difference in slope of the graphs is clear. The increased slope after cure shows the increase in stiffness resulting from the bonding of the carbon fibre plate to the steel. Figure 3 shows the corresponding graphs for specimen 70/1. It can be seen that the slope of the graphs does not change. This was because bond did not develop between the carbon plate and the steel. A similar situation arose with specimen 50/3 although in this case bond started to develop but broke down after 23 hours of curing.

Table 2. Summary of stiffness results from beam bending tests

Specimen	Bending stiffness (kN/mm)	Bending stiffness of control beam (kN/mm)
	First set of tests	
50-4point	damaged	
50/1	38.80	43.20
50/2	40.10	
70/1	32.03*	
	Second set of tests	
25/1	42.60	42.26
34/1	43.57	41.97
42/1	39.80	43.30
50/3	32.27 (37.00)*	39.03
62/1	38.15	39.97

Table 3. Summary of lap-shear test results

Specimen	Failure stress (N/mm^2)	Stiffness (kN/mm)
Beam 50/1		
End 1 mean	12.61	2.64
Centre mean	23.18	3.17
End 2 mean	19.86	2.64
Overall mean	17.97	2.79
Beam 50/2		
End 1 mean	12.30	2.44
Centre mean	20.16	2.72
End 2 mean	15.55	2.38
Overall mean	16.13	2.55
Beam 50/C1		
End 1 mean	15.18	2.58
Centre mean	23.85	2.81
End 2 mean	19.70	2.56
Overall mean	20.13	2.66

Figures 4 and 5 show the development of bending stiffness with time for specimens 34/1 and 42/1 respectively. It can be seen that stiffness increases rapidly at first and the rate of increase slows with time. Figure 4 shows the variability in the test results but it is clear that specimen 34/1 achieves a final stiffness which is the same as that of the control beam while specimen 42/1 achieves a final stiffness lower than that of the control beam. Table 2 compares the final stiffness values of all the specimens with those of the control specimens. The values marked with an asterisk indicate specimens in which there was a bond failure. It is clear that above a maximum cyclic load value of 34kN the specimens did not achieve the full bending stiffness. The slip between the carbon plate and the steel beam during cyclic loading had an effect on the bond developed in the adhesive but only when the slip was above a certain value. Separate calculations have shown that at 34kN the slip was 0.17mm in half the span increasing to 0.21mm at 42kN. A possible explanation is that movement above the limiting value breaks some of the chemical chains forming in the adhesive preventing full cure from occurring. It is not possible at this stage to give a limiting value at which bond will not develop because no bond developed in specimen 70/1, bond did develop in 62/1 and bond developed but then broke down in 50/3. At this stage it is prudent to put a conservative limit to the required shear strength in the adhesive layer of $1N/mm^2$, which corresponds to a maximum cyclic load in these tests of 35kN.

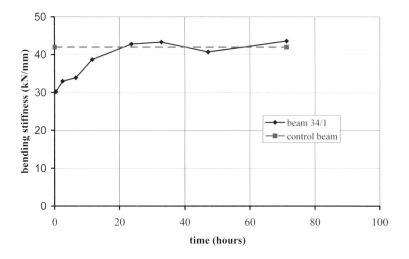

Figure 4. Bending stiffness - time graphs for specimens 34/1 and 34/C

The lap-shear tests revealed further interesting information. It was found that the shear strength was lower at the ends of the beam that at the centre. As shown in Table 3 there was a difference of approximately 25% between values measured at the ends and the centre. The slip would have been greatest at the ends and theoretically zero at the centre during cyclic loading. Interestingly there were similar results in the control beam but the end means were affected by rogue results. An attempt was made during the lap-shear tests to evaluate the shear stiffness of the adhesive layer and values are presented in Table 3. However these values should be treated with some caution because the load-displacement curves obtained from the test machine were difficult to interpret. The lap-shear tests themselves did not conform to the usual test standards mainly because the thicknesses of the adherends were too great. It is not usual to measure displacement during these tests.

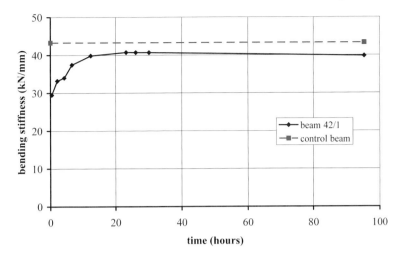

Figure 5. Bending Stiffness - time graphs for specimens 42/1 and 42/C

The second series of specimens were not cut up for lap-shear testing. They have been subjected to further bending tests as reported separately.

CONCLUSIONS
- The tests confirmed the gradual build up of stiffness as the adhesive cures and bond develops between the steel and the carbon plate.
- Cyclic load during adhesive cure will have an effect on the final stiffness of the reinforced beam. In the tests reported when the maximum cyclic load was 34kN there was no reduction but when the load was 42kN and greater there was significant reduction.
- Bond will fail to develop if the deformation of the adhesive during cure is too great. However it is not possible to quote a limiting value because of the variability in the test results.
- It is prudent to limit the shear stress in the adhesive layer to a maximum of 1N/mm^2.
- The shear strength of the adhesive is reduced more at the ends of the beam where the slip during cyclic load is greatest. At the centre of the beam where the slip is theoretically zero there is no reduction in strength compared to manufacturers quoted values.
- Further testing will show the effect on the collapse load of the reduced stiffness of the steel-carbon interface.

REFERENCES
[1]   Moy, S.S.J. Early age curing under cyclic loading - an investigation into stiffness development in carbon fibre reinforced steel beams. Department of Civil and Environmental Engineering Research Report, University of Southampton, 2000.

[2] Moy, S.S.J. Early age curing under cyclic loading - a further investigation into stiffness development in carbon fibre reinforced steel beams. Department of Civil and Environmental Engineering Research Report, University of Southampton, 2001.

[3] Moy, S.S.J. Lap-shear tests on specimens cut from CFRP reinforced steel beams. Department of Civil and Environmental Engineering Research Report, University of Southampton, 2001.

3.2 Flexural behaviour of steel beams reinforced with carbon fibre reinforced polymer composite

S S J Moy and F Nikoukar
University of Southampton, Southampton, UK

INTRODUCTION

Carbon fibre reinforced polymer composite (CFRP) is being used to strengthen metallic and concrete structures and design guides [1, 2] have been written to help the designer. The number of concrete structures strengthened in this manner is now considerable and there is enough evidence to show that the process is robust and reliable. This is not the case with metallic structures. Only a few beams, for example those on the London Underground steel bridge at Acton in West London, have been strengthened using CFRP. In addition very little testing has been carried out on steel structures with CFRP so that the design guidance [2] is based on limited evidence.

The Acton bridge was strengthened with CFRP plate bonded to the tension flange of the steel beams with the objective of reducing live load stresses by about 25%. The bridge layout was such that the beams carried insignificant dead load so that the cyclic imposed load from trains passing over the bridge was a applying a severe fatigue loading to the beams. Reducing live load stresses by 25% would extend the fatigue life considerably. The strengthening was completely successful, and monitoring pre- and post-strengthening showed that the desired stress reduction had been achieved.

The strengthening scheme was designed on the basis of the method of transformed sections. In order to reduce the amount of CFRP it was necessary to use ultra-high modulus material. There were concerns about the effect of cyclic load during adhesive cure on the final strength of the adhesive. Various tests have been undertaken to investigate this and these have been reported elsewhere [3, 4]. However a number of specimens were retained after those tests and these have been tested to failure. This paper presents the results of those tests and shows that there are further considerations which need to be addressed when using CFRP to strengthen metallic structures.

DETAILS OF THE SPECIMENS

The specimens all consisted of 127×76UB13 steel beams 1.2m long. A 7.6mm thick by 76mm wide carbon fibre composite plate was attached to one flange of the beam using epoxy adhesive. The steel was nominally grade 43A ($E = 205kN/mm^2$ and $\sigma_y = 275N/mm^2$) and ultra-high modulus uni-directional K1310 carbon fibres were used in the composite plates ($E = 310kN/mm^2$). The flange of the steel beam was shot-blasted to the SA2½ standard immediately prior to applying the CFRP. Thin, shaped, layers of adhesive were applied to the steel and the CFRP, which were then carefully bonded to exclude air from the adhesive layer. G-clamps were used to lightly hold the surfaces together during adhesive cure. Some of the specimens were subjected to cyclic loading during adhesive cure [3, 4] others were retained as control specimens.

The properties of the reinforced section were calculated using the method of transformed sections and are compared with the properties of the original steel section in Table 1.

Table 1. Beam section properties

Section	Distance of centroid from top (mm)	Second moment of area, I (cm^4)	Section elastic modulus (cm^3)	
			top	bottom
Unreinforced	63.5	473	74.5	74.5
Reinforced	86.8	732	84.3	231.5

The identification of the specimens was determined by the earlier cyclic load tests. Thus specimen 34/1 was the first specimen subjected to 34kN maximum cyclic load and specimen 50/3 was the third specimen subjected to 50kN maximum cyclic load. The control specimens were identified typically by C50, which indicated the control beam corresponding to the specimens subjected to 50kN maximum cyclic load.

TEST DETAILS
The specimens were subjected to three-point bending. This was chosen because it matched the loading used in the cyclic load tests. Several of the specimens were heavily strain gauged on both the steel and the composite. Mid-span vertical deflections were also measured during the test.

The test procedure was identical for each specimen. A modest load was applied and removed to allow the specimen to bed-in. Initial readings were taken on all gauges and load was then increased in increments until failure occurred, at each increment of load all the gauges were scanned. As failure approached the magnitude of the increments was reduced.

TEST RESULTS
Figures 1 and 2 show load-deflection graphs for several of the specimens, Figure 3 shows strain gauge readings at key locations for specimen C42, Figure 4 shows strain distributions across the depth of specimen C42 at various loads and Table 2 gives the bending stiffness (taken as the slope of the linear portion of the load-deflection curve) and the failure load of the specimens. Results are only given for one control beam (C50) because the results for all the control beams were remarkably similar.

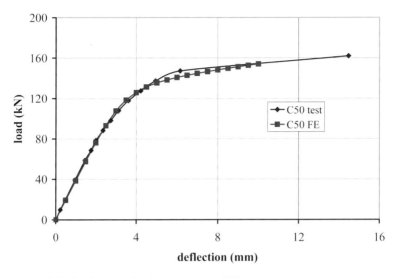

Figure 1. Load-deflection graphs for specimen C50

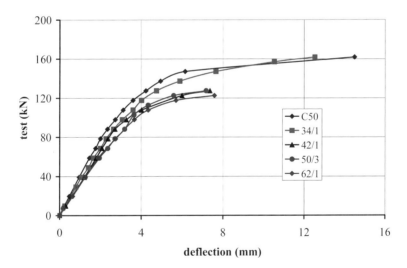

Figure 2. Load-deflection graphs for several specimens

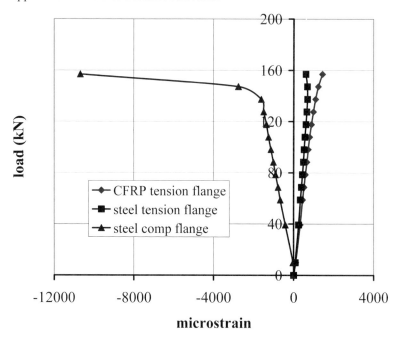

Figure 3. Strain gauge readings at key locations

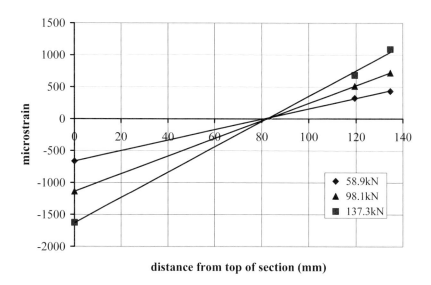

Figure 4. Strain distributions across the depth of the section at various loads

Table 2. Summary of test results

Specimen	Bending stiffness (kN/mm)	Failure load (kN)
C50	39.0	162
25/1	37.7	103
34/1	35.9	162
42/1	32.9	127
50/3	28.2	127
62/1	31.0	122

DISCUSSION OF THE TEST RESULTS

The failure load for beam 25/1 was very low because of an oversight on our part. The bare steel beam is classified as plastic and was also satisfactory against web crushing but no check was made against crushing of the web of the reinforced beam. During the test the beam web crushed giving the low failure load. As a result web stiffeners were welded into the remaining specimens at mid-span and the support positions as shown in Figure 6. There were no further problems with web crushing. Failure was either due to steel yielding accompanied by very large deflections where effectively a plastic hinge had formed at mid-span or to debonding of the CFRP plate from the steel. The rotation capacity of the plastic hinge would have been limited by the strain capacity of the CFRP but the tests were stopped before failure of the CFRP.

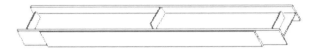

Figure 6. Schematic showing a typical specimen with added stiffeners.

It can be seen from Table 2 that the level of the maximum cyclic loading during adhesive cure had a marked effect on the bending stiffness of the beams. Generally the greater the maximum cyclic load the lower the bending stiffness. It was known from previous tests [3, 4, 5] that the cyclic loading reduced the shear stiffness of the adhesive and this in turn affected the reinforced beam. An analysis based on bending theory for laminated timber beams [6] was carried out to give a theoretical estimate of the effect of the adhesive stiffness on the beam stiffness. Figure 7 shows the results of the analysis. It can be seen that even at very high adhesive shear stiffness there is some reduction in second moment of area, but the reduction is modest until the shear stiffness drops below about $10kN/mm^2$. From the test results it would appear that the adhesive stiffness is about $4kN/mm^2$ which compares well with typical manufacturer quoted values for epoxy adhesive.

It can be seen in Table 2 that the bending stiffness for beam 50/3 did not follow the expected sequence. This was because some debonding had occurred during adhesive cure under cyclic loading which further reduced the average shear stiffness of the adhesive.

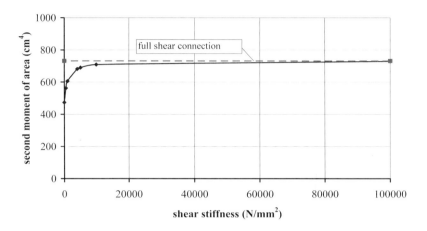

Figure 7. Bending stiffness (second moment of area) versus adhesive shear stiffness

It can also be seen in Table 2 that there was a reduction in failure load with higher levels of cyclic loading. With specimens C50 (and the other control specimens) and 34/1 failure resulted from yielding in the compression extreme fibres spreading into the section, producing eventually a plastic hinge with limited rotation capacity (limited by the strain capacity of the CFRP) as had been predicted in earlier theoretical studies [7]. In the other specimens failure was due to debonding of the CFRP plate. Analyses have been carried out [4] to determine the level of shear stress in the adhesive. It was found that the maximum shear stress in the adhesive during cyclic loading was less than $1.0N/mm^2$ in specimens which later failed by plastic hinge formation, although the shear stress at failure in all specimens was considerably greater than $1.0N/mm^2$.

Figure 4 shows strain distributions across the depth of the section of beam C42 at various loads. It can be seen that the readings lie on or very close to the best-fit straight line as would be expected if there was perfect bond between steel and CFRP. Also the position of the neutral axis in the figure is very close to the predicted centroid of the reinforced section (see Table 1). The maximum load shown (137.3kN) was 87.5% of the failure load. Figure 3 confirms that the behaviour of the reinforced beam was linear up to this load. This behaviour is reassuring because it confirms that the method of transformed sections is good enough for predicting the reinforced section behaviour although it will need to be modified eventually to take account of the adhesive flexibility. This should be possible by using an efficiency factor for the CFRP reinforcement as in the method for timber sections [6].

FINITE ELEMENT ANALYSIS

The test programme was limited to relatively small beams of a particular section size and reinforcement detail. The cost of CFRP means that testing larger beams and other configurations will be very expensive. Consequently the test results are being used to calibrate a finite element model of the bending behaviour of the beams, which can subsequently be used for parametric studies.

The model has to be able to deal with progressive damage in the CFRP as well as non-linear behaviour in the steel. There are five types of damage, fibre breakage, fibre buckling, resin cracking , resin crushing and inter-laminar shear, which are relevant to the unidirectional fibres used and these have been modelled by modifying appropriate material properties in the damaged regions of the CFRP [8]. The adhesive layer has been considered as an unreinforced resin with properties which vary with time to model strength and stiffness gain during cure and also to allow for the deleterious effects of cyclic loading. Figure 8 shows a typical strength-time relationship. The results of the finite element analyses are very promising and as can be seen in Figure 1 the model predicts the load-deflection characteristics very accurately. The load-strain trends are predicted qualitatively, but the accuracy needs to be improved. Lack of space precludes a figure to illustrate this.

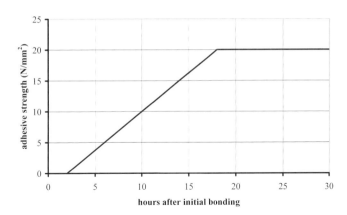

Figure 8. Typical strength-time relationship assumed for the adhesive

CONCLUSIONS

- The higher loads resulting from bonding CFRP reinforcement can cause web crushing which would not arise in the steel beam alone. The reinforced section needs to be checked against this possibility.

- Adhesive cure under cyclic loading can affect the bending stiffness and failure load of the reinforced beam. In some situations the effect is serious enough to change the failure mode of the beam from flexure with a limited rotation capacity plastic hinge to a debonding failure in the adhesive. To prevent this possibility the maximum adhesive shear stress during curing should be limited to $1.0N/mm^2$.

- The method of transformed sections is good enough for determining the properties of the reinforced section although it may need to be modified to allow for adhesive flexibility.

- The bending behaviour of steel beams reinforced with CFRP is predictable, even when the adhesive is cured under cyclic loading as would be the case when vehicle loading is not restricted during the strengthening operation.

REFERENCES

[1] Design Guidance for Strengthening of Concrete Structures using Fibre Composite Materials. The Concrete Society, Technical Note No.55. 1999 ISBN 0 9466 9184 3.

[2] FRP Composites. Life Extension and strengthening of metallic structures. ICE design and practice guide. Editor S.S.J. Moy. Thomas Telford Ltd, July 2001. ISBN 0 7277 3009 6.

[3] S.S.J. Moy. Early age curing under cyclic loading - an investigation into stiffness development in carbon fibre reinforced steel beams. Department of Civil & Environmental Engineering Research Report, University of Southampton, September 2000.

[4] S.S.J. Moy. Early age curing under cyclic loading - a further investigation into stiffness development in carbon fibre reinforced steel beams. Department of Civil & Environmental Engineering Research Report, University of Southampton, July 2001.

[5] S.S.J. Moy. Lap-shear tests on specimens cut from CFRP reinforced steel beams. Department of Civil & Environmental Engineering Research Report, University of Southampton, January 2001.

[6] W. Schelling. Laminated Timber Sections. Translation of paper in German communicated by J. Mallinckrodt, 1990.

[7] S.S.J. Moy. A theoretical investigation into the benefits of using carbon fibre reinforcement to increase the capacity of initially unloaded and preloaded beams and struts. Department of Civil & Environmental Engineering Research Report, University of Southampton, March 1999.

[8] G.S. Pahdi, R.A. Shenoi, S.S.J. Moy, and G.L. Hawkins. Progressive Failure and Ultimate Collapse of Laminated Composite Plates in Bending. Composite Structures. Vol 33, No 11, 1997.

3.3 Mechanical splice of grid CFRP reinforcement and steel plate

K Sekijima[1], T Kuhara[2], S Seki[2] and T Konno[3]
[1]Institute of Industrial Science, The University of Tokyo, Japan,
[2]Shimizu Corporation, Tokyo, Japan,
[3]Asahi Glass Matex Co, Ltd, Kanagawa, Japan

INTRODUCTION

Generally, the substructures in urban areas of Japan have a tendency to become deep and constructed in the soft ground with a high water pressure. The closed type shield tunnelling is an appropriate method to secure this situation. The lateral wall of a vertical shaft used for breaking in or breaking out of a shield tunnelling machine is constructed with a continuous underground reinforced concrete wall, steel sheet-piles and so on.

Recently, a new shield-cuttable tunnel wall system has been developed. The low shear strength of Carbon Fiber Reinforced Polymer (CFRP) reinforcement perpendicular to the carbon fibers was effectively used as the reinforcement in the concrete wall of a vertical shaft. This system enables a shield tunnelling machine to break in or break out by directly cutting the launch or arrival part of the wall with its cutter bits without performing prior ground improvement works and manual demolition [1].

When this system is applied to a column-type diaphragm wall of a vertical shaft, H-shaped steel piles used as the reinforcement in the wall should be replaced by the bit-cuttable mixed-in-place concrete piles reinforced with CFRP reinforcement.

In order to connect the grid CFRP reinforcement and the flange of H-shaped steel, a specially designed mechanical splice has been developed. The splice was made of some pieces of steel plate and resin mortal, and it contained three cross points of the grid CFRP reinforcement.

The tension test of the mechanical splice was carried out to investigate its mechanical behavior. A rupture of the longitudinal bar of the grid CFRP reinforcement outside the splice occurred without a failure of the splice. It was confirmed that the mechanical splice could transmit the tensile capacity of the grid CFRP reinforcement to the steel plate.

GRID CFRP REINFORCEMENT
Specification
The grid CFRP reinforcement was made of PAN type high-strength carbon fibers impregnated with vinylester resin, and it was molded into a grid shape by a pin winding method, which is a kind of filament winding methods. The volume content of carbon fibers was 43 %. Five types of the grid CFRP reinforcement of large diameters were used, and the range of their nominal cross-sectional area was from 100 to 395 mm^2 as shown in Table 1.

Tensile properties
The tension test was carried out in accordance with "Test Method for Tensile Properties of Continuous Fiber Reinforcing Materials" proposed by the committee of the Japan Society of Civil Engineers [2]. The mechanical properties of the grid CFRP reinforcement are shown in Table 1. Each data was an average of five specimens. The ultimate strain and the tensile strength depended on the nominal cross-sectional area, consequently an obvious scale effect was observed.

Table 1. Specification and tensile properties of grid CFRP reinforcement

Bar No.	Nominal cross-sectional area A (mm^2)	Tensile capacity F_u (kN)	Ultimate strain ε_u (%)	Tensile strength f_u (N/mm^2)	Young's modulus E (kN/mm^2)
C16	100	170	1.59	1,700	104
C19	148	223	1.58	1,510	93.6
C25	260	355	1.48	1,360	86.9
C29	320	426	1.39	1,330	64.5
C32	395	522	1.38	1,320	94.2

STRENGTH OF CROSS POINT
Background
The grid CFRP reinforcement has a high strength at cross point, therefore the transverse bar provides a good bond and an anchorage to concrete. A specially designed mechanical splice also made use of the high strength of cross point. The authors have already investigated the strength of cross point of four types of the grid CFRP reinforcement of large diameters [3, 4].

Material and specimen
Material
The specification of the grid CFRP reinforcement is shown in Table 1. This time, the specimens for C19 were newly investigated.

Specimen
The specimens were made by referring to "Test Method for Bond Strength of Continuous Fiber Reinforcing Materials by Pull-out Testing" proposed by the

committee of the Japan Society of Civil Engineers [5]. An example of the
dimensions of the specimens is shown in Figure 1. However, the dimensions of the
specimens were much larger than those of Ref. [5]. The number of the specimens
was three for each type of the grid CFRP reinforcement.

The grid CFRP reinforcement which had only one cross point was arranged in the
centre of a cubic form, and then concrete was placed. The longitudinal bar of the
grid CFRP reinforcement was wound with a thin polyethylene film and a vinyl tape
to remove bond between the longitudinal bar and concrete. A spiral steel
reinforcement hoop was arranged to prevent a splitting failure of concrete. A high
early strength portland cement and coarse aggregate with the maximum size of 20
mm were used for concrete mixture. The compressive strength of concrete was from
42.3 to 45.6 kN/mm². After a steel pipe had been installed at the opposite end of the
longitudinal bar, it was filled with a highly expansive paste [6].

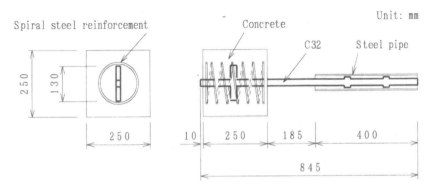

Figure 1. Dimension of specimen for test for strength of cross point (C32)

Test method
Test method was a modification of Ref. [5], too. First, the bearing device was
installed on two H-shaped steel beams and tightened to them with bolts. A centre-
hole jack and a load-cell were installed horizontally, and then the specimen was also
set horizontally. A tensile force was applied to the longitudinal bar of the specimen
by pulling the steel pipe with the jack. The load, the displacement of the free end
and the strain of the longitudinal bar were measured during the test duration.

Test result
Failure mode of cross point
The test results of the strength of cross point are shown in Table 2. Each data was an
average of three specimens. There were three types of failure modes, namely mode
A (two facial shear failure of the transverse bar), mode B (rupture of the longitudinal
bar at cross point) and mode C (mainly rupture of the longitudinal bar at cross point,
but partly slipping out) as shown in Figure 2.

Table 2. Test results of strength of cross point

Bar No.	Maximum load	Strength of cross point	Ratio of strength of cross point and tensile strength	Failure mode
	F_{cu} (kN)	f_{cu} (N/mm^2)	f_{cu} / f_u	
C16	61.5	615	0.36	A, A, A
C19	102	690	0.46	B, A, A
C25	183	703	0.51	B, B, B
C29	203	634	0.48	B, C, C
C32	220	556	0.42	B, A, B

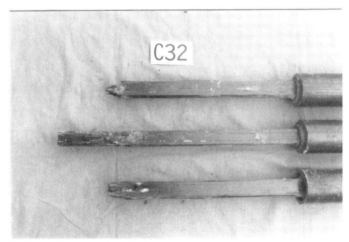

Figure 2. Failure of cross point (C32)

Strength of cross point
In this test, the strength of cross point was defined as the maximum load divided by the nominal cross-sectional area. As shown in Table 2, the strength of cross point of the grid CFRP reinforcement was from 36 to 51 % of the tensile strength.

As for the specimens larger than C19, the strength of cross point became lower as an increase of the nominal cross-sectional area, and consequently, a scale effect was observed. However, the strength of C16 was lower than that of C19, and this phenomenon differed from the tensile strength. As already described, the grid CFRP reinforcement was composed of carbon fibers and vinylester resin, and carbon fibers were laminated alternatively at cross point. In case of C16, the strength of cross point decreased owing to a small number of carbon fiber laminates at cross point.

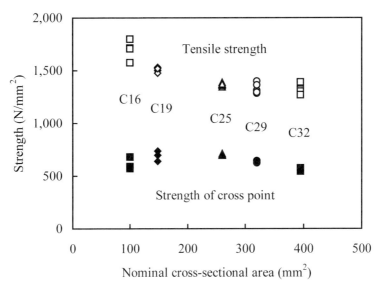

Figure 3. Relationship between strength of cross point and nominal cross-sectional area

MECHANICAL SPLICE OF GRID CFRP REINFORCEMENT AND STEEL PLATE

Material and specimen
Material
The specification of the grid CFRP reinforcement was C16, C19, C25 and C29 shown in Table 1. The specification of the steel plate was SS400 (nominal tensile strength is 400-510 kN/mm^2).

Specimen
As shown in Figure 3, the strength of cross point of the grid CFRP reinforcement was from 36 to 51 % of the tensile strength. Therefore, three cross points would be necessary to transmit the tensile capacity of the grid CFRP reinforcement to the steel plate.

Figure 4 shows the dimension of the specimens for C25 or C29. In case of those for C16 or C19, the interval of the transverse bars was 100 mm. The number of the specimens was three for each type of the grid CFRP reinforcement. The mechanical splice was mainly made of three steel plates. Eight small pieces of steel plate with a circular hole were welded on the surface of the lower steel plate to make a U-shaped groove.

First, the grid CFRP reinforcement which had three cross points was arranged in the groove of the lower steel plate, and second, resin mortar was placed to fill up a gap between the grid CFRP reinforcement and the steel plates. Then, the upper steel plate was tightened to the lower steel plate with bolts and nuts. After a steel pipe had been installed at the end of the longitudinal bar, it was filled with a highly expansive paste [6].

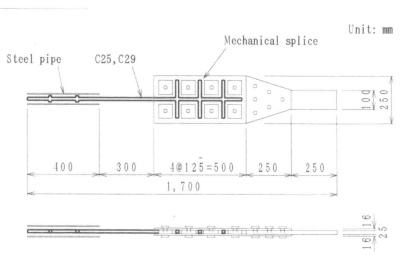

Figure 4. Dimension of specimen for test for mechanical splice (C25 or C29)

Test method
The specimen was set vertically. The steel pipe and the opposite end of the middle steel plate were gripped by the devices of a testing machine, and then a tensile force was applied to the specimen. The load and the strain of the longitudinal bar were measured until a rupture of the specimen.

Test result
Failure mode
The test results are shown in Table 3. Each data was an average of three specimens except C29. In case of two specimens of C29, the longitudinal bar slipped out from the steel pipe (mode S). The other specimens were failed owing to a rupture of the longitudinal bar of the grid CFRP reinforcement outside the splice without a failure of the splice (mode R). The inside of the mechanical splice was observed after the test. A crack occurred on the resin mortar near the end of the splice, however the resin mortar of the other part was not damaged.

Performance of mechanical splice
The ultimate load of each specimen became higher as an increase of the nominal cross-sectional area as well as the tensile capacity of the grid CFRP reinforcement.

The performance of the splice should be express as the ratio of the ultimate load P_u and the tensile capacity F_u. Figure 5 shows the relationship between the performance of the splice and the nominal cross-sectional area. That of one specimen for C16 was a little bit low, however those of the other specimens except two specimens for C29, which were failed owing to a slipping out of the longitudinal bar from the steel pipe, were higher than 90 %. It was confirmed that the splice could transmit the tensile capacity of the grid CFRP reinforcement to the steel plate [7].

Table 3. Test result of mechanical splice

Bar No.	Ultimate load	Ratio of ultimate load and tensile capacity	Failure mode
	P_u (kN)	P_u / F_u	
C16	154	0.91	R, R, R
C19	214	0.96	R, R, R
C25	365	1.03	R, R, R
C29	436	1.02	S, S, R

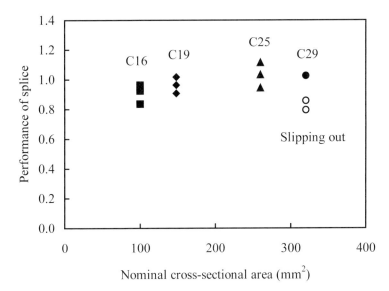

Figure 5. Relationship between performance of splice and nominal cross-sectional area

CONCLUSIONS
In order to connect the grid CFRP reinforcement and the flange of H-shaped steel, a specially designed mechanical splice made of some pieces of steel plate and resin mortal has been developed.

The mechanical splice contained three cross points of the gird CFRP reinforcement, because the strength of cross point was from 36 to 51 % of the tensile strength.

The mechanical splice could transmit the tensile capacity of the grid CFRP reinforcement to the steel plate.

REFERENCES

[1]   Matsumoto, Y., Sonoda, T., Nakamura, M. and Takano, Y., "Novel Material Shield-Cuttable Wall System," Proceedings of an International Conference, Association Francaise des Travaux en Souterrain, Oct. 1993, pp.333-340.

[2]   "Test Method for Tensile Properties of Continuous Fiber Reinforcing Materials," Concrete Engineering Series, No.23, Japan Society of Civil Engineers, Oct. 1997, pp.91-94.

[3]   Sekijima, K., Shinmyo, M., Kuhara, T. and Hayashi, K., "Study on Tensile Strength and Cross Point Strength of Grid Continuous Carbon Fiber Reinforcing Material," Proceedings of the Japan Concrete Institute, Vol.18, No.1, June 1996, pp.1167-1172. (in Japanese)

[4]   Kuhara, T., Sekijima, K., Shinmyo, M. and Konno, T., "Grid Continuous Fiber Reinforcements for Shield-Cuttable Tunnel Wall System," Proceedings of the Third International Symposium on Non-Metallic (FRP) Reinforcement for Concrete Structures, Vol.2, Oct. 1997, pp.599-606.

[5]   "Test Method for Bond Strength of Continuous Fiber Reinforcing Materials by Pull-out Testing", Concrete Engineering Series, No.23, Japan Society of Civil Engineers, Oct. 1997, pp.120-124.

[6]   "Test Method for Tensile Properties Using Highly Expansive Material," Concrete Library, No.88, Japan Society of Civil Engineers, Sept. 1996, pp.337-341. (in Japanese)

[7]   Seki, S., Sekijima, K., Otsuka, Y. and Konno, T., "Study on Mechanical Joint of Grid Continuous Fiber Reinforcing Material and Steel Plate," Proceedings of the 53rd Annual Conference of the Japan Society of Civil Engineers, Vol.5, Oct. 1998, pp.836-837. (in Japanese)

3.4 Repair of steel bridges with CFRP plates

M Tavakkolizadeh[1] and H Saadatmanesh[2]
[1]Jackson State University, USA
[2]The University of Arizona, USA

ABSTRACT

The conventional methods for repair or retrofit of the substandard structures are costly and time consuming. In recent years, several new techniques have examined the possibility of utilizing the superior mechanical and physical properties of Fiber Reinforced Polymer (FRP) composites. This paper presents the results of a study on the behavior of damaged steel-concrete composite girders repaired with Carbon Fiber Reinforced Polymers (CFRP) sheets under static loading. A total of three large-scale composite girders were prepared and tested. The thickness of the CFRP sheet was constant and 1, 3 and 5 layers of these sheets were used to repair the specimen with 25, 50 and 100 % loss of the cross sectional area of their tension flange, respectively. The test results showed that epoxy bonded CFRP sheet could restore the ultimate load carrying capacity and stiffness of damaged steel-concrete composite girders.

INTRODUCTION

Different local and federal agencies have developed programs to rate bridges throughout the United States during the last few decades. It has been found that more than one third of the highway bridges in the United States are considered substandard. Steel bridges comprise more than 34% of the total number of these bridges and were among the most recommended group for improvement based on the National Bridge Inventory report [1]. Many of these bridges need to be repaired due to permanent damage to their critically stressed locations of the tension members. These damages were caused by direct physical impact, corrosion, or growth of small cracks under fatigue loading. The rehabilitation and repair in most cases is far less costly than the replacement. Since there are limited resources available to mitigate the problems associated with substandard bridges, the need for adopting cost-effective techniques and new materials is apparent.

Composite materials possess excellent mechanical and physical properties that make them excellent candidates for repair and retrofit applications. FRPs are made of high strength filaments (tensile strength in excess of 2 GPa) such as glass, carbon and kevlar placed in a resin matrix. CFRPs display outstanding mechanical properties with typical tensile strength and modulus of elasticity of more than 1200 MPa and 140 GPa, respectively. In addition, the CFRP laminates weigh less than one fifth of the steel and are corrosion resistant. CFRP plates or sheets can be epoxy bonded to the tension face of the damaged members to restore the strength and stiffness as shown in Figure 1. The CFRP sheets bridge over the damaged area, and transfer the stresses across. In addition, the stress level in the original member will decrease and that will result in a longer fatigue life. This paper discusses the effectiveness of epoxy bonding CFRP sheets to the tension flange of damaged steel-concrete composite girders to restore their ultimate load carrying capacity and stiffness.

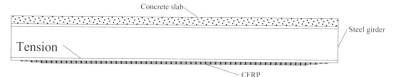

Figure 1. Typical composite girder retrofitted with CFRP

PREVIOUS WORK

The most commonly used techniques for repair of damaged steel bridges include strengthening of damaged members, replacing of members and adding of members. The most common attachment technique for performing these techniques is welding. They usually include welding of a cover plate to the damaged area after performing general clean up and preventative measures. Unfortunately, welded cover-plates pose several problems such as: the need for heavy machinery, sensitivity of the welded detail to fatigue and the possibility of galvanic corrosion between the welded plate, existing members and attachment materials.

The first reported application of epoxy-bonding of steel plates for strengthening of concrete structures dates as far back as 1964 in Durban, South Africa, where the reinforcements in a concrete beam were accidentally left out during construction [2]. A study on adhesive bonding and end bolting of steel cover plates to steel girders at the University of Maryland displayed a substantial increase in the fatigue life of the system. An increase in the fatigue life of more than twenty times, compared to the only welded cover plates was reported [3].

In two separate studies conducted at the University of South Florida and the University of Deleware, the possibility of using CFRP in repair of steel-concrete composite bridges was investigated [4, 5]. It was reported that CFRP laminates could considerably improve the ultimate capacity of intact and damaged composite beams.

In a study conducted by the authors at The University of Arizona the effectiveness of CFRP patch for repairing damaged steel beam was investigated [6]. A total of eight small-scale S 5×10 steel beams were tested. The tension flanges of six beams were cut thoroughly with two different depths and then beams were repaired by CFRP patches. Increases in ultimate load carrying capacity of 145% and 63% for the beams with 80% and 40% loss of tension flange area were reported, respectively. The stiffness of the damaged beam after patching was recovered to 95% the original stiffness. Different modes of failures that were observed in the experiments were end peeling and delamination. They indicated that the drawbacks of this technique included the loss of ductility and the possibility of galvanic corrosion between the CFRP plate and steel, which has been addressed by the authors in a separate article [7].

This paper investigates the effectiveness of epoxy bonding CFRP sheets to tension flange damaged steel-concrete composite girders that lost a portion or their entire tension flange at the most critical location. The experimental results are compared with the conventional analytical models.

EXPERIMENTAL STUDY
The feasibility of epoxy bonding of CFRP sheets on restoring the ultimate load carrying capacity and stiffness of composite girders was examined by testing three large-scale girders repaired with pultruded carbon fiber sheets. In order to observe the effectiveness of this technique, three different damage levels of 25, 50 and 100 % loss of tension flange were considered and different thicknesses of CFRP laminates were used. The tension flanges of the girders were cut with different total depths to simulate 25, 50 and 100% loss of tension flange, respectively. These girders were then strengthened by epoxy bonding of 1, 3 and 5 layers of CFRP sheets to the tension flanges. Concrete slabs with two different compressive strengths were used. The overall lengths of CFRP sheets were identical and the cut-off points for each layer were staggered to prevent premature failure at termination points due to stress concentrations.

Materials
A unidirectional pultruded carbon fiber sheet with width of 7.6 cm and thickness of 1.27 mm was used in this study. A two-component viscous epoxy (tack coat) was used for bonding the laminate to the steel flange surface. This epoxy immediately reached high tack consistency and was ideal for over-head applications in the field. A two-component less viscous epoxy was used for attaching the laminates to each other. This epoxy had a longer gel time and much lower viscosity and was used in between CFRP sheets to insure the least entrapped voids. W14×30 A36 hot rolled sections were used for the experiments. The minimum reinforcement in concrete slabs for temperature and shrinkage was provided by using a 15×15×0.64 cm welded smooth wire mesh. Concrete was ordered from a ready mix plant with the nominal compressive strengths of 15.5 and 27.5 MPa, slump of 10 cm and maximum aggregate size of 1 cm. The mechanical properties of these materials are tabulated in Table 1.

Table 1. Constitutive properties of materials

Concrete (Compressive)			Steel (Tensile)		CFRP (Tensile)	
	Low	High		Web		
Strength (Mpa)	16.6	29.1	Yield Strength (Mpa)	382	Strength (Mpa)	2137
Modulus (Gpa)	14	19	Modulus (Gpa)	178	Modulus (Gpa)	144
Failure Strain	0.0040	0.0035	Poisson's Ratio	0.299	Poisson's Ratio	0.340

Specimen preparation and instrumentation
After cutting the steel sections into 4.9 m long beams, shear studs with diameters of 13 mm and heights of 51 mm were welded to the compression flange in two rows 12.5 cm on centre along the two shear spans. After constructing the forms, the wire mesh was placed in the mid-height of the slab by using 3.8 cm thick chairs. The girders and cylinders were kept moist under a plastic cover for one week. CFRP sheets were cut to the proper length with a band saw and their ends were finished smoothly using grid 150 sandpaper. The surfaces of the sheets were sand blasted with No. 30 sand and then washed with saline solution and rinsed with fresh water. After drying of the sheets, for multiple layer systems, the surfaces of the sheets were covered with thin layers of epoxy and were bonded together.

After the concrete slabs were completely cured, the tension flanges of the girders were cut with 1.27 mm thick blade. The flanges of the girder with the 15.5 MPa concrete strength were cut 4.27 cm deep on both sides at the midspan (50% loss). The tension flanges of the other two girders with higher concrete compressive strength were cut entirely (100% loss) and 2.14 cm deep on both sides at the midspan (25% loss). Just before applying the CFRP sheets, the tension flange of each girder was sand blasted using No. 30 sand, washed with saline solution and rinsed with fresh water.

Upon drying of the steel beam and the CFRP sheets, the tack coat was mixed and applied to the tension flange surface and the sheets. All pieces were covered with uniform and thin layers of tack coat and were squeezed together to force the air pockets out with excess epoxy. The CFRP sheets were secured throughout their lengths using binder clips and 40×40×3 mm aluminium angle bars, while the tack coat was curing.

Strain gauges were mounted on the surfaces of the steel beam, CFRP sheets and the concrete slab. In the midspan, the strain gauges were mounted on the top and bottom of the concrete slab, top flange and web of the steel beam and on the CFRP sheets. In addition, strain gauges were mounted at the end, and at the quarter-length on CFRP sheets and the tension flange. Eight 10×10 cm wooden blocks were cut and tightly fit between the flanges at the supports and under loading blocks to prevent web crippling. Two loading platens on top of the slab were made by quick set grout.

Experimental setup
Four-point bending tests were performed using a 2,200 kN test frame. Loading was applied by an hydraulic actuator. The load was measured by two load cell with a capacity of 500 kN and the deflection was measured by a transducer with a range of ±7.5 cm. Monotonic loading was performed under actuator displacement control with a rate of 0.025 mm/sec. Total of three unloadings were carried out during each test; before steel yielded, after steel yielded and at 500 kN load level. The load, midspan deflection and strains at different points were recorded. The clear span was

478 cm and the loading points were 50 cm apart. The test setup is shown in Figure 2.

EXPERIMENTAL ANALYSIS
Three composite girders were tested in the present study. All three girders were repaired by epoxy bonding of 1, 3 and 5 layers of CFRP sheets to the entire length of their tension flanges. They were subjected to monotonic loading with a few unloadings in the elastic and the plastic regions. The load-deflection plots for three girders are shown in Figures 3 through 5.

Load carrying capacity
Three repaired girders were able to carry ultimate loads higher than the calculated values for the virgin specimens. The girder with 25% loss and 1 layer of CFRP carried 471.8 kN (19.1% increase), while the girder with 100% loss and 5 layers of CFRP carried 434.1 kN (9.6% increase). The girder with 50% loss and 3 layers of CFRP, which was made with the concrete slab with lower strength, withstood 658.5 kN (80.2 % increase).

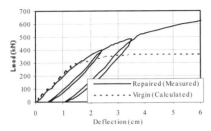

Figure 2. Test set-up

Figure 3. P-Δ plot (25% loss & 1 layer)

The predicted load carrying capacity of the girder with 100% loss and 5 layers of CFRP was 683.8 kN. The girder failed at a load of 434 kN and deflection of 2.82 cm due to early debonding of the CFRP sheet caused by wide crack opening. The edges of the large cut tended to deform unevenly during loading and as a result, normal peeling stresses were developed at the crack edges. This normal stress in addition to the shear stress concentration resulted in progressive debonding that started from the midspan.

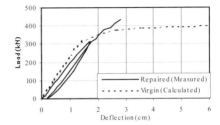

Figure 4. P-Δ plot (50% loss & 3 layers) **Figure 5**. P-Δ plot (100% loss & 5 layers)

Elastic and plastic stiffnesses

The repair technique was not able to recover the loss of the stiffness completely for two girders with higher compressive strength concrete. Average stiffness of the girders were 210.4 (-8.9%) and 199.7 kN/cm (-13.5%) for 25 and 100% loss of flange, respectively. The repaired girder with lower compressive strength (50% loss) recovered the lost stiffness due to the damage completely and displayed the stiffness of 196.7 kN/cm compared to the predicted value of 192 kN/cm (+2.4%). Table 2 summarizes the experimental and theoretical values of the stiffnesses of the virgin and repaired girders.

Table 2. Calculated and Experimental Stiffnesses of Virgin and Repaired Girders

	Measured Stiffness Repaired (kN/cm)		Calculated Virgin Stiffness (kN/cm)	
Repair Techniques	Elastic	Plastic	Elastic	Plastic
25% Loss & 1 Layer	210.4	67.3	231	3.2
50% Loss & 3 Layers	196.7	45.7	192	2.4
100% Loss & 5 Layers	199.7	101.3	231	3.2

Failure modes

The steel-concrete-CFRP system could display several distinct failure modes including concrete crushing, CFRP debonding, CFRP rupture, flange buckling, web crippling and shear stud failure. The wooden blocks prevented web crippling while the shear studs were designed in a way to eliminate their possibility of their failures.

Compression crushing of concrete, as shown in Figure 6-a, was the dominating failure mode for the girder made of 16.6 MPa concrete. The 3-layer repair system for a girder with 50% loss of its tension flange, forced the slab to carry the maximum compressive strain of beyond 0.29% before failure. The progressive debonding of CFRP sheet was limited to a short length of less than 15 cm in the midspan area.

The tension rupture of CFRP sheets, as shown in Figure 6-b, was the distinct mode of failure for composite girder with the 25% loss of tension flange area that was repaired with one layer of CFRP sheet. There was no sign of debonding or concrete compression failure. The rupture of the CFRP sheet was sudden and there was no sign of bond failure between CFRP sheet and the steel flange. The concrete slab barely reached its peak strain and experienced a compressive strain of 0.18% at the time of failure.

The failure of the bond between CFRP laminate and steel flange was the mode of failure in the girder with 100% loss of tension flange, repaired with 5 layers of CFRP sheets. Failure progressed quickly from the first sign of debonding in the midspan (72% of failure load) to the progressive sign of debonding at the end of the sheets (88% of failure load) and eventually the complete failure as shown in Figure 6-c. The web of the girder ruptured as soon as the CFRP sheets debonded. The compressive strain in concrete was 0.16%, well below its capacity.

| a) Compression failure | b) CFRP rupture | c) Plate debonding |

Figure 6. Different modes of failure

CONCLUSIONS

Test results for repair of damaged steel-concrete composite girders by epoxy bonding of CFRP laminates to the damaged area are very promising. For all three damage levels of 25, 50 and 100% considered in this study, this technique improved the ultimate load carrying capacity well beyond that of the virgin girder. The effect of CFRP laminates on the stiffness was also notable. Based on the results of the experimental investigation, the following conclusions are drawn:

- Ultimate load carrying capacities of the girders significantly increased by 20%, 80% and 10% for 25% damaged and 1-layer, 50% damaged and 3-layer and 100% damaged and 5-layer repairing systems.

- The effect of CFRP bonding on the elastic stiffness of the girders was significant. The technique restored the elastic stiffness of the girder to 91%, 102% and 86% of the intact girder for 25% loss and 1-layer, 50% loss and 3-layer and 100% loss and 5-layer repairing systems.

- The effect of the technique on improving the plastic stiffness of the repaired girder was much more pronounced. The plastic stiffnesses were increased more than 20 times compared to the intact girder.

- While, the theory indicates that the ductility of the retrofitted system is less then the virgin girders, repaired girders with small to moderate loss of their tension flange deflected between 5 ~ 7.5 cm which is about 1/100 ~ 1/65 of the clear span.

REFERENCES

[1]   FHWA Bridge Program Group (2001). "Count of Deficient highway Bridges." http://www.fhwa.dot.gov/bridge/britab.htm, The Office of Bridge Technology, The Federal Highway Administration, Washington, D.C.

[2]   Dussek, I. (1980). "Strengthening of the Bridge Beams and Similar Structures by Means of Epoxy-Resin-Bonded External Reinforcement." TRB Record 785, 21-24, Transportation Research Board, Washington, D.C.

[3]   Albrecht, P., Sahli, A., Crute, D., Albrecht, Ph. and Evans, B. (1984). "Application of adhesive to Steel Bridges." FHWA-RD-84-037, 106-147, the Federal Highway Administration, Washington, D.C.

[4]   Sen, R. and Liby, L. (1994). "Repair of Steel Composite Bridge Sections Using Carbon Fiber Reinforced Plastic Laminates." FDOT-510616, Florida Department of Transportation, Tallahassee, Florida.

[5]   Mertz, D. and Gillespie, J. (1996). "Rehabilitation of Steel Bridge Girders through the Application of Advanced Composite Material." NCHRP 93-ID11, 1-20, Transpiration Research Board, Washington, D.C.

[6]   Tavakkolizadeh, M., Saadatmanesh, H. (2001). "Repair of Cracked Steel Girder Using CFRP Sheet." Creative Systems in Structural and Construction Engineering, Proc. ISEC-01, Hawaii, 24-27 January 2001: 461-466

[7]   Tavakkolizadeh, M., Saadatmanesh, H. (2001). "Galvanic Corrosion of Carbon and Steel in Aggressive Environments." Journal of Composites for Construction, ASCE, August 2001, 200-210

3.5 The design of carbon fibre composite (CFC) strengthening for cast iron struts at Shadwell Station vent shaft

A R Leonard
Infraco Sub-Surface Limited, UK

BACKGROUND

The second phase of the East London Railway construction, north of Brunel's brick tunnel below the Thames, was completed in 1876. These works included the construction of large shafts just to the north and to the south of Shadwell Station. The shafts provided ventilation to the deep sections of this steam operated railway. The 15 metre deep ventilation openings are rectangular in plan, see Figure 1. The brick retaining walls forming the sides of the shafts were braced apart by two tiers of cast iron struts.

The struts bracing the south shaft were removed in 1980 when the new station was built. A detailed assessment of the struts in the remaining north shaft revealed that the cast iron struts in the upper tier were stressed above limits normally considered safe for cast iron.

Prior to removal of the cast iron struts during station construction works, they were strain gauged in order to determine the horizontal long term loads imposed by overconsolidated London Clay on subterranean structures. Whilst the struts were being de-stressed, the strain changes were measured. The loads deduced from these strain changes were later used to verify that the upper struts in the remaining vent shaft were overstressed. The more substantial lower struts were found to be relatively lightly loaded.

The Geotechnical Consulting Group (GCG) carried out a detailed finite element analysis using the ICFEP program, developed by Professor Potts of Imperial College London. The output, from the FEA model that most closely matched the likely shaft construction method, gave a reasonable correlation with measured strut loads and confirmed a probable overstress in the upper struts.

Strengthening or replacement of these upper struts was, therefore, considered essential.

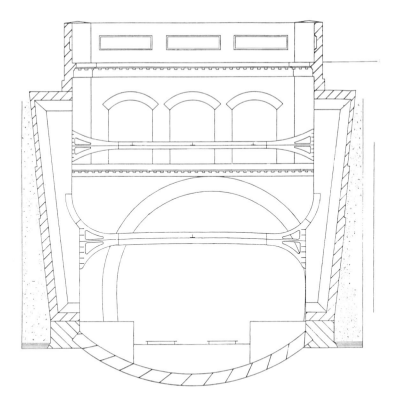

Figure 1 Cross section through vent opening

STRENGTHENING OPTIONS
Several steel and concrete options were proposed but CFC strengthening was adopted because this method of installation had the lowest level of installation risk and caused the least change in load paths through the structure.

OUTLINE OF CONSTRUCTION SEQUENCE
Steel struts were installed immediately above upper tier of cast iron struts and utilised to support working platforms. The steel struts provided a secondary restraint to prevent a catastrophic failure of the retaining walls in the event of a failure of the upper cast iron struts during construction.

Structural Statics Limited (SSL) installed strain gauges to monitor strain changes in the cast iron struts.

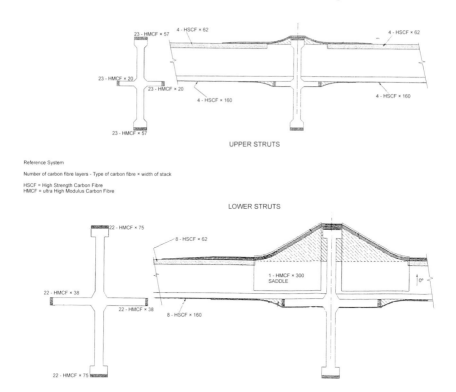

Figure 2 Section to show application of CFC to ends of cruciform arms of struts

AEA Technology (AEAT) used probes to estimate the in situ stress in the cast iron struts. The MAPS equipment, developed by Dr David Buttle of AEAT, uses magnetic pulses to measure several magnetic parameters which are affected by the state of stress in ferrous material.

A light preload was jacked into the steel struts to ensure immediate load take up and to stress relieve upper cast iron members prior to application of CFC strengthening.

Devonport Management Limited (DML) applied the CFC strengthening as shown in Figure 2. DML also provided technical information during the design process and reinforced three test struts to validate the design.

The SSL monitoring confirmed the effectiveness of the CFC reinforcement by measuring the changes in strut behaviour after application during the daily thermal loading cycle.

Finally, the composite struts were painted white to minimise thermal loads and UV degradation.

FAILURE CRITERIA

From a number of bending and compression tests, members have failed when the cast iron failure strain is exceeded. Once cast iron cracks, the CFC will debond along the interface, propagating away from the crack. It follows that the limiting strain in the cast iron is one of the failure criteria for the reinforced member.

In strut tests it was found that CFC could fail in compression without precipitating strut collapse. However, this additional strength reserve has been ignored since the failure of the CFC could be considered a serviceability failure and introduces further complexity into the analysis. Compressive CFC failure could be taken as an early warning of high member loads, warranting further investigation. It follows that the limiting strain in the CFC reinforcement can be taken as another of the failure criteria.

The shear stress along the interface between metal and CFC needs to be limited to prevent a debonding failure, particularly near the ends of the reinforcing stack. Long run-out tapers in the CFC stacks help minimise these stresses.

To prevent any possibility of a buckling failure occurring at less than the material limiting strains, failure is also deemed to have occurred if the transverse deformation due to axial loading exceeds length/100.

Finally, an allowance needs to be made for load term degradation of the CFC. Reduction factors have been taken from the Joint Industry Project document C16100R007 dated 1997. These give a 44% reduction in strength over the life of the struts (taken as 120 years).

DESIGN METHOD

The key design feature for bonded reinforcement is strain compatibility at the interface between cast iron and CFC.

At the time that the CFC is applied to the ends of the arms of the cruciform strut, the CFC is unstressed. The amount by which the CFC can be stressed is limited by the lesser of the total strain capacity of the CFC and the remaining strain capacity of the cast iron. The CFC shares, with the cast iron, any changes of load occurring after application. Provided that the CFC does not fail before the unreinforced cast iron would have reached its maximum capacity, the CFC will enhance the member strength. In compression cast iron has a much larger strain capacity than CFC.

The CFC also stiffens the composite section so that changes in lateral deflection, with increasing axial load, are reduced. Bending effects, resulting from axial load changes, are reduced. This increases the load carrying capacity.

At Shadwell, where the struts are subjected to solar heating, the CFC has further benefits. The CFC has a negative coefficient of thermal expansion. As the struts warm during the day they increase in length, forcing the retaining walls apart and thereby generating an axial resisting load. The CFC reduces the amount of the extension and hence lowers the thermal load effect. The strut that failed at Rotherhithe buckled shortly after midday.

As the strut warms, the CFC tries to shrink, inducing an axial compressive stress in the cast iron. This is beneficial in cases where the strut fails by cast iron reaching its tensile limit. Except for very short struts, this is the most likely failure mode as cast iron is far weaker in tension than compression.

The first step in the design process is to determine the cast iron strains at the time of CFC application, in order to assess the residual cast iron strain capacity and hence the extent to which the CFC can be stressed before the cast iron fails. The strains are estimated by a non-linear interpolation using the working load and calculated failure load as the main parameters.

The working loads have been calculated by GCG using finite element methods.

The failure load is calculated using the Claxton Fidler formula of 1887, which shows a good correlation with cast iron strut test results. The formula defines two failure plots. One is based on compressive failure on the concave side of the strut and the other on tensile failure on the convex face. The lesser load is the failure capacity.

If σ_b is the average stress at strut failure:

If compressive stress governs, $\quad \sigma_b = \dfrac{\sigma_e + \sigma_c - \sqrt{(\sigma_e + \sigma_c)^2 - 4\sigma_e \sigma_c (1-\phi)}}{2(1-\phi)}$

If tensile stress governs, $\quad \sigma_b = \dfrac{\sigma_e - \sigma_t + \sqrt{(\sigma_e - \sigma_t)^2 + 4\sigma_e \sigma_t (1+\phi)}}{2(1+\phi)}$

Euler buckling stress, $\quad \sigma_e = E \dfrac{\pi^2}{(l/r)^2} \quad E = 80{,}000 \text{ N/mm}^2, \ (l/r) = $ slenderness ratio

$\sigma_c = $ characteristic compressive stress $\quad \sigma_t = $ characteristic tensile stress $\quad \phi = 0.4$

The characteristic compressive stress was taken as $1{,}300 \times t^{-0.3}$ N/mm^2 (with an upper bound of 650 N/mm2)

The characteristic tensile stress was taken as $320 \times t^{-0.3}$ N/mm^2 (with an upper bound of 160 N/mm^2),

where t is the effective thickness of the cast iron (measured in millimetres) and can be defined as twice the area of the highly stressed region of the cross section, divided by its exposed perimeter. All other things being equal, the thicker the cast iron the lower the failure stress.

The calculated failure load implies a load eccentricity that is just sufficient to cause the cast iron to reach a failure stress. The lesser of the eccentricities resulting in a tensile or compressive failure is used. The eccentricity at working load is interpolated taking account of initial imperfections and non-linear growth of lateral deformation with load. The stresses in the cast iron are calculated using the working load eccentricity. The associated compressive and tensile strains are deduced from the non-linear cast iron stress-strain relationship. The residual strain to cast iron failure is then used to determine the maximum strain achievable in the CFC reinforcement. If the cast iron failure strains govern then the failure load eccentricity can be approximated to that causing failure in the unreinforced strut, but will be less if the CFC strain limit governs. As a reasonable approximation, the load in the cast iron element of the composite strut will be the assessed failure load of the unreinforced strut and the enhanced buckling capacity will be the additional axial load carried by the CFC reinforcement. The net CFC component of the strut behaves linearly up to failure and the axial load in the CFC can be calculated using the net area and the net elastic modulus (but assessed at its interface with the cast iron). The eccentricity of load acting on the CFC can be taken as that estimated for failure of the unreinforced cast iron strut. A check is then required to confirm that the CFC has not failed prematurely at the extreme fibres.

A more sophisticated analysis has been undertaken for the analysis of the behaviour of the struts tested at East Kilbride. The test analysis method wasn't considered convenient for use in the design office however.

After CFC strengthening, the assessed capacity of the struts exceeded the minimum requirements acceptable to the Chief Civil Engineer of London Underground.

A thorough Category 3 check was carried out by Dr David Morris, Principal Engineer employed by Howard Humphreys, Consulting Engineers and his report confirmed that the strength enhancement requirements had been achieved.

DESIGN VALIDATION

A cast iron strut failed by buckling at Rotherhithe Station vent shaft during demolition of an adjacent road bridge which spanned across the vent opening. Subsequent strengthening works highlighted the risks associated with manoeuvring heavy construction loads adjacent to brittle cast iron members. This influenced the choice of strengthening option at the Shadwell vent opening, see Figure 3.

During the Rotherhithe vent shaft strengthening works, the three remaining cast iron struts were removed intact and were halved in length. Three half lengths were reinforced (by DML) with high modulus CFC applied along the extremities of the strut cruciform arms and three were left unreinforced, as test control pieces. A different thickness of CFC was applied to each of the three reinforced struts. The six half length struts were tested by National Engineering Laboratories (NEL) in their 1,250 ton compression test machine at East Kilbride.

Figure 3 Complexities inherent in repair

The test was designed to simulate one half of the S bow failure expected in the centre braced struts. The load was applied eccentrically at the wall face of the strut and the far end of the strut was pinned. These tests confirmed that CFC reinforcement would significantly enhance the load carrying capacity of the struts. The tests also confirmed that the CFC could fail on the compression side of the strut without precipitating a buckling failure.

TEST ANALYSIS
The induced strains are calculated by applying the working load in small increments to a strut with an assumed sinusoidal imperfection and slight end eccentricity. As the strut deforms laterally bending stresses are induced which eventually exceed the axial stresses in magnitude. The deformations are calculated using finite difference techniques. At each load increment, the load and moment are evaluated at equi-spaced intervals along the strut.

For any given load and moment, the effective tangent modulus of elasticity can be evaluated. The surface plot of tangent modulus relative to load and moment is determined by specifying strains at the extreme fibres of the cast iron and numerically integrating across the section to determine the internal load and moment (taken about the elastic centroid). Plane sections are assumed to remain plane and the non-linear stress-strain relationship for cast iron has been taken from tensile and compressive tests on specimens cut from contemporary struts removed from the Rotherhithe vent shaft. The properties of the carbon fibre were specified by DML following testing. The properties were similar to those provided by the carbon fibre supplier, Sumitomo, who supplied the Mitsubishi carbon fibre. The CFC stress-strain relationship was taken as linear to failure. The tangent modulus of elasticity of the cast iron drops significantly as the load increases. A single function can be defined which approximates the tangent modulus at any given load and moment. Once the tangent modulus is specified at each interval along the strut, the change in deformation due to the next load increment can be evaluated.

The load-moment failure envelope for the unreinforced cast iron strut can be specified as the area bounded by four curves, as shown in Figures 4 and 5. Each curve defines all the possible coexisting loads and moments when an extreme fibre on one side or the other has reached the limiting tensile or compressive strain.

Within the load-moment failure envelope it is possible to plot the curving load-moment trajectory up to the point where it hits the envelope and the strut is deemed to have failed.

The direction of the trajectory is changed after application of the CFC, as the relative increase in moment with load increment is reduced. The addition of CFC to the cast iron will extend the area of the load-moment failure envelope in those areas where the CFC does not fail before the cast iron reaches its limiting strain. Application of CFC to a partially stressed strut increases the length of the perimeter of the load-moment failure envelope that is enhanced by the reinforcement, but reduces the magnitude of the enhancement, as the spare strain capacity in the cast iron is reduced.

The measured load – deformation plots closely followed those predicted by the analysis. The calculated failure load was highly sensitive to orientation and magnitude of initial imperfection relative to direction of the end load eccentricity. It should be noted that assumed initial imperfections should include material variation across the cross section as well as longitudinal curvature. Some calibration, using historical test data, would be needed to decide on reasonable initial imperfection assumptions.

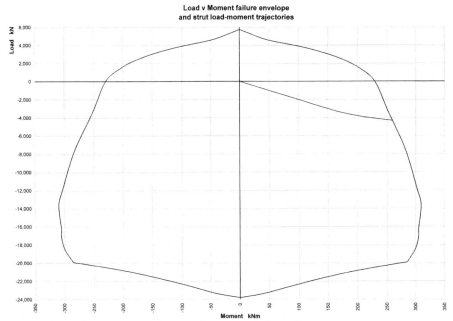

Figure 4 Load – Moment trajectory for an unreinforced strut

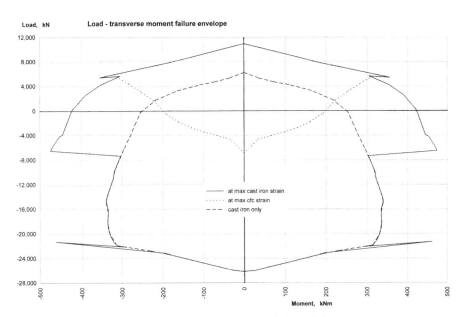

Figure 5 Load – Moment Failure Envelope for Reinforced Strut

Session 4 : **Application of FRP in concrete cylinders and columns**

4.1 Influence of specimen size and resin type on the behaviour of FRP-confined concrete cylinders

L De Lorenzis, F Micelli and A La Tegola,
Innovation Engineering Department, University Of Lecce, Italy

ABSTRACT
One of the most attractive applications of FRP materials is their use as confining devices for concrete columns, which may result in remarkable increases of strength and ductility as indicated by numerous published experimental results. Aim of this study was to investigate the effect of two variables, namely, cylinder size and resin type, on the behaviour of the confined cylinders. Twenty-six specimens were instrumented with strain gages and LVDTs and tested in compression until failure. Results are presented and discussed in this paper, including a comparison with expected theoretical results.

INTRODUCTION AND OBJECTIVE
In the last two decades, the use of fibre-reinforced polymer (FRP) composites for strengthening of concrete structures has been gaining increasing popularity, due to the favourable intrinsic properties possessed by these materials (high strength-to-weight ratio, good corrosion behaviour, electromagnetic neutrality). Among many possible applications of FRP materials, it was seen that columns confinement resulted one of the most effective.

Failure of the confined cylinders occurs when the FRP confining device ruptures in hoop tension. However, many experimental studies on wrapped cylinders have highlighted that the experimental value of the hoop strain in the FRP at tensile failure is usually lower than the FRP ultimate strain in uniaxial tension [1-3]. In the following, the ratio of the measured hoop strain in the confining FRP at tensile failure to the FRP ultimate strain in uniaxial tension is termed "reduction factor". In previous research, number and thickness of overlapping layers, elastic modulus of the wrap and radius of curvature, that is, for cylindrical specimens, radius of the cylinder, were identified as the most significant parameters [1, 4].

Aim of this study was to investigate the effect of two variables, namely, cylinder size and resin type, on the behaviour of Carbon FRP (CFRP)-confined cylinders. Twenty-six specimens were instrumented with strain gages and LVDTs and tested in compression until failure. Results are presented and discussed herein.

EXPERIMENTAL PROGRAM

Prior to starting the compression test series, four coupon specimens of the CFRP laminate were tested in uniaxial tension according to ASTM D3039, in order to determine their tensile strength and modulus of elasticity. Testing was performed in displacement-control mode on a 150-kN universal testing machine, with a cross-head displacement rate of 2 mm/min. Results are reported in Table 1. Failure occurred in all cases by tensile rupture of the laminate in the central part of the coupon, therefore, results were reliable.

Table 1. Mechanical Properties of the CFRP Laminate

Specimen No.	Nominal Thickness (mm)	Tensile Strength (MPa)	Tensile Modulus (MPa)	Ultimate Strain (%)
1	0.15	1057	93050	1.136
2	0.15	1022	90440	1.130
3	0.15	1034	96720	1.069
4	0.15	998	84220	1.185
Average		**1028**	**91108**	**1.130**
Standard Dev.		21.3	4560	0.041
Coefficient of Variation (%)		2.07	5.00	3.66

Twenty-six concrete cylinders were wrapped with one or more layers of Carbon FRP (CFRP) laminate using the wet lay-up technique. Two different epoxy resins, characterized by different degrees of viscosity, were used; they will be indicated in the following as R1 (epoxy resin with lower viscosity) and R2 (epoxy resin with higher viscosity). Cylinders of three different sizes, namely, 55-mm diameter by 110-mm height, 120-mm diameter by 240-mm height and 150-mm diameter by 300-mm height were tested. For each size, three unconfined cylinders were tested to determine the concrete unconfined strength, f'_{co}. The number of layers of CFRP used for wrapping was equal to one for the smallest cylinders, two for the medium, and three for the largest ones. In this way, the cylinders with different sizes had approximately the same values of E_l and p_u, where:

$$E_l = \frac{2\,E_f nt}{D} \tag{1}$$

is the confinement modulus or lateral modulus, which is a measure of the stiffness of the confining device, and:

$$p_u = E_l \cdot \varepsilon_{fu} = \frac{2\,f_{fu} nt}{D} \tag{2}$$

is the maximum value of confinement pressure that the FRP can exert before rupturing in tension. In (1) and (2), E_f, f_{fu} and ε_{fu} are elastic modulus, tensile strength and ultimate strain of the FRP, respectively, n is the number of plies, t the thickness of each ply, and D the diameter of the cylinder. The experimental program is reported schematically in Table 2.

Table 2. Test Matrix

Specimen Code	Diameter (mm)	Number of Plies	Resin Type	Number of Repetitions
D50-UR	55	-	-	3
D50-R1	55	1	R1	2
D50-R2	55	1	R2	3
D100-UR	120	-	-	3
D100-R1	120	2	R1	3
D100-R2	120	2	R2	3
D150-UR	150	-	-	3
D150-R1	150	3	R1	3
D150-R2	150	3	R2	3

Load was applied with a 3000-kN compression machine and measured by means of a pressure transducer. Two LVDTs were used to monitor the relative displacement between the extreme faces of the cylinder, and, from it, the average axial strain at each load level. Four strain gauges, two in the longitudinal and two in the hoop direction, were applied on each cylinder at mid-height (Figure 1).

It is well known that the reason the size effect arises in a compression test is twofold: the partial restrain of the lateral deformations caused by friction between the extreme faces of the specimen and the loading platens, and the localization of deformations in a compressive failure. In order to minimize the first source, a thin Teflon® layer was positioned between the upper and lower faces of the cylinders and the loading platens. However, the second source of size effect will always be present, as confirmed in a Round Robin Test by van Mier *et al* [5].

In order to avoid that the axial load be applied directly on the FRP, the FRP wrap did not cover the full height of the cylinder; instead, a narrow gap was left between the end concrete surfaces and the extreme composite fibres. Direct axial loading of the FRP may result in local buckling of the composite in the axial direction close to the loaded surfaces. The buckling portions expand outward, which results in less contact pressure with concrete. Therefore, direct loading of the FRP should yield a lower confinement effectiveness, as recently shown by Fam and Rizkalla [6].

EXPERIMENTAL RESULTS

During preparation of the specimens, both epoxy resins had been mixed with the respective hardeners in the proportions indicated by the manufacturers. Nevertheless, prior to testing it was noticed that resin R1 had not reached the glass-like consistency characteristic of epoxies after hardening. Instead, it still had a pseudo-rubbery consistency as if the amount of hardener used in mixing had been not enough to allow a complete polymerisation. The specimens were tested to see how the different resins behaviour may affect the effectiveness of FRP-wrapping. It could be noted that an epoxy resin, when correctly hardened, reaches a pseudo-rubbery consistency for temperatures close to its glass transition temperature, T_g. Therefore, the behaviour of cylinders with resin R1 could probably resemble that of wrapped cylinders at a temperature close to the glass transition temperature of the resin used for impregnation.

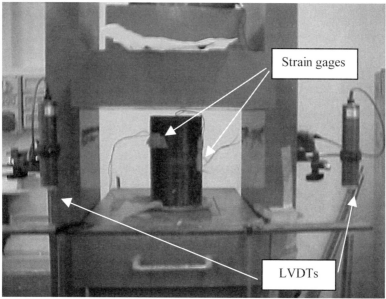

Figure 1. Test Setup

Table 3 presents the experimental results in term of unconfined strength, f'_{co}, confined strength, f'_{cc}, axial strain at peak stress, ε_{cc}, hoop strain at ultimate, ε_{fr}, and failure mode. Also reported is the "reduction factor" $\varepsilon_{fr}/\varepsilon_{fu}$, where ε_{fu} is the ultimate strain of the FRP in uniaxial tension. The values in the table are the average of three (or, in one case, two) repetitions as indicated in Table 2. Figure 2 is a picture of some failed specimens.

Apart from specimens D50-R1, which both failed by debonding at the overlap of the FRP wrap and thus did not show any appreciable increase in strength due to confinement, all the other specimens failed by hoop fracture of the FRP. The increase in compressive strength of the confined specimens with respect to the unconfined ones ranged from 26% for specimens D50-R2 to 77% for specimens D150-R2.

Table 3. Test Results

Spec. Code	f'_{co} (MPa)	p_u/f'_{co}	ε_{co}	f'_{cc} (MPa)	ε_{cc}	ε_{cu}	ε_{fr}	$\varepsilon_{fr}/\varepsilon_{fu}$	Failure mode
D50-UR	43	0.130	0.0041	-	-	-	N/A	N/A	CF
D50-R1	43	0.130	0.0041	43	0.0041	N/A	N/A	N/A	DB
D50-R2	43	0.130	0.0041	54.3	0.0143	0.0149	0.0059	0.52	FR
D100-UR	43	0.120	0.0041	-	-	-	0.0011	-	CF
D100-R1	43	0.120	0.0041	58.5	0.0087	0.0116	0.0071	0.63	FR
D100-R2	43	0.120	0.0041	65.6	0.0082	0.0095	0.0081	0.72	FR
D150-UR	38	0.159	0.003	-	-	-	0.0011	-	CF
D150-R1	38	0.159	0.003	62	0.0071	0.0095	0.0075	0.66	FR
D150-R2	38	0.159	0.003	67.3	0.0124	0.0135	0.0084	0.74	FR

N/A = Not Available
CF = Compressive Failure; DB = Debonding at the Overlap; FR = FRP Rupture

Figure 2. Failed Specimens

The influence of diameter on the strength of the confined cylinders was weak. Specimens with D=150 mm had slightly higher strengths compared with those with D=120 mm (but they also had a higher value of the p_u/f'_{co} ratio), and the latter had in turn higher strengths compared to those with D=50 mm. In all cases, specimens with type of resin R1 had a lower value of confined strength compared with those with resin R2, the difference being 26% for the smallest cylinders, 12% for the medium ones and 8.5% for the biggest cylinders. The higher difference for the smallest cylinders is due to the fact that specimens D50-R1 failed by debonding at the overlap, whereas in D50-R2 the resin performed well and the overlap was efficient in transferring the load up to rupture of the composite wrap.

The reduction factor had a trend similar to that of the confined strength as the specimen diameter and resin type were changed. As the diameter changed from 55 to 120 to 150 mm, the reduction factor for specimens with R2 increased correspondingly from 0.52 to 0.72 to 0.74. This seems to indicate that the "premature" failure was mainly due the curvature of the specimen and was delayed for specimens with smaller curvature. Cylinders with resin R2 had higher values of the reduction factor compared with the R1 ones, due to the pseudo-rubbery behaviour of resin R1 which facilitates asymmetries and local stress concentrations. No conclusions can be drawn on the influence of both parameters on the axial strain at peak, ε_{cc}, due to high scatter in results.

Figure 3 shows, as an example, the axial stress-strain behaviour of specimens with 120-mm diameter. As usual in FRP-confined cylinders, the curves are characterized by a first region (approximately up to f'_{co}) in which the behaviour is similar to that of the unconfined cylinder, followed by a second region in which the confining effect is visible. Due to the low confinement ratio of the specimens (as expressed by p_u/f'_{co}), the peak point does not always coincide with the ultimate point, but is followed in some cases by a descending branch, and especially so for specimens with resin R1. The curves relative to specimens with R2 are always higher and stiffer than those with R1, due to the higher stiffness of the laminate in the latter case.

COMPARISON WITH MODELS

Table 4 compares results of this experimental study (in terms of peak stress and strain, both normalized with respect to the unconfined values) with predictions of some of the available models on FRP-confinement of concrete cylinders. The model by Spoelstra and Monti [7] has not been included as it predicts the ultimate values rather than the peak values of axial stress and strain. For the various models, the errors in predictions of peak stress and strain have been computed inserting into the equations both the nominal and the experimental values of the FRP ultimate strain.

Prediction of the confined strength is rather accurate for all models, and becomes more conservative when the experimental values of the FRP ultimate strains are used. Obviously, since the equations do not account for the resin type and cylinder size, predictions are more conservative for specimens with resin R2 and bigger diameter. Conversely, prediction of the axial strain at peak resulted unsatisfactory for all models, especially with the nominal values of FRP ultimate strain. The best predictions are those of the model by Toutanji and Saafi [8] with the experimental ultimate strain, although the errors are still rather high (ranging from −43.2 to 67.1%).

Table 4. Comparison with Models

Specimen	Exp. f'_{cc}/f'_{co}	Error (Saadatmanesh)		Error (Toutanji)		Error (Saafi)		Error (Saaman)	
		Err' (%)	Err'' (%)	Err' (%)	Err'' (%)	Err' (%)	Err'' (%)	Err' (%)	Err'' (%)
D50-R2	1.263	34.7	11.5	28.3	7.5	10.7	-2.6	16.1	2.6
D100-R1	1.360	21.6	6.2	15.8	2.0	0.7	-8.1	5.8	-3.2
D100-R2	1.526	8.4	-1.9	3.3	-6.0	-10.2	-16.1	-5.7	-11.6
D150-R1	1.632	11.5	-2.6	6.3	-7.1	-9.9	-18.4	-4.6	-13.2
D150-R2	1.771	2.7	-6.9	-2.1	-11.3	-17.0	-22.9	-12.1	-18.0

Specimen	Exp. $\varepsilon_{cc}/\varepsilon_{co}$	Error (Saadatmanesh)		Error (Toutanji)		Error (Saafi)		Error (Saaman)	
		Err' (%)	Err'' (%)	Err'(*) (%)	Err''(*) (%)	Err'(*) (%)	Err''(*) (%)	Err' (%)	Err'' (%)
D50-R2	3.488	29.2	-12.8	-30.6	-43.2	-6.1	-27.9	40.8	-10.9
D100-R1	2.129	100.5	51.3	38.4	16.5	93.6	55.5	124.4	61.8
D100-R2	1.988	114.7	75.3	93.2	67.1	179.3	134.1	140.3	90.4
D150-R1	2.373	114.6	66.3	85.9	54.4	172.6	117.9	183.0	109.3
D150-R2	4.133	23.3	2.6	25.0	8.2	85.7	56.7	62.5	31.1

Err' = computed using the nominal ultimate strain of the FRP
Err'' = computed using the measured ultimate strain of the FRP
(*) with the experimental value of f'_{cc}/f'_{co}

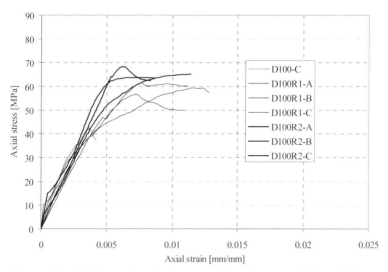

Figure 3. Axial Stress-Strain Curves of Specimens with D = 120 mm

CONCLUSIONS

The effect of cylinder size and resin type on the behaviour of FRP-confined concrete cylinders was investigated. Specimens with larger diameter tended to have higher strengths than the smaller ones, however, the influence of the diameter was weak. In all cases, specimens with type of resin R1 (rubber-like consistency) had a lower value of confined strength compared with those with resin R2 (glass-like consistency). The reduction factor had a trend similar to that of the confined strength as the specimen diameter and resin type were changed. No conclusions can be drawn on the influence of both parameters on the axial strain at peak, ε_{cc}, due to high scatter in results. For some of the available models, the errors in predictions of peak stress and strain were computed using nominal and experimental values of the FRP ultimate strain. Prediction of the confined strength was rather accurate for all models, and became more conservative when using the experimental values of the FRP ultimate strains. Future research is needed in order to measure the influence of significant parameters such as resin properties and size effects on effectiveness of FRP confining techniques.

ACKNOWLEDGEMENTS
Supply of material by EPO GmbH, Leuna-Harze GmbH and Mitsubishi Chemical Co. is gratefully acknowledged.

REFERENCES
[1]   Matthys, S. (2001): Structural behavior and design of concrete members strengthened with externally bonded FRP reinforcement. Doctoral Thesis, Department of Structural Engineering, University of Ghent, Belgium, 345 pp.

[2]   Shahawy, M., Mirmiran, A., and Beitelmann, T. (2000): Tests and modeling of carbon-wrapped concrete columns, Composites Part B: Engineering, No. 31, pp. 471-480, Elsevier Science Ltd., London, UK.

[3]   Rousakis, T. (2001): Experimental investigation of concrete cylinders confined by carbon FRP sheets, under monotonic and cyclic axial compressive load, Research Report, Chalmers University of Technology, Göteborg, Sweden.

[4]   De Lorenzis, L., and Tepfers, R. (2001) : A Comparative Study of Models on Confinement of Concrete Cylinders with FRP Composites, Publication 01 :04, Dept. of Building Materials, Chalmers University of Technology, Gothenburg, Sweden, 81 pp.

[5]   van Mier, J.G.M., et al (1997): Strain softening of concrete in uniaxial compression, Materials and Structures, RILEM, Vol. 30, May 1997.

[6]   Fam, A.Z., and Rizkalla, S.H. (2001): Behavior of axially loaded concrete-filled circular fiber-reinforced polymer tubes, ACI Structural Journal, v. 98, No. 3, pp. 280-289.

[7]   Spoelstra M. R., Monti G. (1999): FRP-Confined Concrete Model. Journal of Composites for Construction, ASCE, v. 3, No. 3, August 1999. pp. 143-150.

[8] Toutanji H., Saafi M. (2000): Behavior of concrete columns encased in PVC – FRP composite tubes. 3rd International Conference on Advanced Composite Materials in Bridges and Structures, 15-18 August 2000. Ottawa, Ontario, Canada. pp. 809-817.

4.2 Performance and application of a design-oriented stress-strain model for FRP-confined concrete

X F Yuan, L Lam and J G Teng
Department of Civil and Structural Engineering
The Hong Kong Polytechnic University, Hong Kong, China

ABSTRACT
Various stress-strain models for fibre-reinforced polymer (FRP)-confined concrete have been developed. This paper is concerned with the performance and application of the recently proposed Lam and Teng model. The paper first provides a brief description of the Lam and Teng model. This is then followed by a comparison of this model with independent test data of CFRP-confined concrete, showing a close agreement between the two, particularly for cases with a significant amount of confinement. Finally, typical results from section analysis employing this model are presented to demonstrate the application of the model.

KEYWORDS
Concrete, columns, FRP confinement, stress-strain models, section analysis, strength, ductility

INTRODUCTION
It is well known that lateral confinement of concrete can greatly enhance both the compressive strength and the ultimate strain of concrete. Fibre-reinforced polymer (FRP) composites have become a popular confining material for concrete in recent years, in both the retrofit of reinforced concrete (RC) columns by the provision of an FRP jacket, and in new construction in concrete-filled FRP tubes. For the reliable design of concrete members with FRP confinement, the stress-strain behaviour of FRP-confined concrete needs to be well understood and accurately modelled. As a result, many stress-strain models have been proposed in the past few years for FRP-confined concrete, which may be classified into two categories: design-oriented models in closed-form expressions and analysis-oriented models employing an incremental numerical procedure. Only models of the former category are suitable for use in direct application in design calculations by hand or spread sheets. Detailed reviews of and comparisons between these models have been presented in Lam and Teng [1, 2] and Teng *et al* [3].

This paper is concerned with the performance and application of the recently proposed Lam and Teng [4, 5] model which is a design-oriented model. This model

ACIC 2002, Thomas Telford, London, 2002

was based on a large test database of a number of key parameters including the ultimate compressive strength and the ultimate compressive strain, and a number of rational assumptions carefully interpreted from test observations. Limited direct comparisons between predictions from this model and test stress-strain curves have also shown this to be the most accurate design-oriented model. In this paper, a brief description of the Lam and Teng model is provided. This is then followed by a comparison of this model with independent test data of CFRP-confined concrete, showing a close agreement between the two, particularly for cases with a significant amount of confinement. Finally, typical results from section analysis employing this model are presented to demonstrate the application of the model.

LAM AND TENG'S STRESS-STRAIN MODEL
Assumptions
The Lam and Teng' model [1, 4, 5] is for concrete which is well confined with FRP, which means that the stress-strain curve is monotonically ascending. In establishing their model, Lam and Teng made the following assumptions based on a careful interpretation of existing test data: (a) the stress-strain curve consists of a parabolic portion followed by a linear portion; (b) the parabolic portion has an initial slope being the elastic modulus of unconfined concrete and meets smoothly with the linear portion; and (c) the linear portion intercepts the stress axis at the compressive strength of unconfined concrete f'_{co} and terminates at a point defined by the compressive strength f'_{cc} and the ultimate axial strain ε_{cc} of confined concrete.

Best-fit version
Based on the above assumptions, the stress-strain curve can be described by the following expressions:

$$\sigma_c = E_c \varepsilon_c - \frac{(E_c - E_2)^2}{4 f'_{co}} \varepsilon_c^2 \quad \text{for } 0 \le \varepsilon_c \le \varepsilon_t \tag{1}$$

$$\sigma_c = f'_{co} + E_2 \varepsilon_c \quad \text{for } \varepsilon_t \le \varepsilon_c \le \varepsilon_{cc} \tag{2}$$

where σ_c and ε_c = axial stress and strain of confined concrete respectively, E_c = elastic modulus of unconfined concrete, ε_t = axial strain at the transition point between the parabolic and the linear portion, and E_2 = slope of the linear second portion. The latter two parameters are given by

$$\varepsilon_t = 2 f'_{co} / (E_c - E_2) \tag{3}$$

$$E_2 = (f'_{cc} - f'_{co}) / \varepsilon_{cc} \tag{4}$$

The compressive strength of FRP-confined concrete is defined by

$$\frac{f'_{cc}}{f'_{co}} = 1 + k_1 \frac{f_l}{f'_{co}} \tag{5}$$

with
$$f_l = \frac{2 f_{frp} t_{frp}}{d} \tag{6}$$

where k_1 = confinement effectiveness coefficient and = 2 (Lam and Teng 2002c), and f_l = lateral confining pressure provided by FRP at rupture. The ultimate axial strain of FRP-confined concrete is defined by

$$\frac{\varepsilon_{cc}}{\varepsilon_{co}} = 2 + k_2 \frac{f_l}{f'_{co}}$$
(7)

where ε_{co} = axial strain at the compressive strength of unconfined concrete, and k_2 = strain enhancement coefficient with k_2 = 15 for CFRP-wrapped concrete and $k_2 = 27(f_l / f'_{co})^{-0.3}$ for concrete filled GFRP tubes.

Due to the limited test results available, expressions for ultimate axial strains were proposed by Lam and Teng [1] only for two types of FRP-confined concrete, but the stress-strain model is generally applicable to other types of FRP once test data allow the establishment of a suitable ultimate axial strain expression. Future fine-tuning of expressions for the compressive strength and the ultimate axial strain can be easily accommodated by the model. The only limitation to the model is that the stress-strain curve has an ascending second portion as may be ensured by a sufficient amount of FRP.

Lower bound version for design use
In addition to the best-fit version described above, Lam and Teng [1, 4, 5] also proposed a lower-bound version of their model for design use, in which the ultimate axial strain is predicted by

$$\frac{\varepsilon_{cc}}{\varepsilon_{co}} = 1.75 + k_2 \frac{f_l}{f'_{co}}$$
(8)

where k_2 = 10 for CFRP-wrapped concrete and k_2 = 22 for concrete filled GFRP tubes, to provide conservative predictions of the ultimate axial strain in most cases. An important advantage of this lower bound model is that it reduces to the idealized stress-strain curve for unconfined concrete adopted by Eurocode 2 [6] for design use, making it particularly attractive for application in design.

PERFORMANCE OF LAM AND TENG'S MODEL
Xiao and Wu [7] carried out compression tests on 27 concrete cylinders which had the dimensions of 152 mm (diameter) by 305 mm (height), and the compressive strengths of unconfined concrete f'_c of 33.7, 43.8, and 55.2 MPa, respectively. These cylinders were wrapped with one to three layers of CFRP. The thickness of CFRP was 0.381 mm per layer, and the tensile strength and elastic modulus of CFRP were 1,577 MPa and 105,000 MPa respectively. Four of the 27 stress-strain curves obtained from these specimens have been used in the assessment of available design-oriented models in Lam and Teng [1]. Here, all the 27 stress-strain test curves are compared with predictions from the best-fit version of Lam and Teng's

model (Figures 1 to 3). The stress-strain curves of Lam and Teng's model were obtained by assuming that the elastic modulus of concrete $E_c = 4,730\sqrt{f'_{co}}$.

It can be seen from these figures that 23 of the 27 test stress-strain curves are in close agreement with Lam and Teng's model, in terms of both the shape of stress-strain curve and the ultimate stress and strain. Four of the 27 test stress-strain curves show a descending branch. Theses curves are for specimens with low amounts of confinement, with the confinement ratio $f_l / f'_{co} = 0.18$ for one specimen (Figure 2a) and $f_l / f'_{co} = 0.143$ for the other three specimens (Figure 3a). This is as expected since Lam and Teng's model is satisfactory for concrete with sufficient FRP confinement so that the stress-strain curve has a monotonically increasing second portion. For cases whose stress-strain curves possess a descending second portion, the gain in compressive strength is generally very small and is thus not of great interest in strength-enhancement applications. Nevertheless, they can be important in ductility enhancement as typically required in seismic retrofit applications. Work is therefore required in the future both to define the lower limit for the FRP confinement ratio for the application of the Lam and Teng model and to extend the scope of this model to cover these low confinement cases.

APPLICATION OF LAM AND TENG'S MODEL IN SECTION ANALYSIS

This section presents typical numerical results from section analysis of eccentrically-loaded RC columns as a demonstration of the application of Lam and Teng's model. Results are presented of both the strength and ductility of circular RC columns wrapped with CFRP. The procedure of analysis is given in Yuan et al [8] and Teng et al [3] and is thus not detailed here.

The example circular RC column has a diameter $d = 1500$ mm with a concrete cover of 50 mm to the longitudinal reinforcement bars. A total of 32 high yield steel bars are evenly distributed around the circumference. The diameter and yield strength of the bars are 40 mm and 460 MPa, respectively. The CFRP has a tensile strength of 3,483 MPa and a thickness of 0.165 mm per layer, with fibres oriented in the hoop direction only. The confinement effect of transverse steel is ignored. The behaviour of unconfined concrete is predicted by the stress-strain curve recommended by Eurocode 2 (ENV 1992-1-1 1991) while the behaviour of the confined concrete is predicted by the lower bound version of Lam and Teng's model. The elastic modulus of unconfined concrete is assumed to be $E_c = 4,730\sqrt{f'_{co}}$.

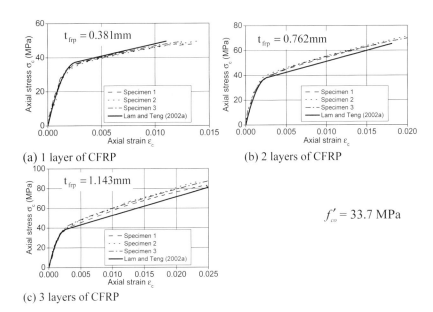

Figure 1 Lam and Teng's model versus test stress-strain curves with $f'_{co} = 33.7$ MPa

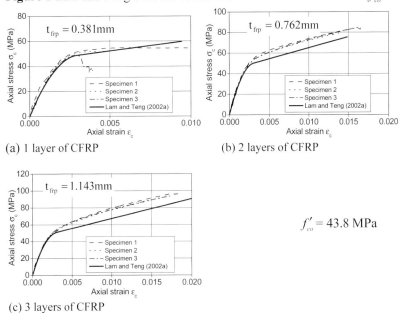

Figure 2 Lam and Teng's model versus test stress-strain curves with $f'_{co} = 43.8$ Mpa

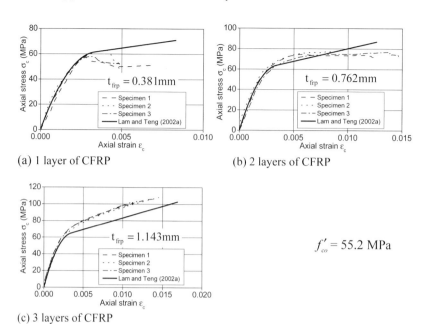

(a) 1 layer of CFRP

(b) 2 layers of CFRP

(c) 3 layers of CFRP

Figure 3 Lam and Teng's model versus test stress-strain curves with $f'_{co} = 55.2$ MPa

Figures 4 and 5 show the effects of different amounts of FRP confinement on the predicted behaviour of confined columns of different un-confined concrete strengths. In these plots, N_u / N_{uo} and M_u / M_{uo} are the ultimate axial load and moment normalised by corresponding values of the original column in pure compression and pure bending respectively. The curvature ductility factor u_ϕ shown in Figures 4 and 5 is defined as the ratio of the curvature of the section at the ultimate ϕ_u to that at the first yield ϕ_y, that is:

$$u_\phi = \frac{\phi_u}{\phi_y} \qquad (9)$$

Figure 4 shows that in general, FRP confinement leads to significant enhancement in the strength (Figure 4a), curvature (Figure 4b) and ductility factor (Figure 4c) of the column, and the degree of enhancement increases with the amount of FRP confinement. The enhancement in strength is seen to be small if bending is the dominant action (Figure 4a).

Figure 5 shows the effect of unconfined concrete strength on the behaviour of confined columns. The three columns have the compressive strengths of $f'_{co} = 33.7$, 43.8 and 55.2 MPa, respectively, and are wrapped with 10 layers of CFRP. It can be seen that an increase in the unconfined concrete strength yields a significant increase in the normalised moment capacity (Figure 5a) at intermediate levels of axial compression, a small reduction in curvature capacity at the same normalised axial load (Figure 5b), but little change in the ductility factor curve (Figure 5c).

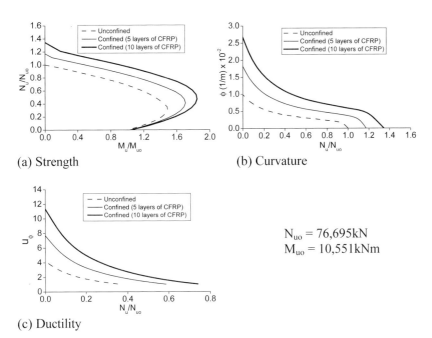

(a) Strength

(b) Curvature

(c) Ductility

N_{uo} = 76,695kN
M_{uo} = 10,551kNm

Figure 4 Effect of the amount of FRP on column behaviour

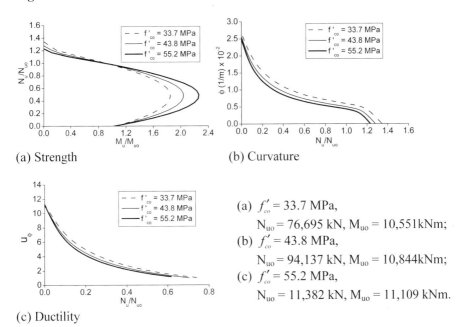

(a) Strength

(b) Curvature

(c) Ductility

(a) f'_{co} = 33.7 MPa,
    N_{uo} = 76,695 kN, M_{uo} = 10,551kNm;
(b) f'_{co} = 43.8 MPa,
    N_{uo} = 94,137 kN, M_{uo} = 10,844kNm;
(c) f'_{co} = 55.2 MPa,
    N_{uo} = 11,382 kN, M_{uo} = 11,109 kNm.

Figure 5 Effect of unconfined concrete strength on column behaviour

CONCLUSIONS

This paper has been concerned with the performance and application of the recent design-oriented stress-strain model for FRP-confined concrete proposed by Lam and Teng (2001, 2002a). Following a description of this model, a comparison between this model and independent test stress-strain curves and typical results from section analysis employing this model were presented. The model has been found to predict closely test stress-strain curves and is thus suitable for use in design, although a definition of the lower limit of FRP confinement ratio should be defined as the model is applicable only to stress-strain behaviour with an ascending second portion. Typical results from section analysis employing this model showed that FRP confinement is effective in providing both strength and ductility enhancement, and illustrated the effects of the amount of FRP confinement and the strength of the unconfined concrete.

ACKNOWLEDGEMENTS
The authors wish to thank The Hong Kong Polytechnic University for its financial support provided through the Area of Strategic Development (ASD) Scheme for the ASD in Advanced Buildings Technology in a Dense Urban Environment and through a postdoctoral fellowship awarded to the second author. They are also grateful to Dr. Y. Xiao of the University of Southern California who provided the test data used in this study.

REFERENCES
[1] Lam L. and Teng J.G. (2002a), "Stress-Strain Models for FRP-Confined Concrete". To be published.

[2] Lam L. and Teng J.G. (2002b), "Stress-Strain Models for Concrete Confined by Fibre-Reinforced Polymer Composites", Accepted for inclusion in the proceedings of 17th Australasian Conference on the Mechanics of Structures and Materials, Griffith University, Australia.

[3] Teng, J.G., Chen, J.F., Smith, S.T. and Lam, L. (2002) FRP-Strengthened RC Structures, John Wiley and Sons Ltd, UK, November, 238 pp.

[4] Lam L. and Teng J.G. (2001), "A Stress-Strain Model for FRP-Confined Concrete", Proceedings of the International Conference on FRP Composites in Civil Engineering, Hong Kong, China, Edited by Teng J.G., pp. 283-292.

[5] Lam L. and Teng J.G. (2002c), "Strength Models for FRP-Confined Concrete", Journal of Structural Engineering, ASCE, 128(5), in press.

[6] ENV 1992-1-1 (1991), Eurocode 2: Design of Concrete Structures – Part 1: General Rules and Rules for Building, European Committee for Standardization, Brussels.

[7] Xiao Y. and Wu H. (2000), "Compressive Behavior of Concrete Confined by Carbon Fiber Composite Jackets", Journal of Materials in Civil Engineering, ASCE, 12(2), 139-146.

[8] Yuan X.F., Lam L., Teng J.G. and Smith S.T. (2002), "Stress-Strain Models for FRP-Confined Concrete in Analysis of Columns under Combined Bending and Compression". To be published.

4.3 Performance assessment of FRP-confinement models – part I: review of experiments and models

L De Lorenzis[1] and R Tepfers[2]
[1]University Of Lecce, Lecce, Italy
[2]Chalmers University, Göteborg, Sweden

ABSTRACT

Despite a large research effort, a proper analytical tool to predict the behavior of FRP-confined concrete has not yet been established. Most of the available models are empirical in nature and have been calibrated against their own sets of experimental data. On the other hand, the experimental results available in the literature encompass a wide range of values of the significant variables. The objective of this paper and its companion one [1] is a systematic assessment of the performance of the existing models on FRP-confinement of concrete columns. In this paper, the experimental data available on FRP-confinement of concrete cylinders are classified according to the values of the significant variables and the existing empirical and analytical models are reviewed.

INTRODUCTION AND OBJECTIVE

One of the most attractive applications of fiber-reinforced polymer (FRP) composites is their use to achieve confinement of concrete columns. Since the early attempts [2-3], a remarkable amount of experiments have been performed to investigate the behavior of concrete columns confined with FRP spirals [4], wraps [5-14] and tubes [15-18]. Despite this large research effort, a proper analytical tool to predict such behavior has not yet been established. Early investigations attempted to extend to FRP confinement the analytical models previously in use for steel [3, 19], but it was soon stated that this operation yielded inaccurate and often unconservative results [20]. Since then, various models specifically suited for FRP-confined concrete columns have been proposed [8-10, 16, 21-23], most of which are empirical in nature and have been calibrated against their own sets of experimental data. On the other hand, the experimental results available in the technical literature encompass a wide range of values of the significant variables and can therefore be used to assess the accuracy of the existing models.

The objective of this work is a systematic assessment of the performance of the existing models on FRP-confinement of concrete columns. For the time being, only the case of concrete cylinders has been considered, due to the limited set of experimental data available for the case of non-circular columns and full-size columns. The assessment has focused on prediction of the ultimate axial stress and strain of the confined concrete, which are the most significant parameters from a design standpoint. In this paper, the experimental data on confinement of concrete cylinders with FRP available in the technical literature are collected and classified according to the values of the significant variables, and the existing empirical and analytical models are reviewed. The comparison between experimental and predicted results is presented in a companion paper [1].

REVIEW OF FRP-CONFINEMENT EXPERIMENTS AND MODELS

Results of about 180 tests from 17 different experimental sets documented in the published literature have been collected. Table 1 shows the range of variation of the variables in each experimental set, for a complete classification see [24]. The values of other variables whose influence is secondary or questionable (e.g. length of the overlap, presence of anti-friction devices, etc.) have been reported elsewhere [24].

The steel-based confinement models extend to FRP the formulations of Richart *et al* [25], Newman and Newman [26] and Mander *et al* [27]. Such formulations are based on ultimate strength surfaces modelled on triaxial test data, and therefore they predict the enhancement in strength of the confined concrete as a function of one value of confining pressure, assumed to be constant throughout the loading history. The other models, except the one by Spoelstra and Monti, are empirically based: they use best fitting of experimental data to correlate the enhancement in strength and ultimate strain of the confined cylinders to the values of the significant parameters. The equations are reported in Table 2. Among these models, the distinct feature of those by Toutanji [10] and Saafi *et al* [16] is that the equations giving f'_{cc} and ε_{cc} as functions of the respective relevant parameters have not been obtained by best fitting of the experimental values of f'_{cc} and ε_{cc}. Instead, they have been found by extrapolating to ultimate conditions (that is, $\varepsilon_l = \varepsilon_{fu}$) the equations giving the *current* axial stress and strain as functions of the *current* lateral strain, such equations being obtained by best fitting of experimental results. This is an expedient through which the ever-increasing confinement pressure exerted by the FRP is accounted for. In other words, the triaxial failure envelope is obtained assuming that the state of the confined concrete core is represented by a point on the envelope at each current value of axial stress and strain. The only difference between the two models is that regression analysis was conducted on experimental results obtained by FRP-encased rather than wrapped cylinders. Different values of the coefficients were found and the discrepancy was attributed to the bond between FRP sheets and concrete being stronger as compared to that of FRP tubes.

The basic idea of the model by Spoelstra and Monti [22] is to apply Mander's confinement model for each current value of lateral pressure exerted by the FRP. Such pressure is related by equilibrium and compatibility to the concrete current lateral strain. This in turn depends on the current axial strain and at the same time on the confinement pressure itself, which prevents the concrete from expanding laterally as much as it would in unconfined conditions. Therefore, the most critical aspect of modeling is the choice of the relationship that links the lateral strain to the axial strain and to the confining pressure. This is accomplished by means of the model in [28]. While the model adopts Mander's failure envelope, it does not assume that the current state of the concrete core is always represented by a point on the envelope. On the contrary, for the current value of axial strain, f'_{cc} and ε_{cc} competing to the corresponding lateral pressure are calculated, and then the current axial stress is derived from Popovics equation [27], therefore, it is always lower than the current value of f'_{cc}. This is true also for the ultimate stress, which is always lower than f'_{cc} corresponding to the maximum FRP confinement pressure. Another distinct feature of the model is that, as opposed to the others, it is able to predict stress-strain curves with a descending branch, and even losses of strength with respect to the unconfined case, which has actually been observed for very low FRP volumetric ratios. From the "exact" formulation of the model, the authors carried out regression analysis to provide "approximate" and more readily applicable expressions of the ultimate axial stress and strain, reported in Table 2.

DISCUSSION
As for the strength enhancement, almost all models relate f'_{cc}/f'_{co} to the ratio p_u/f'_{co}. Most models yield expressions very similar to each other, note for example the similarity not just qualitative but numerical between the equations proposed by Toutanji and Fardis and Khalili (the latter taken from the triaxial tests by Newman and Newman), although derived differently. The ductility enhancement, as expressed by the ratio $\varepsilon_{cc}/\varepsilon_{co}$, appears to be related not just to the strength properties but also to the stiffness of the confining device. This was recognized already in Fardis model, which expresses ε_{cc} as a function of the ratio of the lateral modulus to the unconfined concrete strength E_l/f'_{co}, and confirmed by most of the subsequent formulations.

Many authors have raised towards the steel-based confinement models the objection that they do not account for the profound difference in uniaxial tensile stress-strain behavior between steel and FRP [21-22]. According to these authors, while the assumption of constant confining pressure is still realistic in the case of steel confinement in the yield phase, it cannot be extended to FRP materials which do not exhibit any yielding and therefore apply on the concrete core a continuously increasing inward pressure. However, it should be noted that, in the case of steel confinement and for normal-strength concrete and steel, the peak stress corresponds to yielding of the confining device. Therefore, even though the confining pressure remains constant *afterwards*, the coordinates of the peak points should be predicted on the basis not of the yielding pressure but of a continuously increasing inward pressure due to a passive confinement effect. Hence, the success of the active confinement models in predicting the peak stress of columns passively confined with steel is to be attributed not to the fact that pressure is constant, but rather on the path-independency of the compressive strength of concrete under triaxial stresses. Already Richart *et al* [25], when comparing results of actively and passively confined specimens, found the behavior similar. Imran and Pantazopoulou [29] performing triaxial tests found that the concrete strength was essentially path-independent whereas the deformation behavior was path-dependent. Similar results were found by Lan and Guo [30]. The consequence is that ultimate strength surfaces

Table 1. Summary of Experimental Data

Ref.	Fiber Type	DxH (mm)	nt (mm)	E_f (MPa)	f_{fu} (MPa)	ε_{fu} (%)	f_{co} (MPa)	ε_{co} (%)	E_l (MPa)	p_u (MPa)	p_u/f_{co}	f_{cc} (f_{cu}) (MPa)	f_{cc}/f_{co}	ε_{cc} (ε_{cu}) (%)	ε_{cc} (ε_{cu})/ε_{co}	ε_{fp} (%)	$\varepsilon_{fp}/\varepsilon_{fu}$
Watanabe et al, 1997	CFRP	100x 200	0.14+ / 0.67	223400+ / 611600	2728.5+ / 1562.7	1.221+ / 0.256	30.20	0.25	748+ / 5137	36.56+ / 4.38	1.21+ / 0.14	104.60+ / 41.70	3.46+ / 1.38	4.151+ / 0.575	16.60+ / 2.30	1.000+ / 0.167	0.55+ / 0.98
Miyauchi et al, 1997	CFRP	100x200+ / 150x300	0.11+ / 0.33	230500	3481	1.510	31.20+ / 51.90	0.190+ / 0.219	338+ / 1014	5.11+ / 15.32	0.11+ / 0.49	52.40+ / 104.60	1.31+ / 2.62	0.945+ / 2.013	4.32+ / 10.32	N/A	N/A
Kono et al, 1998	CFRP	100x200	0.167+ / 0.501	235000	3820	1.626	32.30+ / 34.80	0.170+ / 0.234	785+ / 2355	12.76+ / 38.28	0.37+ / 1.19	50.70+ / 110.10	1.46+ / 3.16	0.785+ / 2.490	4.08+ / 10.87	0.960+ / 0.610	0.59+ / 0.38
Toutanji, 1999	GFRP	76x305	0.236	72600	1518	2.091	30.93	0.190	451	9.43	0.30	60.82	1.97	1.530	8.05	1.630	0.78
Toutanji, 1999	CFRP	76x305	0.22+ / 0.33	230500+ / 372800	3485+ / 2940	1.512+ / 0.789	30.93	0.190	1334+ / 3237	20.18+ / 25.53	0.65+ / 0.83	95.02+ / 94.01	3.07+ / 3.04	2.450+ / 1.550	12.89+ / 8.16	1.250+ / 0.550	0.83+ / 0.70
Matthys et al, 1999	CFRP	150x300	0.117+ / 0.235	220000+ / 500000	2600+ / 1100	1.182+ / 0.220	34.90	0.210	343+ / 1567	4.06+ / 3.45	0.12+ / 0.10	46.10+ / 45.80	1.32+ / 1.31	0.900+ / 0.600	4.29+ / 2.86	1.260+ / 0.310	1.07+ / 1.41
Shahawy et al, 2000	CFRP	153x305	1.25+ / 0.36	82700	2275	2.751	19.40+ / 49.00	0.200	1356+ / 390	37.30+ / 10.74	1.92+ / 0.22	33.80+ / 112.70	4.13+ / 1.21	3.560+ / 0.620	17.80+ / 3.10	0.570+ / 0.730	0.21+ / 0.27
Rochette and Labossiere, 2000	CFRP	100x200	0.6	82700	1265	1.530	42.00	N/A	992	15.18	0.36	67.62+ / 73.50	1.61+ / 1.75	1.350+ / 1.600	N/A	0.800+ / 0.950	0.52+ / 0.62
Rochette and Labossiere, 2000	AFRP	150x300	1.26+ / 5.04	13600	230	1.691	43.00	N/A	228+ / 914	3.86+ / 15.46	0.09+ / 0.36	47.30+ / 74.39	1.10+ / 1.73	1.110+ / 1.740	N/A	1.550+ / 1.180	0.92+ / 0.70
Micelli et al, 2001	GFRP	100x200	0.35	72400	1520	2.099	32.00	0.145	507	10.64	0.33	48.00+ / 54.00	1.50+ / 1.69	0.850+ / 1.494	5.87+ / 10.32	0.990+ / 1.500	0.47+ / 0.71
Micelli et al, 2001	CFRP	100x200	0.16	227000	3790	1.670	37.00	0.190	726	12.13	0.33	57.00+ / 62.00	1.54+ / 1.68	0.900+ / 1.090	4.74+ / 5.74	1.070+ / 1.350	0.64+ / 0.81
Rousakis, 2001	CFRP	150x300	0.169+ / 0.845	118340	2024	1.710	25.15+ / 82.13	0.290+ / 0.350	267+ / 1333	4.56+ / 22.80	0.06+ / 0.54	38.75+ / 137.93	1.14+ / 2.79	0.44 (0.46) / +2.45	1.42+ / 7.66	0.095+ / 0.770	0.06+ / 0.45
Saafi et al, 1999	GFRP	152x435	0.8+ / 2.4	32000+ / 36000	450+ / 560	1.406+ / 1.556	35.00	0.250	336+ / 1134	4.72+ / 17.64	0.13+ / 0.50	52.80+ / 83.00	1.51+ / 2.37	1.90+ / 3.00	7.60+ / 12.0	1.650+ / 1.700	1.21+ / 1.09
Saafi et al, 1999	CFRP	152x435	0.11+ / 0.55	367000+ / 415000	3300+ / 3700	0.910+ / 0.892	35.00	0.250	530+ / 2995	4.76+ / 26.71	0.14+ / 0.76	55.00+ / 97.00	1.57+ / 2.77	1.00+ / 2.22	4.00+ / 8.88	1.300+ / 0.900	1.45+ / 1.01
Mirmiran and Shahawy, 1997	GFRP	153x305	1.44+ / 2.97	37233+ / 40749	524+ / 641	1.407+ / 1.573	29.64+ / 31.97	N/A	703+ / 1587	9.90+ / 24.97	0.31+ / 0.84	53.66+ / 114.66	1.74+ / 3.87	2.900+ / 5.330	N/A	1.150+ / 1.940	1.36+ / 0.80
La Tegola & Manni, 1999	GFRP	150x300	4.28+ / 5.9	25250+ / 25450	652+ / 670	2.582+ / 2.633	25.61	N/A	1441+ / 2002	37.21+ / 52.80	1.45+ / 2.06	71.00+ / 110.30	2.77+ / 4.31	4.100+ / 9.900	N/A	N/A	N/A
Fam and Rizkalla, 2000	GFRP	219x438+ / 100x200	3.73+ / 3.08	33400+ / 23000	548+ / 398	1.641+ / 1.730	58.00+ / 37.00	0.200	1483+ / 1129	18.52+ / 24.52	0.32+ / 0.66	90.00+ / 68.00	1.17+ / 2.19	1.350+ / 1.030	6.75+ / 5.15	1.200+ / 1.650	0.73+ / 0.95

Table 2. Summary of Models

Model	Theoretical f'_{cc}	Theoretical ε_{cc}	Range of p_u/f'_{co} used for calibration	"Premature" failure incorp.?
Fardis and Khalili	$\dfrac{f'_{cc}}{f'_{co}} = 1 + 4.1\dfrac{p_u}{f'_{co}}$; $\dfrac{f'_{cc}}{f'_{co}} = 1 + 3.7\left(\dfrac{p_u}{f'_{co}}\right)^{0.86}$	$\varepsilon_{cc} = \varepsilon_{co} + 0.0005\dfrac{E_l}{f'_{co}}$	0.1-0.6	NO
Saadatm. et al	$\dfrac{f'_{cc}}{f'_{co}} = 2.254\sqrt{1 + 7.94\dfrac{p_u}{f'_{co}}} - 2\dfrac{p_u}{f'_{co}} - 1.254$	$\dfrac{\varepsilon_{cc}}{\varepsilon_{co}} = 1 + 5\left(\dfrac{f'_{cc}}{f'_{co}} - 1\right)$	-	NO
Miyauchi et al	$\dfrac{f'_{cc}}{f'_{co}} = 1 + 3.485\dfrac{p_u}{f'_{co}}$	$\dfrac{\varepsilon_{cc}}{\varepsilon_{co}} = 1 + 10.6\left(\dfrac{p_u}{f'_{co}}\right)^{0.373}$ (f'_{co}=30 MPa)	0.1-0.5	N/A
Kono et al	$\dfrac{f'_{cc}}{f'_{co}} = 1 + 0.0572 p_u$	$\varepsilon_{cc} = 1 + 0.280 p_u$	0.37-1.19	YES
Saaman et al	$\dfrac{f'_{cc}}{f'_{co}} = 1 + 6.0\left(\dfrac{p_u}{f'_{co}}\right)^{0.7}$	$\varepsilon_{cc} = \dfrac{f'_{cc} - f_0}{E_2}$; $f_0 = 0.872 f'_{co} + 0.371 p_u + 6.258$	0.31-0.84	NO
Toutanji	$\dfrac{f'_{cc}}{f'_{co}} = 1 + 3.5\left(\dfrac{p_u}{f'_{co}}\right)^{0.85}$	$\dfrac{\varepsilon_{cc}}{\varepsilon_{co}} = 1 + (310.57\varepsilon_{fu} + 1.90)\left(\dfrac{f'_{cc}}{f'_{co}} - 1\right)$	0.3-0.83	NO
Saafi et al	$\dfrac{f'_{cc}}{f'_{co}} = 1 + 2.2\left(\dfrac{p_u}{f'_{co}}\right)^{0.84}$	$\dfrac{\varepsilon_{cc}}{\varepsilon_{co}} = 1 + (537\varepsilon_{fu} + 2.6)\left(\dfrac{f'_{cc}}{f'_{co}} - 1\right)$	0.13-0.76	NO
Spoelstra and Monti (approx)	$\dfrac{f'_{cu}}{f'_{co}} = 0.2 + 3\left(\dfrac{p_u}{f'_{co}}\right)^{0.5}$	$\dfrac{\varepsilon_{cc}}{\varepsilon_{co}} = 2 + 1.25\dfrac{E_{co}}{f'_{co}}\varepsilon_{fu}\sqrt{\dfrac{p_u}{f'_{co}}}$	-	NO
Xiao and Wu	$\dfrac{f'_{cc}}{f'_{co}} = 1.1 + \left(4.1 - 0.75\dfrac{f'^{2}_{co}}{E_l}\right)\dfrac{p_u}{f'_{co}}$	$\varepsilon_{cc} = \dfrac{\varepsilon_{fu} - 0.0005}{7\left(\dfrac{f'_{co}}{E_l}\right)^{0.8}}$	0.14-0.7	NO

based on triaxial strength data should be valid and successful for FRP-confined concrete as they are for steel confined concrete. This is actually the case for Mander's model [1], but not for the expressions by Richart *et al* and Newman and Newman. The reason is probably that such expressions overestimate the effectiveness of high confining pressures and especially so for medium to high strength concretes [29]. This becomes particularly evident in the comparison with FRP-confinement tests, where the confining pressure can attain unusually high values (up to about twice the unconfined strength) due to the high tensile strength of the FRP. Yet, the formula by Newman and Newman is very similar to that by Toutanji, the latter being developed on the basis of FRP confinement tests (Table 2).

For the computation of the ultimate strain of the confined concrete, path dependency cannot be ignored any longer and the passive confinement mechanism should be taken into account. While Fardis and Khalili proposed an empirically based equation, Saadatmanesh *et al* continued to follow Mander's model in the computation of both the strain at peak stress and the ultimate strain, not accounting for the experimental evidence that in FRP-confined specimens these two values coincide. This approach is evidently inappropriate [15, 22].

Figures 1 illustrates the prediction curves of the various models for f'_{cc}/f'_{co} and $\varepsilon_{cc}/\varepsilon_{co}$, respectively. The comparison can only be partial since not all models introduce the same significant variables, especially in computation of the peak strain. As for the prediction of f'_{cc}/f'_{co}, the highest curve is that based on Richart's formula. The models by Fardis and Khalili (based on Newman and Newman), Miyauchi *et al* and Toutanji yield very similar curves, due to the similar value of the coefficient multiplying the ratio p_u/f'_{co} and to the weak influence of the exponents, close to the unity, in the respective formulae. Rather close to each other are also the curves by Saadatmanesh (Mander), Saafi *et al*, and Spoelstra and Monti (approximate), the first one predicting higher values of strength at lower pressures and lower values at higher pressures than the others. The equations by Saaman *et al* and Kono *et al* have been plotted for two different values of unconfined strength, namely, 20 and 60 MPa. The first model predicts a lower confinement effectiveness for higher strength concrete at a given p_u/f'_{co} ratio, whereas the opposite trend is suggested by Kono's model.

In prediction of $\varepsilon_{cc}/\varepsilon_{co}$, the models show more different trends. When plotting the equations by Toutanji and Saafi *et al*, since $\varepsilon_{cc}/\varepsilon_{co}$ is also related to f'_{cc}/f'_{co}, the latter has been calculated according to the respective models. In this way, the curves appear to be very similar, irrespective of the value of FRP ultimate strain. However, due to the remarkable difference between the two models in predictions of f'_{cc}/f'_{co}, very different results would be obtained if $\varepsilon_{cc}/\varepsilon_{co}$ had to be computed with the same set of f'_{cc}/f'_{co} values. Both models yield lower values of $\varepsilon_{cc}/\varepsilon_{co}$ for FRP with lower ultimate strain, that is, for a given strength, with higher modulus of elasticity. The model by Kono *et al* is very sensitive to the concrete unconfined strength, and predicts higher strains for higher concrete strengths at a given p_u/f'_{co} ratio. Same trend is displayed by Miyauchi´s model at high values of p_u/f'_{co}, although with less sensitivity. The model by Saadatmanesh (of which the expression of the strain at peak has been considered, using the f'_{cc}/f'_{co} value predicted by the model itself) is rather close to Miyauchi´s curves, and predicts lower values of strains if compared to the models by Saafi $et\ al$ and Toutanji.

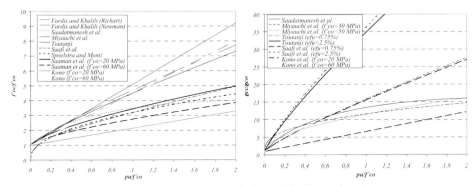

Figure 1. Comparison of the models in prediction of f'_{cc}/f'_{co} and $\varepsilon_{cc}/\varepsilon_{co}$

CONCLUSIONS

It has often been remarked that steel-based models are not able to capture the behavior of FRP-confined concrete due to the fact that the FRP confining device exerts a continuously increasing pressure as opposed to steel in the yield phase. However, attainment of the confined strength corresponds in both steel- and FRP-confined concrete to the end of the linear-elastic phase, with the different behavior coming afterwards. Therefore, triaxial strength curves based on a unique value of confining pressure should be neither valid for steel nor for FRP. The reason they are (and, hence, they are for both) is to be attributed to the path independency of the strength properties of concrete under triaxial stresses already documented in the literature. This explains the high similarity observed in this study between empirical strength curves developed for concrete under triaxial stresses, steel-confined and FRP confined concrete: compare the formulae by [26] with that by [10] and the predictions of the confined strength by [27] and by Spoelstra and Monti "exact" [22]. Different is the case of the deformation properties, for which path-dependency is observed and hence the stiffness characteristics of the confining device become crucial.

REFERENCES

[1] De Lorenzis, L., and Tepfers, R. (2002): Performance Assessment of FRP-Confinement Models – Part II: Comparison of Experiments and Predictions, Proc. ACIC 2002, Southampton, UK.

[2]   Kurt, C.E. (1978): Concrete filled structural plastic columns, J. of the Struct. Div. ASCE, v. 104, No. ST1, pp. 55-63.

[3] Fardis, M.N., and Khalili, H. (1981): Concrete encased in fiberglass-reinforced-plastic. J. of the Am. Concr. Inst. Proceedings, v. 78, No. 6, pp. 440-446.

[4] Nanni, A., and Bradford, N.M. (1995): FRP jacketed concrete under uniaxial compression, Constr. and Build. Mat., v. 9, No. 2, pp. 115-124, Elsevier Ltd., UK.

[5] Harmon T.G., Slattery K.T. (1992): Advanced composite confinement of concrete. Proceedings ACMBS-I, 1992. Sherbrooke, Québec, Canada. pp 299-306.

[6] Picher F., Rochette P., Labossiére P. (1996): Confinement of concrete cylinders with CFRP. Proceedings ICCI'96, Tucson, Arizona. pp.829-841.

[7] Watanabe K., Nakamura H., Honda Y., Toyoshima M., Iso M., Fujimaki T., Kaneto M., Shirai N. (1997): Confinement effect of FRP sheet on strength and ductility of concrete cylinders under uniaxial compression. Proc. of FRPRCS-3, Sapporo, Vol 1, pp. 233-240.

[8] Miyauchi K., Nishibayashi S., Inoue S. (1997): Estimation of strengthening effects with carbon fiber sheet for concrete column. Proc. of FRPRCS-3, Sapporo, Vol 1, pp. 217-224.

[9] Kono S., Inazumi M., Kaku T. (1998): Evaluation of Confining Effects of CFRP Sheets on Reinforced Concrete Members. Proceedings of ICCI'98, Tucson, Arizona. pp.343-355.

[10] Toutanji H. (1999): Stress-Strain Characteristics of Concrete Columns Externally Confined with Advanced Fiber Composite Sheets. ACI Mat.J., v. 96, No. 3, pp.397-404.

[11] Matthys S., Taerwe L., Audenaert K. (1999): Tests on Axially Loaded Concrete Columns Confined by Fiber Reinforced Polymer Sheet Wrapping. Proc. FRPRCS-4, Baltimore, MD, pp. 217-228.

[12] Shahawy, M., Mirmiran, A., and Beitelmann, T. (2000): Tests and modeling of carbon-wrapped concrete columns, Composites Part B: Eng., No. 31, pp. 471-480, Elsevier Ltd, UK.

[13] Micelli, F., Myers, J.J., and Murthy, S. (2001): Effect of environmental cycles on concrete cylinders confined with FRP, Proceedings of CCC2001 Porto, Portugal.

[14] Rousakis, T. (2001): Experimental investigation of concrete cylinders confined by carbon FRP sheets, under monotonic and cyclic axial compressive load, Research Report, Chalmers University of Technology, Göteborg, Sweden.

[15] Mirmiran A., Shahawy M. (1997): Behavior of Concrete Columns Confined by Fiber Composites. J. of Struct. Eng., ASCE, v. 123, No. 5, pp. 583-590.

[16] Saafi M., Toutanji H.A., Li Z. (1999): Behavior of Concrete Columns Confined with Fiber Reinforced Polymer Tubes. ACI Mat. J., v. 96, No. 4, pp. 500-509.

[17] La Tegola A., Manni O. (1999): Experimental Investigation on Concrete Confined by Fiber Reinforced Polymer and Comparison with Theoretical Model. Proc. FRPRCS-4, Baltimore, MD, pp. 217-228.

[18] Fam A.Z., Rizkalla S.H. (2000): Concrete – filled FRP tubes for flexural and axial compression members. Proceedings ACMBS-3, Ottawa, Canada, pp. 315-322.

[19] Saadatmanesh H., Ehsani M.R., Li M.W. (1994): Strength and Ductility of Concrete Columns Externally Reinforced with Fiber Composite Straps. ACI Struct. J., v. 91, No. 4, pp. 434-447.

[20] Mirmiran, A., and Shahawy, M. (1996): A new concrete-filled hollow FRP composite column, Composites Part B: Engineering, v. 27B, No. 3-4, pp. 263-268, Elsevier Ltd., UK.

[21] Samaan M., Mirmiram A., Shahawy M. (1998): Model of Concrete Confined by Fiber Composites. J. of Struct. Eng., ASCE, v. 124, No 9, pp. 1025-1031.

[22] Spoelstra M. R., Monti G. (1999): FRP-Confined Concrete Model. J. of Comp. for Constr., ASCE, v. 3, No. 3, pp. 143-150.

[23] Xiao Y., Wu H. (2000): Compressive Behavior of Concrete Confined by Carbon Fiber Composite Jackets. J. of Mat. in Civ. Eng., ASCE, v. 12, No. 2, pp. 139-146.

[24] De Lorenzis, L. and Tepfers, R. (2001) : A Comparative Study of Models on Confinement of Concrete Cylinders with FRP Composites, Publ. 01 :04, Dept. of Building Materials, Chalmers University of Technology, Göteborg, Sweden, 81 pp.

[25] Richart, F.E., Brandtzaeg, A., and Brown, R.L. (1928): A study of the failure of concrete under combined compressive stresses, Univ. of Illinois. Eng. Exp. Station Bulletin No. 185.

[26] Newman, K., and Newman, J.B. (1972): Failure theories and design criteria for plain concrete, Proceedings, International Engineering Materials Conference on Structure, Solid Mechanics and Engineering Design, Southampton, Wiley Interscience, Part 2, pp. 963-995.

[27] Mander, J.B., Priestley, M.J.N., and Park, R. (1988): Theoretical stress-strain model for confined concrete, J. of Str. Eng., ASCE, vol. 114, No. 8, pp. 1804-1826.

[28] Pantazopoulou S.J., Mills R.H. (1995): Microstructural Aspects of the Mechanical Response of Plain Concrete. ACI Mat.J., v. 92, No. 6, pp. 605-616.

[29] Imran, I., and Pantazopoulou, S.J. (1996): Experimental study of plain concrete under triaxial stress, ACI Mat. J., v. 93, No.6, pp. 589-601.

[30] Lan, S., and Guo, Z. (1997): Experimental Investigation of Multiaxial Compressive Strength of Concrete under Different Stress Paths, ACI Mat. J., v. 94, No. 5, pp. 427-434.

4.4 Performance assessment of FRP-confinement models – part II: comparison of experiments and predictions

L De Lorenzis[1] and R Tepfers[2]
[1]University Of Lecce, Lecce, Italy
[2]Chalmers University, Göteborg, Sweden

ABSTRACT
This paper follows up to a companion one [1] in which the experimental data on FRP-confinement of concrete cylinders available in the technical literature were classified according to the values of the significant variables and the existing empirical and analytical models were reviewed. In this paper, the whole set of available experimental results is compared with the whole set of analytical models.

PERFORMANCE ANALYSIS OF THE EXISTING MODELS
In the following, the experimental confined concrete strength, f'_{cc}, and axial strain at peak stress ε_{cc}, both divided by the respective unconfined values, are compared with the predictions of the models reviewed in [1] to assess their accuracy. The accuracy of the models was quantitatively evaluated computing the average absolute percent error. Also, the comparison was graphically made by plotting experimental versus theoretical values. Due to limited space, these graphs are not reported herein, for the complete set see [24].

Prediction of f'_{cc}
Figure 1 summarizes results by showing the average absolute error of the various models in the prediction of f'_{cc} for specimens confined by FRP wraps and tubes.

- The models by [3], [8], [10], and [23] largely overestimate the strength of most confined cylinders. The error is larger for cylinders with heavy confinement, that is, for those with the highest values of the ratio p_u/f'_{co}. This may be due to the use of formulae that overestimate the effectiveness of high lateral pressures [3] or to the empirical calibration being conducted on specimens with low ultimate confinement pressures [8,10,23]. For these models, the prediction is less accurate for the tube-encased cylinders, which typically have higher values of the ratio p_u/f'_{co} due to the greater thickness of the tube. The experimental set with the highest p_u/f'_{co} values is that by [17], with p_u/f'_{co} ranging from 1.45 to 2.06. If this set is excluded, the average absolute error decreases approaching the value relative to the wrapped specimens.

- The remaining models give better predictions of the confined strength, with no appreciable difference between wrapped and encased specimens. It is interesting to note that the models by Spoelstra and Monti "exact" and by Saadatmanesh (Mander) yield very similar results: the ratio between f'_{cc}/f'_{co} computed by the two models has an average value of 0.9637 with a standard deviation of 0.0347 (coefficient of variation 3.6%). This supports what observed in [1].

- The three most accurate models are those by Saaman *et al*, Saafi *et al*, and Spoelstra and Monti "approximate", whose average absolute errors on the whole set of specimens are 13.4%, 13.8% and 15.2%, respectively.

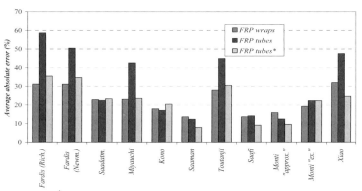

*Excluding data from (La Tegola and Manni, 1999).

Figure 1. Average absolute error in the prediction of f'_{cc}/f'_{co}

Prediction of ε_{cc}

Figure 2 shows the average absolute error of the various models in the prediction of ε_{cc} for specimens confined by FRP wraps (or filaments) and tubes. It appears that:

- The errors in the prediction of ε_{cc} are much larger than those on the prediction of f_{cc}. In most cases, the error results from overestimating the strain at peak stress. This conducts to overestimate the ductility of the confined cylinder.

- Unlike the prediction of f'_{cc}, that of ε_{cc} is less accurate for wrapped than for encased cylinders. This is due to the fact that, in wrapped cylinders, the experimental value of the lateral strain in the FRP at tensile failure is mostly lower than the FRP ultimate strain in uniaxial tension. This discrepancy affects the accuracy in the prediction of ε_{cc}, in fact, the models that mostly show a difference in accuracy between wrapped and encased specimens are the most sensitive to the value of the FRP ultimate strain (or stress). Such discrepancy is not registered in encased specimens, as shown later.

- The model by [19] (Mander) has been applied for the prediction of the strain at peak stress, which in most cases coincides with the ultimate strain of the FRP confined cylinder. However, Saadatmanesh intended to compute the ultimate strain with the energy balance approach suggested by Mander, which is not appropriate for FRP confinement. Since the expression of ε_{cc} suggested by Mander is related to the ratio f'_{cc}/f'_{co}, the comparison has been made both with the theoretical and experimental values of f'_{cc}/f'_{co}. In the latter case, the error is smaller. A similar approach has been adopted for the models by Toutanji and Saafi, that also relate $\varepsilon_{cc}/\varepsilon_{co}$ to the ratio f'_{cc}/f'_{co}. Both models are inaccurate in predicting ε_{cc} for wrapped cylinders. The model by Toutanji gives a better prediction when the experimental f'_{cc}/f'_{co} are used (the error decreases from 118% to 50%), whereas the error of the Saafi model remains approximately the same (decreases from 126% to 120%). This because the Saafi model is more accurate in predicting the confined strength, and thus the difference between experimental and theoretical f'_{cc}/f'_{co} is small and does not influence much the error in the prediction of $\varepsilon_{cc}/\varepsilon_{co}$. Conversely, the error of the Toutanji model in the prediction of $\varepsilon_{cc}/\varepsilon_{co}$ is in large part a consequence of its overestimating f'_{cc}/f'_{co}. Summarizing, of the two models, the Saafi one is more accurate in the strength prediction whereas the Toutanji model is more accurate in predicting the strain at peak.

- The three models that perform best in the prediction of f'_{cc} do not perform equally well when it comes to ε_{cc}. It appears that no model is able to predict the strain at peak stress of the confined cylinders with reasonable accuracy.

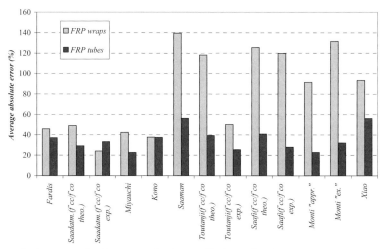

Figure 2. Average absolute error in the prediction of $\varepsilon_{cc}/\varepsilon_{co}$

EFFECTIVE FRP ULTIMATE STRAIN

As previously mentioned, failure of the confined cylinders occurs when the FRP confining device ruptures in hoop tension, but, in wrapped cylinders, the experimental value of the hoop strain in the FRP at tensile failure is usually lower than the FRP ultimate strain in uniaxial tension. Object of this section is to formulate some observations on this phenomenon based on the available experimental database. In the following, the ratio of the measured hoop strain in the confining FRP at tensile failure ("effective ultimate strain") to the FRP ultimate strain in uniaxial tension ("reference ultimate strain") is termed "reduction factor". The experimental reduction factor is available for 3 sets of experiments regarding FRP-encased specimens and 9 sets with wrapped specimens.

For FRP-encased cylinders, Table 1 in [1] shows that the reduction factor oscillates around the unity value, the average of all specimens being 1.073 with a coefficient of variation of 18%. The fact that values greater than one are reached may be attributed to scatter in the FRP tensile strength and in the strain measurement. This result seems to indicate that FRP-encased specimens do not display "premature" tensile failure of the confining device but are able to exploit the full strength of the composite material. The situation is very different for wrapped specimens, with the reduction factor ranging from more than 1 to less than 0.1.

In order for the "premature" tensile failure to be predicted, the parameters that exert an influence on it must be identified. The possible reasons for this phenomenon have been suggested in [11]: the multiaxial stress state existing in the FRP; non-homogeneous deformations in the cracked concrete at high load levels; the curved shape of the wrapping reinforcement; the quality of the execution; size effect when applying multiple layers. The different behavior between encased and wrapped cylinders seems to suggest that the reduction factor is mostly affected by those of the above mentioned causes that are typical of wrapped rather than encased specimens. For example, the application of FRP wraps by hand lay-up raises the problem of the quality of execution. Local misalignments or wavings may lead to unequal stretching of the fibers, so that the most stretched ones reach failure before the hoop strain is globally equal to the ultimate tensile strain of the laminate. Rupture of some fibers would initiate failure in the second most stretched ones, and so forth in a progressive fashion that finally leads to catastrophic failure of the specimen. This phenomenon is witnessed by the progressive crackling noise that is typically heard during tests of confined cylinders. For a given difference in fiber stretching, high-modulus wraps should display higher differences in load taking, and therefore progressive failure is likely to be more pronounced for high-modulus fibers [31]. Conversely, FRP tubes are manufactured with automated filament winding technique, and therefore a nearly equal stretching of all fibers should be expected. Size effect due to overlap of more layers can be intended in two ways: from the standpoint of the strain measurement, the hoop strain measured on the external surface of the specimen is somewhat lower than the strain in the innermost FRP layer; from the standpoint of the execution quality, the overlap of more layers is detrimental due to its manual difficulty and to the fact that, for example, at each protruding or wavy spot on the first layer will correspond a larger defect on the second layer and so forth. The first type of size effect, if present in the wrapped specimens, is a fortiori present in the encased ones due to the greater thickness of the tubes, whereas the second source of size effect is typical of the hand lay-up technology and therefore of wrapped specimens. The multiaxial stress state in the FRP confining device is common to both wrapped and encased specimens. Even if load is not applied directly on the FRP, part of the axial load acting on the concrete is transmitted to the FRP by means of bond stresses at the contact surface. However, the bond between wraps and concrete is ensured by the impregnating resin whereas that between the tube and the concrete surface is essentially based on friction and might be affected by shrinkage of the inner concrete. Finally, the radius of curvature of the FRP should have an influence on its tensile strength for both wraps and tubes. Results in [32] indicated that the tensile strength of a CFRP wrap decreased with decreasing corner radius, and was approximately equal to 66% of the reference strength for the highest radius investigated, equal to 51 mm.

The previous considerations seem to indicate that the most important parameters affecting the "premature" tensile failure of the confining wraps are the number of overlapping layers, the elastic modulus of the wrap and the radius of curvature, that is, for cylindrical specimens, the radius of the cylinder. However, other parameters may be influential [24]. In an attempt to find a quantitative correlation between the three chosen parameters and the reduction factor, the available experimental data have been plotted versus the product of the FRP elastic modulus by the wrap thickness for the three most frequent specimen diameters: 50, 100 and 150 mm (Figure 3). Although in some cases considerable scatter exists, the consistency in trend is evident. While the wide scatter could be interpreted as due to other parameters exerting an influence and not being considered, equally wide scatters exists among identical specimens tested by the same author indicating that a high degree of randomness is inherent to the phenomenon. It appears that, for a given specimen diameter, the reduction factor decreases for increasing stiffness of the wrap. Best fitting interpolation conducted to the exponential-type equations shown in the figures. When comparing graphs competing to different diameters, it appears that both the starting value of the interpolating curve (that is, the value of the reduction factor for zero stiffness of the wrap) and its slope decrease as the specimen diameter increases. In other words, the reduction factor is smaller for big diameters at low stiffnesses and for low diameters at high stiffnesses. A possible explanation is the following. For a given stiffness of the wrap, specimens with small diameter are better confined than the bigger ones, since the confinement ratio is higher. This delays the phase in which uneven deformations in the concrete and high local stresses in the FRP are developed. On the other hand, this beneficial effect is counteracted by the "corner radius" effect, which causes the opposite tendency: sheets with higher radius of curvature perform better in tension than those with small radius. The observed behavior might result from the combination of these two opposite tendencies, the first one prevailing on the second at low stiffness values and vice versa at high stiffnesses.

It can be noticed [24] that the reduction factor is higher in the case of AFRP with respect to CFRP confined cylinders with the same diameter (150 mm) and similar stiffness. This is probably due to the higher toughness that characterizes aramid fiber composites as opposed to carbon and glass.

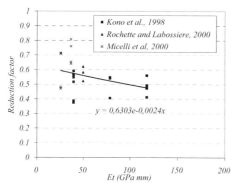

Figure 3a. Reduction factor vs. FRP stiffness (specimens with D=100 mm)

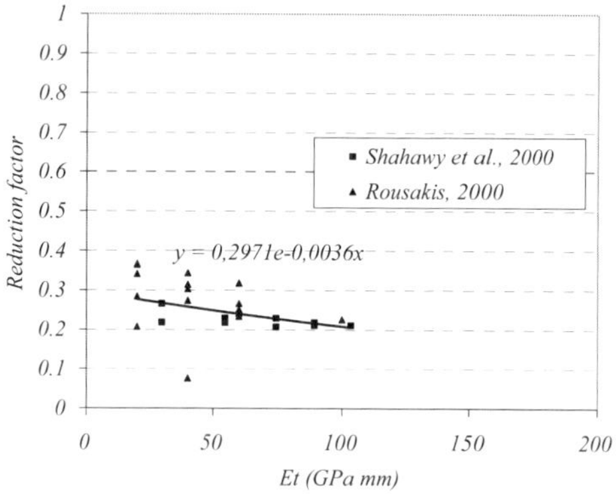

Figure 3b. Reduction factor vs. FRP stiffness (specimens with D=150 mm)

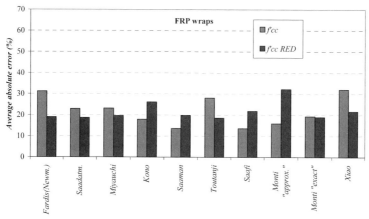

Figure 4. Average absolute errors in the prediction of f'_{cc}/f'_{co} based on the uniaxial and on the effective FRP rupture strain (FRP-wrapped specimens)

Prediction of f'_{cc} with the effective FRP ultimate strain

Figure 4 shows the average absolute error of the various models in predicting f'_{cc} for wrapped specimens when the effective ultimate strain and the reference ultimate strain are introduced into the equations. The following conclusions can be drawn:

- For FRP-wrapped specimens, since using the effective FRP ultimate strain results in lower theoretical values of f'_{cc}, the average absolute error decreases (as compared to that obtained with the reference ultimate strain) for the models previously found to be unconservative. Conversely, the models previously found to be the most accurate become too conservative and their absolute error increases. The data sets that mostly influence results are those by [12] and [14], both characterized by very low values of the reduction factor, and partly those by [5] and [9].

- For FRP-encased specimens, since the reduction factor is close to the unity, no appreciable differences are noticed.

- If the effective rupture strain of the FRP wrap was known in advance, predictions of the strength enhancement by the existing models would not necessarily be more accurate than they are using the reference ultimate strain. Based on the overall average absolute percent error, the most accurate models are now those by Saaman *et al*, Saafi *et al*, Spoelstra and Monti "exact" and Saadatmanesh *et al*, with average errors 16.9%, 18.2%, 19.0% and 19.5%, respectively.

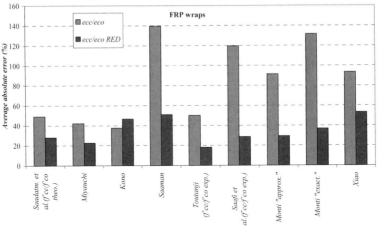

Figure 5. Average absolute errors in the prediction of $\varepsilon_{cc}/\varepsilon_{co}$ based on the uniaxial and on the effective FRP rupture strain (FRP-wrapped specimens)

Prediction of ε_{cc} with the effective FRP ultimate strain

Figure 5 shows the average absolute error of the various models in predicting ε_{cc} for wrapped specimens with both the ultimate and effective FRP hoop strains. It appears that:

- For wrapped specimens, the average absolute error of all models shows a remarkable decrease when the FRP effective strain is inserted in the equations. This confirms what anticipated previously, that the accuracy in the prediction of ε_{cc} is influenced by the premature rupture of the FRP. Such influence is much stronger than for the prediction of f'_{cc}, probably because of the higher sensitivity of the equations giving $\varepsilon_{cc}/\varepsilon_{co}$ to the value of p_u and because of the wider range of percent variation of $\varepsilon_{cc}/\varepsilon_{co}$ when compared to f'_{cc}/f'_{co} (see Figure 1 in [1]). Since all models overestimate ε_{cc} when using the reference ultimate strain of the FRP wrap, using the effective one always results in decreasing error.

- For encased specimens, since the reduction factor is close to the unity, only small differences are noticed due to the different databases used for the two predictions.

- If the effective rupture strain of the FRP wrap was known in advance, the model by Toutanji would be the most accurate in predicting ε_{cc}, with an average absolute error of about 18%. However, this model has been applied using the experimental value of the ratio f'_{cc}/f'_{co} rather than that calculated by the model itself.

CONCLUSIONS

In most experimental studies, the FRP confining wraps have been observed to rupture at a tensile strain lower than the uniaxial ultimate tensile strain. On the basis of the experimental database collected in this study, this phenomenon has been analysed. It appears that it mostly interests FRP wraps, whereas FRP tubes seem to fail at values of tensile strain close to the predicted ones. However, the database involving encased specimens is more limited than that related to wrapped cylinders, and is not large enough to draw final conclusions. Number, thickness and elastic modulus of overlapping layers and radius of curvature were identified as the most significant parameters and used to develop a tentative equation for the reduction factor.

When using the nominal values of p_u, the strength of the confined cylinders was rather accurately predicted by some of the existing models, particularly those by Saaman et al, Saafi et al, and Spoelstra and Monti "approximate". Using the reduced values of p_u improved the accuracy for models found unconservative with the nominal p_u and it made the others more conservative and less accurate. Based on the overall average absolute percent error, the most accurate models with the reduced p_u were those by Saaman et al, Saafi et al, Spoelstra and Monti "exact" and Saadatmanesh et al (Mander).

Using the nominal values of p_u and ε_{fu}, the errors in the prediction of the strain at peak stress of the confined cylinders were larger than those on the prediction of f_{cc} and in most cases on the unconservative side. No model was able to predict ε_{cc} with reasonable accuracy, especially so for wrapped specimens and for models particularly sensitive to the value of ε_{fu}, due to the premature failure phenomenon. When the FRP effective strain was inserted into the equations, the average absolute error of all models showed a remarkable decrease for wrapped specimens, whereas for encased cylinders only small differences were noticed since the reduction factor was close to the unity. If the effective rupture strain of the FRP wrap was known in advance, the model by Toutanji with the experimental value of the ratio f'_{cc}/f'_{co} would be the most accurate in predicting ε_{cc}.

Accuracy of the models on FRP confinement of concrete still needs to be improved. Furthermore, the premature failure phenomenon of FRP wraps confining concrete cylinders still needs investigations from both a qualitative and quantitative standpoint.

REFERENCES
[1] De Lorenzis, L., and Tepfers, R. (2002): Performance Assessment of FRP-Confinement Models – Part I: Review of Experiments and Models, Proc. ACIC 2002, Southampton, UK.

[31] Tepfers, R. (2001): Compatibility related problems for FRP and FRP reinforced concrete, International Workshop "Composites in Construction: A Reality", Capri, Italy, July 20-21.

[32] Yang, X., Nanni, A., and Chen, G. (2001): Effect of corner radius on the performance of externally bonded FRP reinforcement", Proceedings of FRPRCS-5, Cambridge, UK.

For the other references, see the companion paper 4.3.

For references [2]-[30] see the companion paper 4.3.

4.5 Repair of a damaged circular column using carbon fiber sheet (Reno system)

S Lee, T C Wong, S C Ng, L Lee and M L Liu
Advanced Engineering Materials Facility
The Hong Kong University of Science and Technology
Clear Water Bay, Hong Kong

ABSTRACT

Bridge columns designed before 1971 in the United States contained very little transverse reinforcement. As a consequence of the damage observed after the 1971 San Fernanado earthquake, some major problems were identified [1, 2]: 1), the moments and lateral forces induced by seismic loads results in large shear forces, 2) a concrete column is subjected to a large range of varying lateral loads during an earthquake, 3) a column/footing joint suffers from a large shear stress, leading to failure of the column in the plastic hinge region. A column was designed and constructed as part of a research project to consider the effects of seismic loads and the configuration of the column is presented in Figure 1. This column was subjected to a constant compressive load and a series of lateral loads (push and pull action) to simulate the seismic conditions. Because of the strong footing, the tested column exhibited damage on the bottom of the column, especially within the plastic hinge region, see Figure 2.

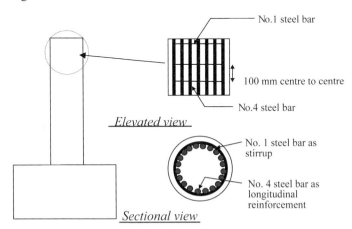

Figure 1. Schematic of a RC column

ACIC 2002, Thomas Telford, London, 2002

Figure 2 Damaged concrete column from previous seismic experiment (extensive damage in the plastic hinge region)

Application of carbon fiber reinforced plastic (CFRP) composites in infrastructure repair has received a lot of attention in the past decade, both for retrofitting and strengthening of deteriorated reinforced concrete structures. CFRP has superior properties such as high specific tensile strength and modulus and high corrosion resistance. In column applications, the hoop wrapped carbon fiber composite material is used for circumferential confinement, greatly increasing the compressive strength of the column. In this study, the strength and the ductility of a damaged column repaired with CFRP were investigated. The specific objectives of this study were as follows:

a) to gain practical experience in handling a Reno CFRP material system
b) to establish repair procedures for damaged columns
c) to determine the maximum lateral load and the maximum deflection

The following sections describe in detail the design, testing and evaluation of the performance of a damaged concrete column repaired with carbon fiber reinforced plastic. The material system was a Reno carbon fiber sheet (ASMT Co.'s product)

DESIGN REQUIREMENTS FOR THE REPAIR OF A COLUMN USING ADVANCED COMPOSITE MATERIALS

The column specimen obtained for this experiment had been subjected to testing under simulated seismic cyclic loads in a previous experiment. The test results are presented in [3]. The results of the previous experiment revealed that most of the crack damage was concentrated at the bottom of the specimen especially in the plastic hinge region.

Basic repairs to this substantially damaged column specimen are necessary before the strengthening. The basic repairs provide an opportunity to carry out remedial treatments following typical local construction site practice to the damaged portion of the column. Cement mortar with the conventional repair method was used to treat the damaged area. The mortar also served as a solid background on which the carbon fiber sheet would be applied and the existing exposed reinforcements were also protected with anti-rust painting.

The procedure simply reflected a typical remedial treatment to damaged concrete structures after a corrosion attack. Since the corrosion attack to the concrete structures is a major concern in Hong Kong, any repair procedures developed for the application are documented in this study so that the performance of concrete structures repaired with composite materials could be optimised.

Since the damage on the tested column specimen was in the plastic hinge region, the strengthening must be concentrated in that region. After the column specimen received basic remedial treatment, three unidirectional carbon fiber sheets were normally laid onto the surface of the column in the hoop direction and performed a confining action. The carbon fiber sheet was wrapped over the plastic hinge region of the column specimen and up to two-thirds of the height of the column.

EXPERIMENTAL PROGRAM: CONFIGURATION, CONSTRUCTION AND MATERIAL PROPERTIES OF THE COLUMN SPECIMEN

The test specimen was a circular concrete column footing measured 300 mm in diameter and 1500 mm in height. The column was constructed with 19 no. 4 steel bars reinforced longitudinally and with 12mm. no.1 steel bars were used as stirrups spaced at 100 mm from centre to centre. The column was embedded in a rectangular footing block (800 x 500 x500 mm) containing 10 no. 4 steel bars. The concrete strength was about 30 MPa. A schematic diagram is presented in Figure 1.

The previous experiment on the column specimen induced significant crushing damage concentrated at the bottom especially along the plastic hinge region (Figure 2). The actual damaged area covered both sides of the column and extended almost around the whole perimeter. The height of the damaged area was approximately 200 mm. Concrete cracking extended behind the longitudinal bars. Significant amounts of loose concrete particles could be seen from the damaged area and were behind the longitudinal reinforcements and at least 50 mm deep. The exposed reinforcement bars were observed to be rusty. The exposed reinforcements were also buckled, The buckling indicated that the yielding occurred as a result of the previous test.

PREPARATORY WORK AND PROCEDURES

After cleaning and removal of all lose particles from the damaged area, the concrete substrate behind the buckled reinforcement was easily seen at the bottom of the column specimen. The substrate was then moistened to ease the water suction. A cement sand mortar (1:2 with a water cement ratio of 0.5) was applied by hand directly to the substrate of the damaged area through the reinforcement bars. Each coat of mortar was controlled to be approximately 10 mm thick to ensure proper bonding between the different coats.

Pressure was required when applying each of the cement patches onto the column so as to ensure reasonable compaction and even thickness of the patches. The third (final) mortar coat was given a smooth finish so that the surface of the coat would be in line with the original curved shape at the bottom of the column. The edges of this final coat were smoothed in line with the original column surface.

After the cure, the repaired cement mortar surface was further polished. Slight bulging of the hardened mortar over the repaired area was observed during the surface polishing (Figure 3). It was not easy to carry out the polishing work at this stage, particularly sanding or trimming the bulged surface close to the column profile.

Figure 3. Bulging area in column

APPLICATION OF CARBON FIBER SHEET (RENO CARBON FIBER SHEET SYSTEM)

To wrap the column to two thirds of its height and to achieve a 100 mm overlap for anchorage when the column was wrapped, a single layer of carbon fiber sheet would need a length of 900 mm long.

The sequence of laying carbon fiber sheets is 1) primer 2) resin 3) carbon fiber sheet 4) resin. Steps 2 to 4 were repeated for 1st, 2nd and 3rd ply of carbon fiber. The ratios of primer to hardener and resin to hardener are both 100:35. A carbon fiber sheet was laid on after the primer. The placement of carbon fiber sheet was started on the bottom of the column. One coat of resin was brushed onto the column and up to a height approximately equal to a width of the carbon fiber sheet. Again, full and even coverage of resin over the column portion was important to achieve good adhesion.

The 1st ply of the sheet was positioned with the fiber-side in contact with the column. Care was taken to ensure that the fibers were aligned perpendicular to the axial direction of the specimen in order to maximize the confining action of the fibers. The fiber sheet should run horizontally and adhere to the column. The backing paper was then removed after the fibers were pressed by hand and a roller to remove excess resin as well as pushing air out from under the sheet.

Figure 4. A repaired column

The 2nd and 3rd ply of fibers were applied on the previous fiber following the above procedures. Approximately 90 minutes was allowed for the resin to gel before the above procedure was repeated for the next ply. The same as with the primer, the application of the resin and fibers were completed within 1 hour while the workability of resin could be maintained.

There were 3 plies of fibers wrapping around the column from the bottom to two thirds of the column height, representing the plastic region of the column under lateral load (Figure 4). A final coat of resin on the 3rd ply of fiber for the whole wrapped portion was applied to ensure that the outer fibers were completely impregnated with resin.

The above procedures were generally considered successful. A practical problem regarding the fiber distortion and wrinkles was first noted when applying the 1st ply of fibers at the bottom of the column. The main reason was the uneven mortar surface underneath. To minimize the problem, the separated fibers were then removed and reapplied in small strips (approximately 1/6 of the original width). The finish was acceptable.

TEST SETUP AND TESTING

The overall setup was designed to test the column and footing specimen repaired with carbon fiber material under cyclic seismic conditions. The loading used in the test was a combination of constant axial load and incremental cyclic lateral load. An axial load of 80 kN was applied stimulating the dead load of an actual structural system. The axial load pushed against the strong floor and was applied by two high strength steel rod. The axial load was maintained through a hydraulic vertical load follower. The cyclic lateral load was applied to the top end of the column by an MTS-400 kN hydraulic actuator with a 100 mm stroke (positive and negative direction), reacting against the reaction wall. The setup is illustrated in Figure 5. During the test, load levels were monitored with the actuator and the vertical load system through load cells.

 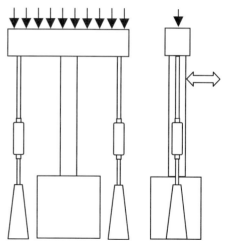

Figure 5. A test setup

Testing was conducted in displacement control with three cycles at each displacement level. The lateral loading particularly corresponding to the applied displacements was measured and monitored. Testing was stopped between sets of displacements. Then visual inspection for cracking and damage was carried out and photographic records were taken if necessary.

The loading sequence was conducted in displacement control in three cycles (0.01 Hz) at displacement stroke (mm) of 2.5 increment from 2.5 to 22.5 mm. Then a stroke of 5.0 increment was continued from 27.5 to 47.5 mm. Finally an increment of 10.0 was applied from 47.5 to 67.5 mm. Test results and data on vertical load, lateral load and control stroke were recorded and stored in the computer.

GENERAL OBSERVATIONS, TEST RESULTS AND DISCUSSION

Testing was conducted in the Structural Laboratory of the Civil Engineering Department at the Hong Kong University of Science and Technology. At displacements 7.5 and 10.0 mm, snapping sounds were heard but no visible cracks, splits or damage of the small strip of the wrap on the edge of the footing were noted. Snapping sounds were heard again at the displacement of 22.5 mm and a horizontal hairline crack was observed at the interface between the column base and the upper surface of the footing. The crack was concentrated on the push side facing the strong wall. This showed a separation between the concrete or repaired mortar and the parent concrete at the interface. At the displacement of 27.5 mm, the hairline crack continued progressing and loose cement particles were also observed near the crack. Snapping was again heard. At the displacement of 32.5 mm, bulging patch (approximate 50 x 50 mm) was observed on the footing surface on the pull side away from the strong wall. The above hairline crack propagated from the displacements of 32.5 to 42.5 mm. The width of the crack was increased to approximately 1mm. Before the end of the experiment, the crack ran around half of the column - footing interface at the push side facing the strong wall. The above bulging patch became enlarged and measured approximately 100 x 50 mm at the displacement of 57.5 mm.

Constant lateral loading was attained at the displacement of 57.5 mm. The failures were observed at the interface between the base and the footing regardless of the wrapped column portion. During the test, the column was loaded under a displacement control. Therefore, the lateral loading corresponding to the applied displacements was measured and monitored. The loading sequence was conducted in displacement control with three cycles (0.01 Hz) at displacement stroke (mm) ascending from 2.5 up to 67.5 mm at the end of the experiment. Test results and data on the vertical load, lateral load, control stroke and strain response were updated and stored in the computer.

A maximum lateral load was attained at the displacement of 57.5 mm and the failures were primarily concentrated at the interface between the base and the footing regardless of the wrapped column portion.

A relatively higher lateral loading of 40 kN was recorded at the displacement of 57.5 mm. This result was compared with that of the same previous conventional reinforced column before being wrapped (max. loading 35 kN and max. displacement 40 mm) and showed a significant improvement in its ductility due to the confinement of the carbon fibers to the damaged column, see Figure 6.

CRACKING AND FAILURE MECHANISM

The link between the column bottom and the footing could be a weak area as the column itself was reinforced with carbon fiber and the footing was originally designed to be very strong. No visible crack or distress was detected in the wrapped column during the experiment and this indicated that the flexural action was

Hysteresis Loop Carbon Fiber Reinforced Column

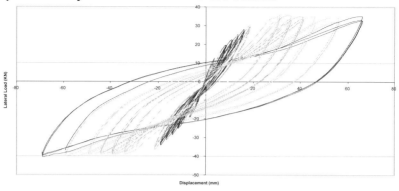

Hysteresis Loop Conventional Reinforced Column

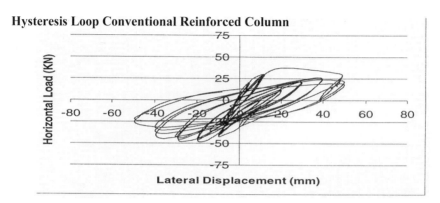

Figure 6. Load and displacement curves

primarily being concentrated at the concrete interface at the base of the column away from the protected plastic hinge region.

The results also showed interfacial cracking and bulging were concentrated at the pull side of the column facing the strong wall. A summary of the lateral loads and the corresponding displacements is presented in Table 1. It can be seen that the lateral loading, under the displacement control, applied to the pull side was consistently higher than that on the push side. This explains the concentration of cracks and bulging subject to higher lateral loading on the pull side.

Table1: A summary of lateral loads and corresponding displacements

Displacement(mm)	5.0	10.0	15.0	20.0	32.5	42.5	57.5
Push side(kN)	12.2	19.2	25.1	29.5	35.1	34.8	34.2
Pull side(kN)	18.3	24.6	30.9	36.0	38.4	38.5	40.0

At the displacement of 32.5 mm, the bulging patch (approximately 50 x 50 mm) was observed on footing surface on the pull side away from the strong wall. No such bulging patches were observed in the bare column specimen. This phenomenon indicates that the buckled reinforcement within the vertical repaired area had been well strengthened so that the next weakest point subject to the lateral loading would be near the bulging patches. At this point, the lowest and turning horizontal part of longitudinal reinforcement connecting the column and footing was expected to be deformed against the applied lateral force and push towards the surface of the footing. The bulging patch enlarged and measured approximately 100 x 50 mm at the displacement of 57.5 mm. At this point, further deformation of the longitudinal reinforcement connecting column and footing would cause more damage to the connecting area.

LOAD-DISPLACEMENT CHARACTERISTICS
The column specimen had been symmetrically reinforced with respect to the positive and negative side. The resulting positive and negative portions of the specimen's hysteresis loops in Figures 6 should be theoretically symmetrical and identical in value. However, due to the limitation of the hydraulic actuator, the maximum strokes reached in the positive and negative directions were slightly different, and this difference could affect the symmetries of the loops.

CONCLUSIONS
- Experience in handling Reno CFRP material system and repairing a concrete column has been gained including the preparatory work and actual fiber application work. With an irregular concrete surface, the ease of application and confining ability of the carbon fibers on concrete structural elements could be adversely affected to a large extent.

- Emphasis should be placed on the surface quality of the concrete background on which the Reno carbon fiber sheet will be applied. The backing material used for repairing damage in structures should be a quality surface as close as possible to the parent concrete itself.

- A maximum lateral load of 40 kN was attained at the displacement of 57.5 mm. This result was compared with the lateral loads on conventionally reinforced column before fiber wrapping (max. loading 35 kN and max. displacement 40 mm) and showed significant improvement.

- The failure occurred at the link between the wrapped column bottom and the footing as the column itself was reinforced with carbon fiber and the footing was strong.

- Based on the test results, three plies of carbon Reno material system for retrofitting a column could be excessive since the failure occurred in the footing section, which was outside of the repaired region.

ACKNOWLEDGEMENT

The research presented in this report is funded by the Hong Kong University Grant Council, under the Cooperative Research Centre program- (CRC98/01.EG14) and the industrial sponsor is ASMT Co. of Lit Cheong Power Engineering Ltd. The authors would like to thank Dr. Chris. Leung and Mr. K.S. Iu, Mr Ivan H.M Chan and Mr. K.K. Chung for their technical discussions and support.

REFERENCES

[1]   H. Saadatmanesh, M.R. Ehsani, and Limin Jin "Seismic Strengthening of circular Bridge Pier models with fiber composites", ACI Structural Journal/ November-December 1996.

[2]   D. Innamorato, F. Seible, G. Hegemier, M.J.N. Priestly, "Carbon Shell Jacket Retrofit Test of Circular Flexural Column with Lap Spliced Reinforcement" Report NO. ACTT-95/18, Division of Structural Engineering, University of California, San Diego, November 1995.

[3]   Yao Jun, "Advanced Composite Reinforced Concrete" Master Thesis, 00, Department of Civil Engineering, the Hong Kong University of Science and Technology, 2000.

Session 5 : **Application of FRP to walls and pipes**

5.1   Composite wall anchors

M Beigay[1], D T Young[2] and J Gergely[2]
[1]*Odell and Associates, Charlotte NC, USA*
[2]*UNC Charlotte, Charlotte NC, USA*

ABSTRACT
Composite materials can be used to upgrade the load-carrying capacities of unreinforced masonry walls. The strength of the upgraded wall, however, is often limited by failure that occurs near the bottom of a wall due to a low capacity for load transfer between the wall and the foundation. The present paper provides information on an experimental investigation of a composite anchoring system on unreinforced concrete masonry walls. This system provides high-capacity load transfer between the fibre reinforced polymer (FRP) laminate retrofitted wall and the foundation.

Six masonry wall-concrete footing assemblages were built, strengthened with FRP laminates, and tested. On all walls, a composite anchoring system provided the only connection between concrete foundation and masonry wall. Three walls (1219 mm x 1219 mm) had composite laminate applied in edge strips to both faces and were subjected to in-plane shear and a 22.24 kN axial load. The capacity of these walls reached more than 120 kN (lateral load) before failure, which resulted from extensive cracking of the masonry units. The other three walls (406 mm x 1016 mm) had the tension surface fully covered with CFRP laminate and were tested in tension (bending) reaching a maximum load of 32 kN before failing due to masonry unit crushing. In each loading case, the anchoring system provided sufficient strength allowing the masonry units to fail before the CFRP laminate or composite anchors failed.

The results of this research reveal that strength gains due to FRP laminate upgrades on unreinforced concrete masonry walls can be fully realized through the use of a composite anchoring system.

INTRODUCTION AND OBJECTIVES
Strengthening existing masonry walls and repairing damaged masonry walls can be accomplished utilizing any of a number of popular techniques. For many years these methods have included surface overlays of various coatings including shotcrete; grout injection with and without reinforcing bars in the cells; and external stiffening including steel and timber bracing elements (FEMA 1997). The popularity of these techniques results from designers, contractors, and building

officials being familiar with the behaviour of these materials and with the governing specifications.

In the past 10-15 years, another type of retrofit and repair method has gained acceptance. This method involves the use of fibre-reinforced polymers (FRP's), or, as described in this paper, carbon fibre reinforced polymers (CFRP's), as a strengthening overlay on structural components. Extensive research has been performed on these materials applied to masonry walls with early research conducted by Seible *et al* [1] on FRP retrofit methods for unreinforced masonry (URM) walls. His results along with those from masonry research conducted by Al-Chaar & Hassan [2], Ehsani & Saadatmanesh [3], Gilstrap & Dolan [4], and Marshall *et al* [5] revealed that shear and flexural strength in masonry walls can be increased through the use of FRP composites externally attached to these masonry elements.

In most applications, strengthening and repair methods on masonry walls require the increased capacity or the wall be accompanied by an increased capacity of the connection between the wall and its supporting member or foundation. In the traditional methods described above, the connection is strengthened by anchoring the repair materials to the foundation. These anchors often are provided by drilling, grouting, injecting epoxy adhesives, clamping with steel angles or channels, or adding additional concrete or masonry material. Currently, if strengthening or repairs to masonry walls are provided by CFRP laminates, then the connection is strengthened using these same traditional anchoring methods. Alternatively, anchoring techniques utilizing the CFRP material itself need to be developed.

SPECIMEN AND TEST DETAILS

The objective of this study was to experimentally investigate unreinforced hollow CMU walls retrofitted with CFRP composite laminates using one lamination schedule and subjected to in-plane shear (shear specimens) and out-of-plane bending (bending specimens) loads. The focus of the investigation was the composite anchoring system used to connect the CFRP laminates to the foundation.

Three identical 1219 mm by 1219 mm masonry walls were constructed and tested in shear, and three identical 406 mm by 1016 mm masonry samples were testing in bending. Each specimen was first retrofitted with CFRP laminate, on the tension face only for the bending samples (Figure 1) and on both faces for the shear specimens (Figure 2). Type S mortar and CMU blocks having a width of 406 mm and a height and thickness of 203 mm were used in each specimen. The average compressive strength of the masonry units was 10.3 Mpa. In each specimen, the CFRP composite laminate was fully anchored to the base as shown in Figure 3. Details of the composite anchor are shown in Figure 4.

The three shear specimens were loaded with in-plane loads, and the three bending specimens were subjected to out-of-plane bending. The loads were applied at the tops of the walls. Displacement transducers were used to monitor the in-plane and out-of-plane deformations of the specimens at several locations throughout the masonry walls. Strain gauges were attached to the FRP laminates to record the stress level in the composite material on the specimen surface.

Figure 1. Retrofitted bending wall specimen

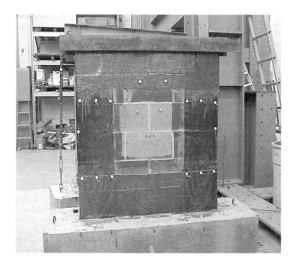

Figure 2. Retrofitted shear wall specimen

As shown in Figures 1 and 2, the boundary condition for the shear and bending test setups modelled cantilever walls with fixed conditions at the bottom. The wall in the bending setup was grouted and dowelled in the bottom course. In order to create a more realistic condition, an axial of 22.24 kN of axial load was applied to the shear samples. No axial load was applied to the bending specimens.

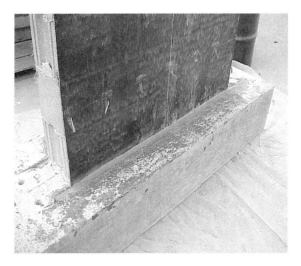

Figure 3. Composite anchor into wall foundation

STRENGTHENING DETAILS AND TEST RESULTS
Baseline Values

Shear and bending capacities of unretrofitted wall specimens were previously measured in the laboratory by Young and Gergely [6]. The unretrofitted, cantilevered shear specimens developed horizontal cracks along the mortar bed joint at the walls' lower section, while simply supported bending specimens cracked at the specimens' mid-height. Although the walls in these previous tests were not the same sizes as the specimens in this project, the failure stresses at failure should be same. Also, the unretrofitted shear walls and bending walls failed at ultimate stresses values consistent with ACI 530/ASCE 5/TMS 402 [7]. Calculated failure loads were 9.34 kN for in-plane cantilever shear and 1.67 kN for out-of-plane cantilever bending.

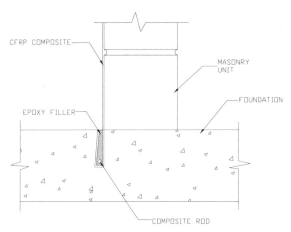

Figure 4. Details of the composite anchor

Retrofitted Specimens

Inherently brittle masonry units and mortar provide masonry walls, which react to loads in a brittle manner. As a result, they fail catastrophically with little or no warning. Using linear materials, such as composites, to strengthen these non-ductile structural elements, shifts the primary focus of the wall's behaviour from its lack of ductility to its increased strength. However, this shift in emphasis is reasonable only if the increased strength is continued from the wall to its foundation. Thus, a proper laminate anchorage system is required.

In a masonry wall CFRP retrofit design, attention must be given to height-to-thickness and height-to-width ratios of the wall, the level of axial load, and the capacity and drift demand. Similarly to any orthotropic material, however, the effectiveness of the composite retrofit also depends on the orientation of the fibres. A 45-degree [+-45] layout is the most effective (although not always the most practical) to carry shear forces in a shearwall. A laminate aligned with the height of the wall (i.e. 0-degree $[0_2]$) is optimal for out-of-plane, vertical bending loads. Finally, a 0-degree + 90-degree [0/90] laminate will provide an efficient method to strengthen walls supported on all four edges and subjected to two-way bending.

The traditional [+-45] layout for shear walls provides a system analogous to cross-bracing, and it behaves quite effectively. To investigate an alternative, shear wall specimens were strengthened with CFRP laminates around the perimeter of the wall (horizontal and vertical CFRP strips on the edges), thereby modelling a stiffened beam-column arrangement. Bending specimens were retrofitted with composite in a full-surface vertical alignment $[0_2]$ on the tension surface of the wall only.

For the in-plane retrofitted shear specimens, strain gauges were positioned in the maximum stress zones, and were aligned in the direction of the fibres. The shear walls, with composite laminate around the wall perimeter, reached a peak lateral load of 120 kN and a maximum horizontal deformation of 13.5 mm (see Figure 5). Failure occurred due to extensive shear stresses in the masonry wall combined with localized damages to the load-applicator beam at the top of the wall. No damages to the CFRP laminate or the composite anchor were observed. The peak strain in the composite laminate reached 1.0%. At 2/3 of the composite's ultimate strain of 1.5%, this laminate strain includes tension strain (uplift) as well as shear strain and is representative of only partial coverage of the wall surface by CFRP laminates.

The load-deformation curve for the wall subjected to cantilever bending and having a $[0_2]$ composite retrofit is shown in Figure 6. The peak lateral load was 32 kN with a maximum displacement of 47 mm. The maximum strain readings in the composite fibre reached 1.2%, approximately 80% of its ultimate capacity. Failure mode for the retrofitted bending specimens was due to extensive shear damage in the masonry wall. No damages were observed in the CFRP laminate or in the composite anchor. Ultimately, the masonry units inside the wall crumbled, and suddenly lost its lateral load carrying capacity (see Figure 7).

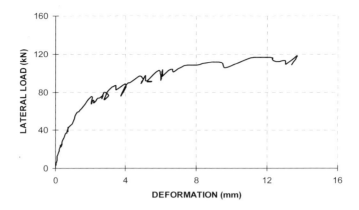

Figure 5. Load versus deformation results for retrofitted walls in shear

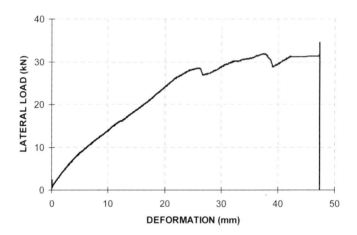

Figure 6. Load versus deformation results for retrofitted walls in bending

CONCLUSIONS

The experimental results revealed that a composite anchoring system can be utilized effectively with CFRP laminates to significantly increase the in-plane shear and out-of-plane bending capacity of unreinforced hollow masonry walls. Walls retrofitted with CFRP laminates and composite anchors and subjected to in-plane shear and out-of-plane bending realized 1285% and 1916% increases in capacity, respectively, when compared to the capacities of unretrofitted walls.

Figure 7. Failure of retrofitted wall subjected to out-of-plane bending

ACKNOWLEDGEMENTS
The authors would like to acknowledge the financial support provided by Mr. Karl Gillette of Edge Structural Composites and by the William States Lee College of Engineering at the University of North Carolina at Charlotte.

REFERENCES

[1] Seible, F., G. Hegemier, N. Priestley, G. Kingsley, A. Igarashi, and A. Kurkchubasche. 1990. Preliminary Results from the TCCMAR 5-story Full-Scale Reinforced Masonry Research Building Test. The Masonry Society Journal, 12(1):53-60.

[2] Al-Chaar, G.K., and H.A. Hassan. 1999. Seismic Testing and Dynamic Analysis of Masonry Bearing and Shear Walls Retrofitted with Overlay Composite. ICE '99, Cincinnati, May 10-12, 1999.

[3] Ehsani, M.R., and H. Saadatmanesh. 1996. Seismic Retrofit of URM Walls with Fiber Composites. The Masonry Society Journal, 14(2):63-72.

[4] Gilstrap, J.M., and C.W. Dolan. 1998. Out-of-plane Bending of FRP-reinforced Masonry Walls. Composites Science and Technology, 58:1277-1284.

[5] Marshall, O.S., S.C. Sweeney, and J.C. Trovillion. 1999. Seismic Rehabilitation of Unreinforced Masonry Walls. Fourth International Symposium on Fiber Reinforced Polymer Reinforcement for Reinforced Concrete Structures, Baltimore.

[6]   Young, D.T. and Gergely, J, In-Plane and Out-of-Plane Behavior of Unreinforced CMU Walls Retrofitted with Composites, Proceedings of the 2001 Second International Conference on Engineering Materials, 2001, San Jose, CA, 2:465-471.

[7]   ACI 530/ASCE 5/TMS 402, 1999, Building Code Requirements for Masonry Structures, The American Concrete Institute, The American Society of Civil Engineers, and The Masonry Society.

[8]   Federal Emergency Management Agency. 1997. NEHRP Guidelines for the Seismic Rehabilitation of Buildings. FEMA 273, Washington, DC.

5.2   GFRP pipe joints reinforced using composite pins

J Lees and G Makarov
University of Cambridge, UK

ABSTRACT

While many of the material properties of glass fibre reinforced polymers (GFRPs) are extremely desirable, in the design of composite pipeline infrastructure, the major challenges are mechanical. In particular, the key problem that prevents the more widespread use of advanced composite pipes is the joint. The joints between FRP pipe sections often represent the weakest link in the system and the structural performance is reduced accordingly. This investigation addresses this issue and focuses on the development of a novel connection system for advanced composite pipelines. The main objectives of the work are to develop a combined bonded / mechanical joint with a longitudinal tensile capacity equal to the strength of the pipe itself and to ensure the water- and air-tightness of the proposed joint.

INTRODUCTION

Composite glass reinforced epoxy (GRE) pipe systems are generally manufactured in sections using filament winding. For the most part, lengths of pipe are joined in the field using adhesive bonding and a variety of joint designs. A typical adhesive bonded joint is a rigid taper/taper connection that consists of a very slightly conical socket and spigot end or two spigots ends with coupler between them (see Figure 1).

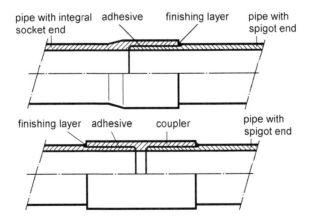

Figure 1. Adhesive bonded joint for composite pipeline sections [1].

ACIC 2002, Thomas Telford, London, 2002

A number of researchers have investigated the strength of bonded tubular joints under tensile loading both theoretically [2-4] and experimentally [5, 6]. The results of the theoretical analyses and numerical simulations of the stresses in tubular lap joints show that the interface shear and normal stresses are a maximum at the end of the joint. In particular, there is a high shear and transverse normal stress gradient in this region, where the stresses increase from zero at the free surface to some maximum value over a very short distance [7]. The principle of strength enhancement consists of embedding relatively small diameter reinforcing pins in the pipe walls in the joint region (see Figure 2.)

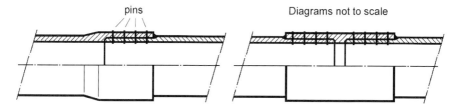

Figure 2. Adhesive bonded joint with reinforcement pins.

In heavily loaded structures such as internally pressurized pipelines used in the oil/gas or marine/offshore industry, mechanical reinforcement of the adhesive joint can effectively prevent the spreading of debonding and delaminations originating from local damage or initial defects. In addition, even in undamaged joints the presence of reinforcement can significantly reduce the stress peaks in the adhesive near the joint ends. In spite of the general opinion [8] that combining bonding and bolting will not improve the joint strength over that of well-designed undamaged adhesive joint (because the load is carried by the adhesive layer almost entirely and there is no load sharing between adhesive and fasteners), our preliminary results show a significant difference in the fracture process and ultimate load capacity of single-lap joints reinforced using a relatively small number of rigid pins.

EXPERIMENTAL PROCEDURES
To investigate the influence of pin reinforcement on the strength of a pipe joint, a number of bonded single-lap specimens were prepared and tested. The pipes used in the investigation were Wavistrong® pipe system EST32 manufactured by Future Pipe Industries B.V. These glass fibre reinforced tensile resistant epoxy pipe sections have a design working pressure of 32 bar. All the experiments were undertaken on 80 mm internal diameter pipe with 4 mm wall thickness (the mechanical properties of these pipes can be found in the Table 1). The pipes were fabricated by the filament winding of glass fibres impregnated with epoxy resin with a winding angle of ± 55°. The adhesive used for bonding was a room temperature-cure two-part Araldite® epoxy resin. The reinforcement pins were cut from unidirectional carbon-fibre reinforced plastic (CFRP) composite rods and steel. The mechanical properties of the pins can be found in the Table 2. Four different

diameters of CFRP pins were used: 0.5 mm, 1.0 mm, 1.5 mm and 2.0 mm. In the case of the steel pins, only 1.2 mm diameter pins were considered.

Table 1. Pipe material properties

Property	Unit	Value
Axial tensile stress	MPa	75
Axial tensile modulus	GPa	12
Hoop tensile stress	MPa	210
Hoop tensile modulus	GPa	20.5
Shear modulus	GPa	11.5
Poisson's ratio (hoop)	-	0.65
Poisson's ratio (axial)	-	0.38
Glass content (by volume)	%	52±7

Table 2. Pin material properties

Property	Unit	CFRP	Steel
Tensile strength	GPa	2.07	0.5 (yield)
Tensile modulus	GPa	138	210
Ultimate tensile strain	%	1.5	0.2
Ultimate shear strength	MPa	41.3	135
Fiber content (by volume)	%	62	-

Single lap-shear tests

The single lap-shear test specimen configuration is illustrated in Figure 3. The original GRE pipes were cut onto eight equal segments along the longitudinal axis and these strips were used for single-lap shear coupon tests. The length of the free arm was 160 mm and the overlap length was 40 mm. This length was chosen because it corresponded to the overlap length of an adhesive joint with a coupler for the 80 mm internal diameter Wavistrong® EST32 pipe. The adherend thickness ranged from 3.5 to 4.0 mm, and the adhesive thickness was 0.1-0.2 mm. The number of reinforcing pins varied from 4 to 35 and different diameters and pin materials were considered. Testing was carried out using a 150 kN Instron tensile machine. During the test, one end of the sample was held fixed by a lower grip and the other end was held by an upper grip that moved at a constant velocity of 1.0 mm/min. A digital extensometer was used to measure the actual displacement between the two points located a distance of 60 mm apart and 30 mm from the middle of the sample. In addition, the total displacement between the grips was measured to observe the crack initiation and propagation. The grips of the testing machine were perfectly rigid. Although the specimens were well aligned at the beginning of each test, the bending moment induced by the lap-shear geometry caused a significant re-alignment that could be seen visually.

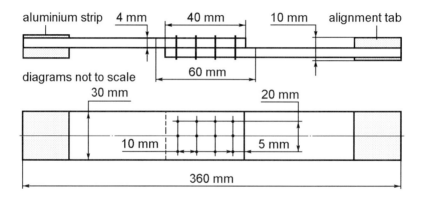

Figure 3. Single lap-shear test sample with 12 reinforcement pins.

Control specimens
In order to identify the potential strength enhancement as a result of the inclusion of mechanical reinforcement in the joint, a set of GRE specimens without joints and a set of adhesive bonded specimens without pin reinforcement were tested. The tensile specimens made from GRE pipe had a length of 200 mm. For both sets of specimens the specimen width was 25 mm. The average tensile strength of the single-lap adhesive joint without any reinforcement pins was 4.3±0.25kN (or 43±2.5MPa), while the strength of the GRE tensile specimens (25 mm width) was 6.8±0.25kN (or 68±2.5MPa). Typical diagrams of the load-deformation (where the displacement was measured over a 60 mm gauge length) behaviour are presented in Figure 4.

Reinforced specimens
Tests were carried out where the number of pins and the pin material were varied. Specimens with either CFRP or steel pins were considered. The combinations tested and the experimental values of fracture load can be found in Tables 3 and 4. As these specimens were wider than the control specimens (30 mm), the approximate fracture stresses calculated using the gross cross-sectional area are shown for comparison purposes.

CFRP pins
For the specimens with 12 CFRP 0.5 mm diameter pins, fracture occurred in the joint area. However, all the other CFRP specimens fractured in the free arm region. The load displacement behaviour of specimens with 12 CFRP pins and 35 pins is compared in Figure 5 (the displacement is measured between the grips of the testing machine).

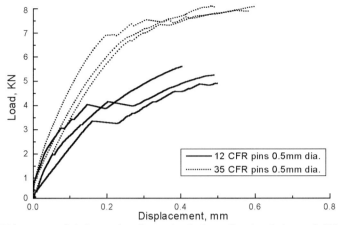

Figure 4. Diagrams of deformation for a single lap adhesive joint and GRE coupon samples

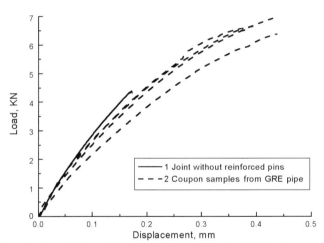

Figure 5. Adhesive joints reinforced using composite CFRP pins.

Table 3. Maximum fracture load of samples with CFRP reinforcement pins

Number of pins	12	12	12	35
Diameter, mm	0.5	1.0	1.5	0.5
Fracture load, kN	5.18±0.55	6.96±0.25	7.29±0.15	8.00±0.15
Fracture stress, MPa	43.16±4.58	58±2.08	60.75±1.25	66.67±1.25

Table 4. Maximum fracture load of samples with steel reinforcement pins

Number of pins	4	6	6*	9
Diameter, mm	1.2	1.2	1.2	1.2
Fracture load, kN	7.53±0.45	7.55±0.7	8.51±0.45	7.79±0.25
Fracture stress, MPa	62.75±3.75	62.92±5.83	70.92±3.75	65±2.08

6* - end joint orientation of pins

Steel pins
The load-displacement diagrams for the joints reinforced using steel pins (three specimens were tested for each combination of pins) are presented on Figure 6. All the specimens with steel pins remained intact until the carrying capacity of material was reached and the fracture of the coupon occurred. A slight difference in behaviour was observed for the specimens with four pins.

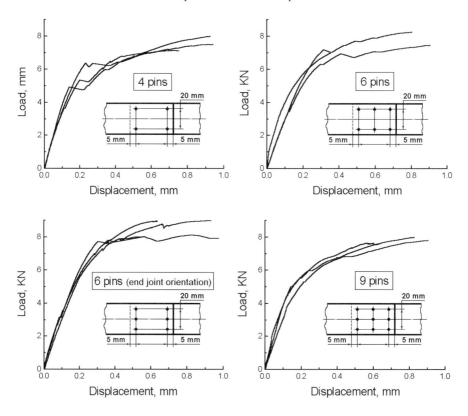

Figure 6. Adhesive joints reinforced using 1.2 mm diameter steel pins.

DISCUSSION
The results show that, in general, a significant enhancement in capacity can be obtained by reinforcing an adhesive joint with small diameter pins. In addition, the

capacity enhancement is greater than would be expected by considering the ultimate shear resistance of the pins.

The pin material, diameter and number are important in ensuring an improvement in the behaviour and, for example, no improvement was noted in the joint with 12 CFRP 0.5 mm diameter pins. In these samples, failure occurred at a load that was similar to that of an adhesive joint without pins. In particular, it appeared that as the capacity of the adhesive was reached, a crack propagated from the end of the joint and the load was transferred to the pins. However, the CFRP pins could not resist this additional force and the pins pulled out of the holes and partially fractured in shear. This transfer process can be identified in the load-displacement behaviour (see Figure 5) where at a load of about 4 kN, a distinct increase in displacement was observed while the load remained fairly constant (the lower curve in this figure is rather spurious, this was perhaps a measurement error). However, when the pin diameter was increased to 1.2 mm or the number of 0.5 mm diameter CFRP pins was increased to 35, the specimens failed in the pipe material.

Using the manufacturers' data for the ultimate shear strength of the pins (see Table 2), the resistance of 12 and 35 CFRP 0.5 mm diameter pins would be 0.097 kN and 0.283 kN respectively. On this basis, it is clear that the incremental improvement in strength over that of an adhesive joint without pins is not solely due to the shear resistance of the pins. Instead it appears that the pins must carry forces in axial tension (as proposed by Cox [9]) and delay the crack propagation and the onset of peeling. The joint is thus able to carry a higher load before failure.

A similar increase in joint strength was noted in the specimens with steel pins. Assuming a yield stress in shear of approximately 135 MPa, then 4, 6 or 9, 1.2 mm diameter pins would have a shear resistance of 0.611 kN, 0.916 kN and 1.374 kN respectively. Yet in all the experiments with steel pins, the incremental increase in capacity was far greater than these numbers would suggest (see Figure 6). Even when only 4 steel pins were used, the specimens failed in the pipe material in the free arm rather than in the joint. It is of interest that for the specimen with 4 steel pins, again, steps in the load displacement diagram were noted at around 4-5kN. Yet, unlike the CFRP specimen with 12 CFRP pins where the pins failed, the steel pins maintained their integrity even after the load transfer took place. The steel pin tests also enable a comparison of the influence of the location of the pins. The tensile loading of a single lap joint causes bending and the peak normal and shear stresses would be expected at the end of the joint. Hence by locating the reinforcing pins close to the end of the joint the pins work like an additional barrier, which prevents the crack propagation through the adhesive layer from one end, and creates a "bridging effect" for the loaded adherends.

On the basis of the single lap test results, three stages of behaviour can be identified: an initial elastic stage where the joint behaves in a similar manner to that of an adhesive joint; a crack propagation/load transfer stage where the force is transferred from the adhesive to the pins; and final failure. Final failure may be due to a failure of the coupon material, a local failure of the pin due to damage within the matrix, the pull-out of the pin or the global failure of the pin. Two possible design philosophies also emerge. On one hand, it would be possible to reinforce the joint locally with discrete pins located in regions of high stress. On the other hand, if the density of pins are increased and distributed throughout the joint, a new 'pinned' material is created and the bridging behaviour can be considered using averaged traction laws [9].

Although the preliminary experiments have been carried out on single lap shear tests, the results will now be extended to consider the behaviour of tubular pipe joints. Of interest will be whether the pins will enhance the capacity of a tubular pipe joint in uniaxial tension with an axi-symmetrical load distribution. Experiments where the pipe joints are subjected to multi-axial loading conditions will also be carried out. In addition, pressure testing to confirm the water-tightness of the joint and further analytical studies are planned.

CONCLUSIONS
The reinforcing of a single lap adhesive joint using relatively small diameter rigid pins resulted in a significant increase of the maximum fracture load that the joint could carry under tensile loading conditions. In particular, the presence of the pins ensured that the joint remained undamaged until the strength of the parent GFRP material was exceeded. This work will now be extended to consider the behaviour of tubular pipe joints.

ACKNOWLEDGEMENTS
This project was supported by funding from the Engineering and Physical Sciences Research Council (GR\R09770).

REFERENCES
[1]   Wavistrong® Epoxy Pipe Systems. Engineering Guide for Wavistrong Filament Wound Epoxy Pipe Systems. Future Pipe Industries B.V., Hardenberg, Netherlands.

[2]   Lubkin J.L., Reissner E. Stress distribution and design data for adhesive lap joints between circular tubes. Trans. ASME. 1956, Vol. 78, pp. 1213-1221.

[3]   Adams R.D., Peppiatt N.A. Stress analysis of adhesive bonded tubular lap joints. Adhesion. 1977, Vol. 9, pp. 1-18.

[4]   Griffin S.A., Pang S.S., Yang C., Strength model of adhesive bonded composite pipe joints under tension. Polymer Eng. Science. 1991, Vol. 31, No. 7, pp. 533-538.

[5]   Hashim S.A., Cowling M.J., Lafferty S. The integrity of bonded joints in large composite pipes. Int. J. of Adhesion & Adhesives. 1998, Vol. 18, No. 6, pp. 421-429.

[6] Knox E.M., Lafferty S., Cowling M.J., Hashim S.A. Design guidance and structural integrity of bonded connections in GRE pipes. Composites Part A. 2001, Vol. 32, pp. 231-241.

[7] Adams R.D., Peppiatt N.A. Stress analysis of adhesive bonded lap joints. Strain Analysis. 1974, Vol. 9, No. 3, pp. 185-196.

[8] Clarke J.L. Structural Design of Polymer Composites. EUROCOMP Design Code and Handbook. London: E & FN Spon, 1996.

[9] Cox B.N. Constitutive Model for a Fiber Tow Bridging a Delamination Crack. Mechanics of Composite Materials and Structures 1999, Vol 6, pp. 117-138.

5.3 A critique of the EUROCOMP simplified design method for bolted joints in long fibre reinforced polymer composite materials

G J Turvey and P Wang

Engineering Department, Lancaster University, Bailrigg, Lancaster, LA1 4YR

ABSTRACT

Three shortcomings with the EUROCOMP simplified design method for bolted joints in composite materials are identified. They are: inadequate definition of materials, stress distributions that violate equilibrium and inadequate specification of the bolt/hole interface conditions. Pultruded GRP material properties are used in Finite Element (FE) analyses of stress distributions for six load cases which are central to the method. The stress distributions obtained from the analyses confirm that equilibrium is violated in some instances by the normalised stress curves given in [1]. Moreover, it is shown that for the two cases of bolt/hole contact, with and without clearance, it is not valid to use a single normalised stress curve for design, as stresses are dependent on the magnitude of the clearance and the applied boundary stress.

INTRODUCTION

Interest in the use of long fibre reinforced composite materials for infrastructure applications has increased rapidly over the past five to ten years. Even so, relatively few all-composite structures have been built. One of the more frequently cited reasons why engineers appear reluctant to design structures made of long fibre reinforced polymer composite materials is the lack of a design code, accepted by the regulatory authorities.

In 1991 the EUROCOMP project was initiated to address this issue. It was supported by the EU and undertaken by European engineers with backgrounds in consulting, materials supply and academia. The principal deliverable was the EUROCOMP design code and handbook [1] - the first design code for these materials, based on limit state principles.

The code addresses many aspects of structural design, one of the most important being the design of joints. Rules and guidance are included for the design of mechanically fastened (hereinafter referred to as bolted) and adhesively bonded joints. This information was strongly influenced by knowledge and experience of composites joint design for aerospace applications.

The code describes four methods for the design of bolted joints. One of them, known as the *simplified* method, is amenable to hand calculation and is the focus of attention here. It may be used to design concentrically or eccentrically loaded joints, provided the bolts are in shear or tension.

The simplified method is attractive, because it requires minimal material stiffness and strength data. Moreover, the design of multi-bolt joints reduces essentially to checking critical stresses adjacent to the most highly loaded bolts using dimensionless graphs. Despite its apparent simplicity the scientific basis of the method has neither been assessed rigorously, nor tested against independent test data.

Here the scientific basis of the simplified method is evaluated. First, the method is described. Then its shortcomings are identified, viz. inadequate definition of material properties, use of stress distributions that violate equilibrium and inadequate definition of contact conditions at the bolt/hole interface. Detailed analyses of the stress distributions around the edge of the hole and along the net tension and shear out planes, undertaken with the ANSYS FE software [2], are presented to highlight and explain the deficiencies with the method. It is concluded that the inability of the simplified method to account for the effect of hole clearance on critical stresses is a serious deficiency, as large clearances are routinely used with bolted joints in infrastructure applications.

SIMPLIFIED DESIGN METHOD FOR BOLTED JOINTS
As explained in [1], the simplified design method for multi-bolt joints in composite materials is only valid when the bolts are subjected to shear or tension and the loading is concentric or eccentric. A large number of practical joints comply with these restrictions. Figure 1 shows a concentrically loaded tension joint in which the bolts are subjected to shear along each of the interfaces between the inner and outer plates.

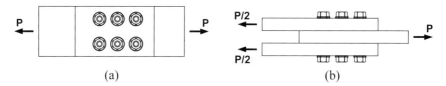

(a)   (b)

Figure 1. Concentrically loaded multi-bolt joint: (a) plan view and (b) side view

The first step in the simplified method is to determine the force(s) carried by each bolt in the joint due to the external loading. For eccentrically loaded joints, they are calculated using the procedures established for bolted joints in structural steelwork (see, for example, [3]). However, for concentrically loaded joints, the bolt forces may be determined from a table of load factors given in [1].

Having established the forces acting on each bolt, they are resolved into one or more of six basic load cases, in which stresses and/or a bolt force act on a square composite material domain surrounding a single open or filled hole. Three load cases, 1,3 and 5, are shown in Figure 2. The other three cases, 2, 4 and 6, are obtained by reversing the directions of the external stresses and the bolt force.

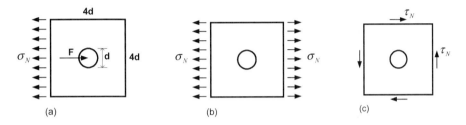

(a)   (b)   (c)

Figure 2. Basic load cases for which dimensionless stress distributions have been determined: (a) case 1, (b) case 3 and (c) case 5.

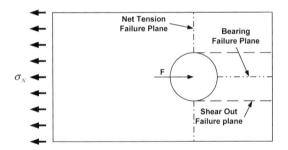

Figure 3. Sketch showing the three principal failure planes intersecting the edge of the bolt hole.

In the simplified method joint failure is assumed to be in one of four modes: bearing, net tension and shear out of the composite and bolt failure. The three composite failure modes are shown in Figure 3 for the case of an orthotropic composite in which the loading, material and geometric principal axes are coincident. (It should, however, be appreciated that the simplified method is claimed to be valid for the general anisotropic case.) Therefore, in [1] dimensionless graphs of the radial and tangential direct stress distributions along the hole edge and the shear stress distribution along the shear out plane are provided for load cases 1 – 6. Stresses derived from these graphs are compared with the material strengths at various locations adjacent to the hole to determine whether or not they exceed the material strengths associated with the three failure modes.

SHORTCOMINGS WITH THE SIMPLIFIED METHOD

As indicated in the Introduction, three principal shortcomings with the simplified design method have been identified. They are: inadequate specification of the composite materials, stress distributions that do not satisfy equilibrium and inadequate specification of the bolt/hole contact conditions for load cases 1 and 2. All of these inadequacies impact on the dimensionless stress distribution graphs given in [1], which are central to the application of the method. It is convenient to discuss each of the shortcomings in turn.

COMPOSITE MATERIALS SPECIFICATION

In [1] three dimensionless stress distributions - two for uni-directional laminates (0° and 90°) and one for a quasi-isotropic laminate - are given for each of the six load cases. For each laminate two materials are identified, eg. A1 and A2 for the 0° laminate. Unfortunately, it is not clear what the differences are between the two materials. Moreover, it is not clear whether the 0° direction is parallel to the stress, σ_N, and the force, F, shown, for example, in Figure 2(a). Given the fact that some of the stresses for the basic load cases show significant differences, such uncertainty scarcely promotes confidence in the method.

VIOLATION OF EQUILIBRIUM FOR SPECIFIC LOAD CASES

The six load cases for which dimensionless stress distributions are given in [1] are in fact three pairs of equal and opposite loads. Thus, for example, load case 2 is the compression version of tension load case 1. Likewise, load case 4 is the compression version of tension load case 3 and load case 6 is the reverse of shear load case 5 (see Figure 2).

An examination of the dimensionless tangential stress distributions around the hole given in [1] for load cases 3 and 4 reveals an anomaly. Not only are the shapes of the distributions dissimilar, i.e. they are not mirror images of each other with respect to the tangential direction, but the magnitudes of corresponding stresses are also dissimilar. This is unexpected, because superposition of the load cases 3 and 4 yields the *stress free* state. Clearly, because of the aforementioned differences in the distribution and magnitude of the tangential stresses for load cases 3 and 4, a stress free state is impossible.

In order to check these observations it was decided to undertake an analysis of load cases 3 and 4 using the ANSYS FE software [2]. The material selected for the analysis was EXTREN® 500 series pultruded GRP plate 6.4mm in thickness. EXTREN® was chosen because it is orthotropic (comprising discretely spaced longitudinal rovings sandwiched between layers of continuous filament mat) and has been used extensively in bolted joint tests [4 & 5]. For the purposes of the analysis, the rovings are assumed to run in the σ_N direction. The elastic property values are: longitudinal modulus (E_L) = 16.2GPa, transverse modulus (E_T) = 11.3GPa, shear modulus (G_{LT}) = 2.9GPa and major Poisson's ratio (v_{LT}) = 0.3. The dimensionless stress distributions obtained from the FE analysis using these elastic properties were expected to correspond to the unidirectional 0° laminates, A1 and A2, given in [1].

Four node solid elements were used to carry out the linear elastic FE analysis for load cases 3 and 4 and symmetry was exploited. About 800 elements were used in the FE model. Figure 4 shows the dimensionless tangential stress distribution around the edge of the hole and the shear stress distribution along the shear out plane [Note: the angle is measured anticlockwise from the intersection of the bearing plane with the hole edge (see Figure 3)]. It is clear that the graphs of both stress distributions are symmetric with respect to their abcissas and that corresponding stress values are equal in magnitude but opposite in sign. These characteristics prove that superposition of the two load cases produces a *stress free* state and that, unlike the stress distributions given in [1], they satisfy equilibrium.

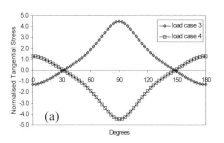 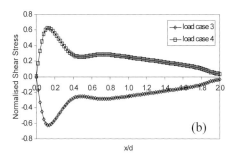

Figure 4   Stress distributions for load cases 3 and 4 normalised with respect to σ_N: (a) tangential stress around the hole edge and (b) shear stress along the shear out plane.

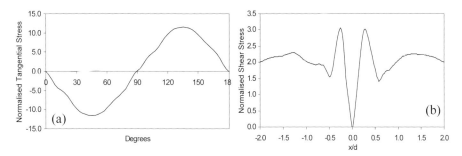

Figure 5   Stress distributions for load case 5 (normalised with respect to τ_N): (a) tangential stress around the hole edge and (b) shear stress along the shear out plane

 Similar linear elastic FE analyses, using the same EXTREN® 500 plate properties have been undertaken for load cases 5 and 6. Again, it was found that the *stress free* state was obtained when the stress distributions for the two cases were superimposed, thereby confirming that equilibrium was satisfied. Figure 5 shows two stress distributions, normalised with respect to τ_N, for load case 5. Comparing the tangential stress distribution (Figure 5(a)) with the corresponding stress distribution for the uni-directional 0° laminates in [1], the magnitudes and shapes of the stress distributions differ substantially, whereas only the shape of the shear stress distribution along the shear out plane is markedly different from those shown in [1], which appear to be almost independent of the laminate lay-up.

INADEQUATE DEFINITION OF THE BOLT/HOLE CONTACT CONDITIONS

The contact conditions at the bolt/hole interface are not defined in [1] for the simplified design method. It is well known that clearance holes are often specified for bolted joints used in infrastructure applications. Hence, it was decided to subject load cases 1 and 2 to closer examination to see if the normalised stress distributions in [1] are valid. Therefore, ANSYS FE analyses were carried out using clearances, c = 0 and 0.2mm, for a nominal 10mm diameter bolt. Because of symmetry only the upper half of the domain was modelled. PLANE 82 solid elements with mid-side nodes were used at 3° intervals around the edge of the hole. In order to reduce the overall size of the analysis four-node PLANE 42 elements were used remote from the hole. Contact between the rigid bolt and the flexible hole was modelled using the CONTACT 172 and TARGET 169 elements. About 1000 elements were used in the FE model for these load cases. The friction coefficient between the bolt and the hole was taken as 0.3.

Accounting for contact and friction makes the analysis nonlinear. Therefore, it was decided to carry out analyses for three values of the stress σ_N, viz. 2, 4 and 10 MPa. The normalised stress distributions obtained with the FE analysis for c = 0 and 0.2mm are shown in Figure 6.

In [1] the magnitude of the hole clearance is not specified and a single normalised curve is given for each of the three laminates (0°, 90° and quasi-isotropic). This contrasts sharply with the present FE results, which show that the stress distributions are not only dependent on the applied load, but also on the magnitude of the clearance. Figure 6(a) shows the normalised radial stress distribution along the hole edge for c = 0mm. The maximum radial stress occurs at an angle of about 40° and loss of contact is at an angle of about 87° (consistent with [1]). When c = 0.2mm, the maximum radial stress occurs at about 3° and the angle at which contact ceases varies from about 18° to 38° according to the magnitude of the applied stress, σ_N (see Figure 6(b)). For this case, the maximum radial stress is much more sensitive to the magnitude of the applied stress. Normalised tangential stresses along the hole edge are shown in Figs.6(c) and 6(d) for c = 0 and 0.2mm respectively. For c = 0mm the maximum normalised tangential stress is insensitive to the applied stress and occurs at 90°, i.e. at the net tension plane. On the other hand, for c = 0.2mm the maximum normalised tangential stress occurs at an angle of 84°. Finally, the normalised shear stress distributions along the shear out plane for c = 0 and 0.2mm are shown in Figs.6(e) and 6(f) respectively. Figure 6(e) shows that the peak normalised shear stress is relatively insensitive to the magnitude of the applied stress, σ_N, though its position varies from x/d = 0.2 to 0.4. However, when c = 0.2mm the maximum normalised shear stress is less than when c = 0mm and is independent of the applied stress. The peak value also occurs at x/d = 0.2. The shape of the shear stress distribution for zero clearance, shown in Figure 6(e) is similar to the curves for the three lay-ups given in [1].

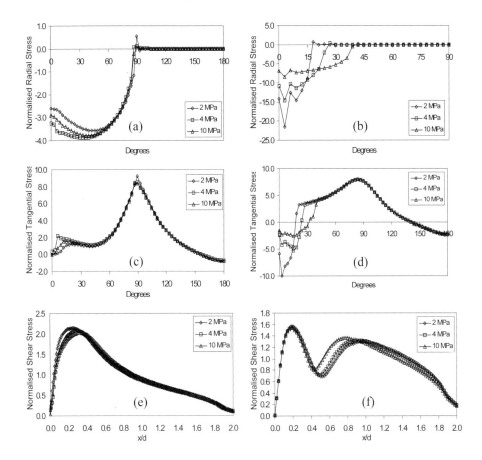

Figure 6. Stress distributions for load case 1 normalised with respect to σ_N: (a) radial stress along hole edge (c = 0mm), (b) radial stress along hole edge (c = 0.2mm), (c) tangential stress along hole edge (c = 0mm), (d) tangential stress along hole edge (c = 0.2mm), (e) shear stress along shear out plane (c = 0mm) and (f) shear stress along shear out plane (c = 0.2mm).

Consideration of the FE analysis results depicted in Figure 6 suggests that, when contact at the bolt/hole interface is accounted for, the normalised stress distributions are dependent on the applied stress and that a single normalised curve representation of the radial, tangential or shear stress, as advocated in [1], is not generally valid. It is presumed, therefore, that the *single curve* stress distributions in [1] have been derived from an assumed contact stress distribution, eg. cosine distribution, applied at the hole edge.

CONCLUDING REMARKS
An assessment of the scientific basis of the simplified method of bolted joint design, as described in the EUROCOMP design and handbook [1], has been presented. It has been noted that the laminates, for which design stress curves are given, have not been adequately defined and, moreover, that some of the stress data for the six basic load cases do not satisfy equilibrium. The major shortcoming with the simplified method has been identified as the inadequate definition of the bolt/hole interface conditions. Through FE analysis, it has been shown that stress distributions are dependent on the applied external stress and the magnitude of the hole clearance. Therefore, it is not valid to represent each stress distribution by a single curve for a given laminate, especially as bolted joints in infrastructure applications routinely use clearance holes.

ACKNOWLEDGEMENTS
The authors wish to express their appreciation to EPSRC for funding this research under grant GR/N11346.

REFERENCES

[1] Clarke, J.L. (ed.), Structural Design of Polymer Composites: EUROCOMP Design Code and Handbook, E. & F.N. Spon, London, pp.751, 1996.

[2] ANSYS Inc., Southpointe, 275 Technology Drive, Canonsburg, PA 15317.

[3] MacGinley, T.J., Structural Steelwork Calculations and Detailing, Butterworths, London, 1973.

[4] Cooper, C. and Turvey, G.J., 'Effects of Joint Geometry and Bolt Torque on the Structural Performance of Single Bolt Tension Joints in Pultruded GRP Sheet Material', Composite Structures, Vol.32, Nos.1-4, pp.217-226, 1995.

[5] Turvey, G.J. and Wang, P., 'Single Bolt Tension Joints in Pultruded GRP Material – Effect of Temperature on Failure Loads and Strengths and Joint Efficiency'. In Strain Measurement in the 21st Century, R.A. Tomlinson (Ed.), British Society for Strain Measurement, pp.20-23, 2001.

5.4 Study and application on strengthening the cracked brick walls with continuous carbon fibre sheet

T Zhao, C J Zhang and J Xie
Tianjin University, China

INTRODUCTION

Masonry buildings are of wide application in China because of less investment and great demand. As brick masonry is a comparatively brittle material, it is generally necessary to design and construct for higher quality than using other materials. Brick masonry buildings must be meticulous in design and construction. Otherwise the masonry structures will not be able to sustain the design loads. Compared with the existing methods, repairing and strengthening of the unreinforced masonry walls by continuous carbon fibre sheet (CFS) has the main advantages of excellent mechanical strength and deformation performance, convenience to use, light weight, no dimension increasing, and immunity to corrosion. This paper carries out the experiment on the strengthening of the cracked brick walls with continuous carbon fibre sheet. And a cracked brick wall in a building was also strengthened with the continuous carbon fibre sheet.

EXPERIMENTAL INVESTIGATION

The brick walls are 240mm in thickness, 1400mm in width and 1000mm in height, connected by two reinforced concrete beams at the top and the bottom, respectively. During the experiment, the vertical compressive stress keeps 1.2MPa. Figure 1 shows the geometric dimensions of the specimens.

Before strengthening with CFS, the specimen Wall-2 should be made cracked ahead. When the load was up to 212kN, Wall-2 cracked suddenly with the cracking displacement being 2.0mm as shown in Figure 2. The maximum width of the diagonal crack was 0.25mm. Then Wall-2 was unloaded and some cracks closed partly. The maximum width of cracks at that time was 0.1mm.

In order to strengthen the specimen, the strengthening area should be cleaned and a surface of cement plaster was employed. After twenty-eight-day curing, the cement plaster area of the specimen was sanded to obtain a rough surface, and then the surface was cleaned. A coat of epoxy consisting of thoroughly mixed resin and hardener was applied. The CFS, previously cut to the required dimensions, was stuck to the prepared surface, and then a resin coat was applied. Using a thin and soft plastic plate, the CFS was gently pressed along the fibre direction to remove any air bubbles. Then anchors were positioned by the bolts at the bottom and top of the strengthened walls. After the CFS was bonded, seven-day resin curing was needed.

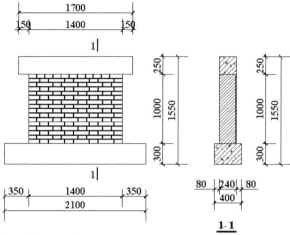

Figure 1. Details of the specimens

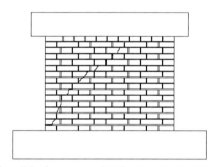

Figure 2. Cracked wall before strengthening

The strengthening conditions of the specimens are given in Table 1 and illustrated in Figure 3. The Arabic numerals in Figure 3 indicate the number of strain gauges. The mechanical properties of CFS and epoxy are listed in Table 2 and Table 3, respectively. Figure 4 is the schematic drawing of loading system in the experiment.

Each specimen was subjected to a number of lateral loading cycles while maintaining constant axial loads (430.2kN). The first loading cycle of Wall-1 was added to 30% of calculating ultimate load (P_u). The following cycles were used to determine the cracking displacement by adding 20% of P_u to Wall-1. Then multiples of the cracking displacement were applied to control all subsequent cycles. When load was down to 85% of the maximum load, the experiment was finished. Wall-2a was loaded to the cracking force directly. Wall-2b was loaded to the ultimate force and then loaded by cracking displacement.

Table 1. Strengthening conditions of the specimens

No.	Strengthening conditions
Wall-1	Control specimen
Wall-2	Strengthened with 300mm width of CFS

Table 2. Mechanical properties of CFS

Calculating thickness	Tensile strength	Modulus of elasticity	Percent elongation at fracture
0.1035mm	2100MPa	2.8×10^5MPa	1.4%~1.5%

Table 3. Mechanical properties of epoxy

Compressive strength	Tensile strength	Shear strength
50~60MPa	15~18MPa	20MPa

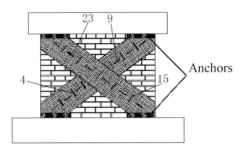

Figure 3. Strengthening conditions of the specimen

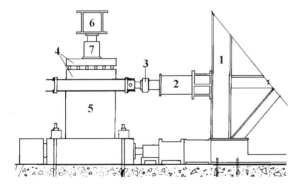

1. Horizontal loading reaction frame
2. Hydraulic jack
3. Load sensor
4. Trimmer
5. Specimen
6. Vertical loading reaction frame
7. Jack

Figure 4. Loading system

EXPERIMENTAL RESULTS AND DISCUSSION
The main experimental results are listed in Table 4.

Table 4. Experimental results

No.		f_1 (MPa)	f_2 (MPa)	P_{cr}, P_u (kN) Value	Increase	Δ_c (mm) Value	Increase	ε_{max} (10^{-6}) Warp	Woof
Wall-1	A	11.64	16.89	224	—	3.07	—	—	—
Wall-2	P	11.64	16.89	276	23.2%	8.0	160.6%	4323	1519
	N			368	64.3%	11.4	271.3%	3392	545

Note:
1) f_1 —Compressive strength of brick; f_2 —Compressive strength of mortar; P_{cr}—Cracking load; P_u—Ultimate load; Δ_c—Cracking displacement
2) P—Positive force; N—Negative force; A—Average value of P and N
3) "Increase" means the comparison of the specimens with CFS to control specimen Wall-1.

Failure characteristics
Experimental results indicate that all the specimens under cyclic loading are failed in shear failure mode and the main failure characteristics are described as follows:

Control wall (Wall-1)
During the early stage of loading, with the horizontal load less than 100kN, the specimen was in elastic stage. The P-Δhysteresis curve presented linear property in this loading period. With the increase of loading, the P-Δdiagram began to curve. When the load was added to the cracking load, the wall cracked abruptly and a main crack appeared along the diagonal direction. The maximum width of crack at this time was about 1mm. During the subsequent loading procedure, the crack was wider obviously and extended to the loading points. At the top of the specimen, some bricks were crushed and extruded, even fallen off. At the bottom of the wall, some bricks were also crushed. At the failure moment, the crack widened quickly. Along horizontal mortar bed joints between the eleventh and twelfth layers of bricks from the bottom, a 2cm slippage appeared. The cracks of Wall-1 are illustrated in Figure 5(a) and Figure 6(a) shows the experimental hysteresis curves of this specimen.

Strengthened wall (Wall-2)
During the early stage of positive loading, the specimen was also in elastic stage. When the specimen began to show some plastic properties, the old crack kept developing and the strain of CFS, especially near the crack, increased remarkably. When the specimen was loaded to the ultimate load (276kN), a new crack appeared. When the wall cracked, the strain of CFS was enlarged abruptly. Then the maximum strain of CFS could be up to 4323 $\mu\varepsilon$. The cracks of Wall-2 are illustrated in Figure 5(b).

When the specimen was loaded negatively, the strengthened wall exhibited the same behaviour as the control wall. When the specimens began to show some plastic properties, even fine cracks did not exist in the area without CFS. At the cracking loading point, the strengthened wall cracked suddenly and a diagonal crack appeared close to the outside of strengthening area quickly. The crack was fine and there were some secondary cracks near the main crack. These cracks could be closed when the specimen was unloaded. When the wall cracked, the strain of CFS was enlarged abruptly. Then maximum strain of CFS could be more than 3400 $\mu\varepsilon$. Finally, the experiment finished by the crush of bricks. Figure 6(b) shows the experimental hysteresis curve of the specimen. The broken curve was measured in the process of making Wall-2 cracked.

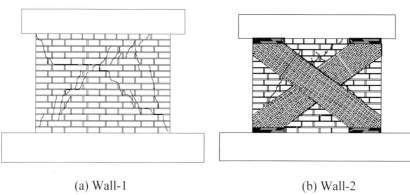

(a) Wall-1   (b) Wall-2
Figure 5. Cracks of specimens

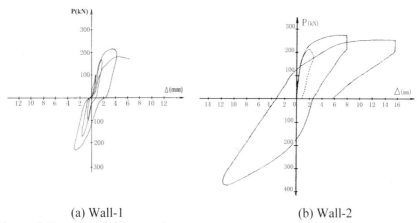

(a) Wall-1   (b) Wall-2
Figure 6. Experimental hysteresis curves

Cracking load and ultimate load
Because the specimens in the experiment reached its ultimate load as soon as it reached its cracking load, the two values could be taken as the same. From Table 4,

it is found that CFS can increase the ultimate load of cracked wall. Compared with Wall-1, the positive and negative ultimate loads are increased by 23.2% and 64.3%, respectively. Moreover, whether the wall cracks or not will affect the strengthening effect. This should be considered when strengthening the uncracked walls.

Deformation behaviour
From Table 4, it is also found that besides increasing the cracking load (ultimate load) of brick walls, CFS can improve the deformation behaviour of brick wall at the same time. The cracking or ultimate displacements of Wall-2 in positive and negative loading process are increased by 160.6% and 271.3% compared with Wall-1, respectively.

STRAIN OF CFS
Figure 7 illustrates the strain development of warp and woof of CFS in Wall-2. The strain gauges are shown in Figure 2 (b). From Figure 7 (a), it can be seen that when CFS is in tension, the strain is small in the early stage of loading and increases abruptly when the load is up to the cracking load. Figure 7 (b) also shows that the strain of CFS woof is increased with the load adding. In the beginning of loading, the strain increases linearly. When the load reaches the cracking load, the strain of CFS woof increases obviously and the strain-stress relationship expresses a turn.

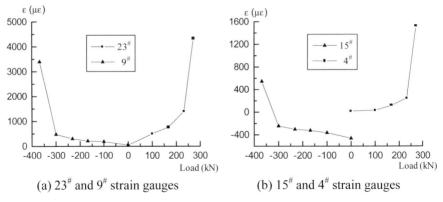

(a) $23^{\#}$ and $9^{\#}$ strain gauges   (b) $15^{\#}$ and $4^{\#}$ strain gauges

Figure 7 Strain development of CFS

CALCULATING METHOD
Based on the experimental result analysis, it is found that the action of CFS is similar to that of the diagonal braces in the truss model. But the compression strut action of CFS is too inconspicuous to be ignored. The shear force contributed by CFS is strongly dependent on the action of tension bar.

Because the CFS used in this experiment has both warp and woof, the tension effect of CFS would be divided into two parts, T_{cf1}, tensile force of warp in tension bar, and T_{cf2}, tensile force of woof in compression shut. The calculating model is illustrated in Figure 8.

In order to make convenient to design, this paper takes the strain of warp and woof of CFS as constants. Based on the equilibrium analysis of the bar forces, the following simple equation is obtained to calculate the shear force of the strengthened brick wall contributed by CFS.

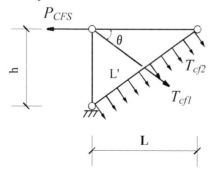

Figure 8. Calculating model

$$P_{CFS} = n\alpha_{cfs}Et'(\varepsilon_1 bL\sin\theta + \frac{1}{2}\varepsilon_2 L^2)/h \tag{1}$$

in which, P_{CFS} is the shear force contributed by CFS. E, t', b and L' are referred to the modulus of elasticity, calculating thickness, width and length of CFS, respectively. L is the length of wall and n is the layers of CFS. Strain values ε_1 and ε_2 mean the strain of warp and woof of CFS. θ is the angle between the warp of CFS and horizontal line. α_{cfs} is the shear bearing coefficient of CFS, which has the relationship with strength of brick, the quality of sticking CFS and using anchors or not. It should carry out further researches on how to reasonably determine the coefficient α_{cfs}. It is taken as 1.0 in this paper temporarily.

Based on the strain measured in the experiment, while considering the uneven coefficient, this paper takes ε_1 and ε_2 as 3000 $\mu\varepsilon$ and 600 $\mu\varepsilon$, respectively.

ENGINEERING APPLICATION
Because of the large deformation of the beam in a frame system, the wall of an office building in Tianjin cracked with an obvious diagonal crack. After several traditional repairing and strengthening, the crack tended to develop more and more. So, we decided to use the new technique to strengthen the wall with continuous carbon fibre sheet. Through calculating, the strengthening program was designed as sticking two pieces of CFS along the diagonal direction of the wall (Figure 9). The width of CFS was 600mm. After a long time observation, the crack has been already controlled.

Figure 9. A cracked brick wall strengthened with continuous CFS

BIBLIOGRAPHY

[1] Tong Zhao, Ming-guo Liu, Jian Xie and Jing-ming Zhang, Investigation on Application of Continuous Carbon Fiber Sheet to Improve Ductility of Reinforced Concrete Columns. Earthquake Engineering and Engineering Vibration 20:4, 66-72.

[2] Jian Xie, Tong Zhao, Ming-guo Liu and Hiroyuki Kawamura, Experimental Study on the Mechanical Properties of Continuous Carbon Fiber Sheet. New Carbon Materials 16:1, 63-66.

[3] Tong Zhao, Jian Xie and Zi-qiang Dai, Experimental Study on Shear Strength of Reinforced Concrete Beams Strengthened with Continuous Carbon Fiber Sheet. Building Structure 30:7, 21-25.

[4] N.Avramidou, M.F.Drdacky, P.P.Prochazka. Strengthening Against Damage of Brick Walls by Yarn Composites. Proceedings of the 6th International Conference on Inspection, Appraisal, Repairs & Maintenance of Buildings & Structures, 15-17 Dec. 1999: Melbourne, Australia, 51-58.

5.5 Structural performance of brick walls strengthened with composite laminates

A Mosallam
California State University, Fullerton, California, USA

ABSTRACT

This paper presents the results of an experimental study on the structural behaviour of full-scale red brick walls retrofitted with composites and subjected to out-of-plane bending. Four 105 x 105 x 3.7 inch (2.67 m X 2.67m X 9.4 cm) brick walls were tested. The experimental program evolved an as-built wall, two retrofitted walls using carbon/epoxy laminates: one wall retrofitted by two layers of unidirectional laminates [0°]2 and the other one retrofitted by one layer of unidirectional laminate in each direction [0°/90°]1. The fourth wall was retrofitted using three unidirectional layers of Eglass/epoxy laminates [0°]3. The laminates were applied to the tension surface of the brick wall specimens. The experimental results showed that significant strength and deformation capacity increases could be achieved. An analytical model was developed to predict the ultimate load of the tested specimens. The analytical modelling is based on deformation compatibility and force equilibrium using simple section analysis procedure. A good agreement between the experimental and theoretical results was achieved.

EXPERIMENTAL PROGRAM

General

In this program, a total of four full-scale wall specimens were subjected to out-of-plane loading and unloading cycles. The overall dimensions of the brick walls were 105" (2.67m) x 105" (2.67m). This represents a half-scale of a large typical wall in building construction. In order to simulate the action of the inertia forces resulting from the earthquake action, a uniform distributed out-of-plane loading condition is desirable. For this reason a special water-bag testing rig was used in evaluating the efficiency of the composite system. The sampling and testing procedures followed the ICBO, ES AC125 requirements.

Control (As Built) Test

At first, an as built, unstrengthened brick wall specimen was tested to provide a baseline for comparison with other FRP strengthened specimens. No external composite reinforcement is used (except of two 3" (76.2 mm) cross single laminates adhered to the compression surface to avoid failure while transporting the specimen to the test rig).

Initially, the "as-built" wall specimen exhibited a near-linear behaviour up to 2.5 tons, after which the behaviour became non-linear up to failure. The ultimate load capacity of this specimen was 4.92 tons. After reaching the ultimate load, rapid stiffness and strength degradations were observed, and the ultimate deflection at the total collapse was about 3 inches (76.2 mm). The failure was initiated around the mid-span in the form of a line crack in the mortar that was propagated across the width of the specimen. As the load increased, the size of the crack increases and a total collapse occurred.

Unreinforced Brick Walls Retrofitted with Carbon/Epoxy Laminates:
Two unreinforced/undamaged specimens fabricated at the same time, same materials, and by the same contractor as for the as-built specimen, were tested to failure under the same out-of-plane uniformly distributed loading condition. Two fibre architectures were evaluated; i) two unidirectional laminates covering the entire wall $(0°)_2$, and ii) two, single cross plies $(0°/90°)_1$. The reason of using the second fibre architecture is that in general retrofit cases, the wall will be exposed to both in-plane as well as to out-of-plane seismic loads, and it is not obvious what is the effect of this layup on the overall performance of the retrofitted wall.

Ultimate Behavior of [0] $_2$ Carbon/Epoxy Strengthened Brick Wall Specimen:
The wall specimen was subjected to several loading/unloading cycles up to a load level of 30 tons, after which, a ramp load (1.5 psi/minute or 10.34 kPa/min) was applied up to the ultimate failure of the wall specimen. A minimal permanent deformation was developed during the cycling phase of 30 tons. However, the first crackling sound was heard at a load level of about 22 tons. This can be attributed to fibres stretching due to of the unevenness of the brick wall surface and the mortar lines.

The behaviour was linear from 30 tons up to failure. The ultimate load was 58 tons as compared to 4.92 ton ultimate load capacity of the unstrengthened wall (about 1,060% increase in the ultimate capacity), with a corresponding deflection and tensile strain values (at the mid span) of 3.417", and 0.714%, respectively. The average increases in the initial stiffness exceeded 100% as compared to the as-built wall specimen. The ultimate failure load was a combination of compression failure of the bricks followed by a cohesive failure as shown in Figure 1.

Ultimate Behavior of [0°/90°] $_1$ Carbon/Epoxy Strengthened Brick Wall Specimen:
The motivation behind the selection of this fibre architecture is although the wall is resisting the out-of-plane uniform pressure loading in one-way action, in the common field application both in-plane and out-of-plane reinforcements for an unreinforced wall are used typically. For this reason, it was decided to use this specific layup to i) investigate the one-way, out-of-plane flexural response of the brick wall specimen when strengthened with one single "effective" laminate of carbon/epoxy system, and ii) to evaluate the contribution of the presence of the 90° laminate. From the first glance, one may expect that the 90° laminate may not contribute in both the stiffness and the strength of the one-way loading resisting

path, however, test results indicated that this scheme may alter the ultimate mode of failure by suppressing the expected longitudinal separation of the "effective" laminates (in the 0°-direction) and force these laminate to work together. In reality, the 90° laminates work as cross supporting members. The contribution of the 90°-ply was shown to be effective and a contributing factor in determining the ultimate failure mode of this specimen. The behaviour of this specimen is considered to be an ideal performance, with respect to the corresponding ductile failure that was observed for this specimen. Test results indicated that, even with a single "effective" laminate of carbon/epoxy *system (laid up perpendicular to the direction of the expected inertia force)*, an appreciable increase in both strength and stiffness was achieved. For example, the ultimate load capacity of this specimen was about 47 tons (refer to Figure 2), as compared to 4.92 tons, and 58 tons mate load capacity of the unstrengthened wall, and the wall specimen strengthened with two unidirectional layers of carbon/epoxy composites.

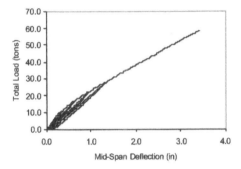

Figure 1. Half-Cyclic Performance of Unidirectional $(0)_2$ Carbon/Epoxy Strengthened Brick Wall Specimen

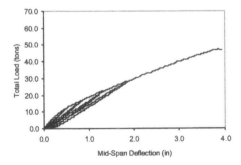

Figure 2. Half-Cyclic Performance of Cross-ply $[0°/90°]_1$ Carbon/Epoxy Strengthened Brick Wall Specimen

This can translated into over eight times the ultimate load capacity of the control specimen. The increase in the initial average stiffness of the wall was about 66% increase in the original wall stiffness as compared to the *as-built* brick wall specimen.

Unreinforced Brick Walls Retrofitted with E-glass/Epoxy System:
In this case, a total of three laminates of Glass/Epoxy were applied to the tension surface to the brick wall specimen. A similar behaviour was observed. The ultimate loading capacity was about 55.58 tons (refer to Figure 3), as compared to 4.92 tons of the as-built specimen with an increase in the flexural strength of over 10 times the original flexural loading capacity. The corresponding average linear stiffness increase of this wall as compared to the as-built specimen was about 60% increase over the original flexural capacity of the as-built wall. Figure 3 shows the load/mid-span deflection relations of this wall. The ultimate load of failure of this specimen is shown in Figure 4.

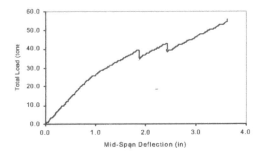

Figure 3. Ultimate Behavior of Brick Wall Strengthened with three Unidirectional E-glass/Epoxy Laminates $[0°]_3$

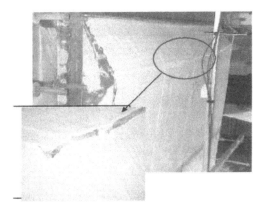

Figure 4. Ultimate Failure Mode of the E-glass Epoxy Retrofitted Specimen

THEORTICAL MODELING

The analytical modelling used in this study is based on very simple section analysis procedure identical to that used in the analysis of concrete beams. However, new parameters have been adopted for the masonry wall based on available experimental data [2,5]. The first part of the analysis is to define the material properties. The stress-strain curve for brick-mortar blocks under compression is as shown in Figure 5. The curve consists of two distinct regions: a parabolic relationship up to the maximum compressive strength, f/ m, and a linear descending branch up to the ultimate ε_{mu}. The first region of the stress-strain curve is assumed polynomial in the form:

$$f_m = A\varepsilon_m^n + B\varepsilon_m + C \tag{1}$$

The four unknowns in Equation (1) are determined from the following boundary conditions:

$$
\begin{array}{lll}
\text{i.} & f_m = 0.0 & \text{at} \quad \varepsilon_m = 0.0 \\
\text{ii.} & f_m = f_m^l & \text{at} \quad \varepsilon_m = \varepsilon_{mo} \\
\text{iii.} & df_m/d\varepsilon_m = E_m & \text{at} \quad \varepsilon_m = 0.0 \\
\text{iv.} & df_m/d\varepsilon_m = 0.0 & \text{at} \quad \varepsilon_m = \varepsilon_{mo}
\end{array}
\tag{2}
$$

The equations of the stress-strain curve have been determined to be:

• For $0 < \varepsilon_m < \varepsilon_{mo}$:

$$f_c = E_m \varepsilon_m \left[1 - \frac{1}{n} \left(\frac{\varepsilon_m}{\varepsilon_{mo}} \right)^{n-1} \right] \tag{3}$$

$$n = \frac{E_m \varepsilon_{mo}}{E_m \varepsilon_{mo} - f_m^l} \tag{4}$$

For $\varepsilon_{mo} < \varepsilon_m < \varepsilon_{mu}$:

$$f_m = f_m^l - E_d (\varepsilon_m - \varepsilon_{mo}) \tag{5}$$

$$E_d = \frac{0.5 f_m^l}{(\varepsilon_{mu} - \varepsilon_{mo})} \tag{6}$$

The parameters of the previous stress-strain curve are given as [2,5]: $f/m = 4500$ psi (31 MPa); $e_{mo} = 0.002$; $e_{mu} = 0.0035$; $E_m = 2.8 \times 10_6$ psi (19.3 GPa) ; $f_{mf} = 0.5 f/m$.

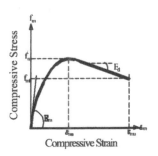

Figure 5 Stress-Strain Model for Brick-Mortar Blocks in Compression

Based on experimental evidences, it is appropriate to assume the FRP composites to be linear elastic up to failure as shown in Figure 6. The properties for both carbon/epoxy and E-glass/epoxy systems are as shown in Table 1.

Table 1 Mechanical Properties for FRP Laminates

FRP System	Thickness/Layer, inch (mm)	Tensile Modulus, E_j ksi (GPa)	Tensile Strength, f_{ju}, ksi (GPa)
Carbon/Epoxy	0.023 (0.58)	15,060 (103.84)	180.7 (1.24)
E-glass/Epoxy	0.045 (1.14)	2,679 (18.47)	61.6 (0.43)

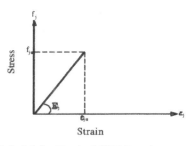

Strain

Figure 6. Stress-Strain Model for Typical FRP Laminates

Section analysis procedure are based on the following assumptions: i) Tensile strength of the brick-mortar blocks is ignored, ii) Tensile resistance of the FRP laminates can be neglected in the transverse direction, iii) The area of the FRP laminates is enough for the failure of the specimen to be due to masonry crushing rather than fibre fracture, and iv) Plane section before bending remains plane after bending, and hence a linear strain distribution can be assumed along the section.

In order to proceed with section analysis, it is necessary to develop the parameters of the equivalent rectangular stress block shown in Figure 7. By integrating the stress-strain curve for brick-mortar blocks in compression, these parameters can be determined as follows:

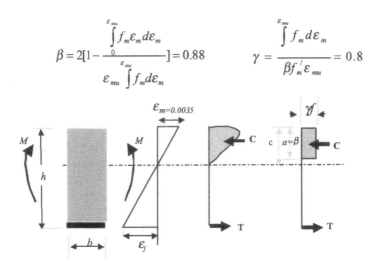

$$\beta = 2[1 - \frac{\int_0^{\varepsilon_{mu}} f_m \varepsilon_m d\varepsilon_m}{\varepsilon_{mu} \int f_m d\varepsilon_m}] = 0.88 \qquad \gamma = \frac{\int_0^{\varepsilon_{mu}} f_m d\varepsilon_m}{\beta f_m' \varepsilon_{mu}} = 0.8$$

Figure 7. Stress and Strain Distribution for Section Analysis

The steps for calculating the ultimate moment and the maximum load for the retrofitted specimen with two unidirectional layers of carbon/epoxy system are detailed below:

Calculation of neutral axis depth:
h = 3.75 + 0.046/2 = 3.773 inch
a = βc = 0.88c
C = γf_m/ab = 0.8 x 4.5 x a x102 = 367.2a
T = Ajfj = AjEjε_j = 102 x 0.046 x 15060 x ε_j = 70661.52ε_j
From strain compatibility,
ε_j = 0.0035(h/c −1) = 0.01162/a – 0.0035
T = 821.146/a – 247.31532
From equilibrium:
C = T
Thus,
367.2a = T = 821.15/a – 247.32
from which;
a = 1.196 inch

Check of FRP strain:
$\varepsilon_j = 0.006215$
$\varepsilon_{ju} = f_{ju}/E_j = 0.012 > \varepsilon_j$ ok, the failure will be due to masonry crushing rather than fibre fracture.
Calculation of ultimate moment and maximum load:
$M_u = \gamma f_m$
$/ab(h-a/2) = A_j f_j(h-a/2)$
$M_u = 1{,}394.46$ kip-inch
$w_u = 8M_u/L_2 = 1.0723$ kip/inch
P_u (for wall) $= 1.0723 \times 102 = 109.37$ kips $= 49.6$ tons
P_u (theoretical) $= 49.6$ tons
P_u (experimental) $= 58.0$ tons
P_u (theoretical)/ P_u (experimental) = 0.86
A summary of theoretical results for all retrofitted specimens is indicated in Table 2.

Table 2 Summary of Theoretical Analysis for Retrofitted Specimens

Specimen Configuration	Experimental Maximum Load, P_{u-exp} (tons)	Theoretical Maximum Load, P_{u-theo} (tons)	P_{u-theo}/P_{u-exp}
$[0^o]_2$ Carbon/Epoxy	58.0	49.6	0.86
$[0^o/90^o]_1$ Carbon/Epoxy	47.0	38.95	0.83
$[0^o]_3$ E-glass/Epoxy	55.8	40.53	0.73

ACKNOWLEDGEMENTS
This study is a part of an evaluation study of Edge Structural Composites repair and rehab systems under a general contract with the California State University at Fullerton.

REFERENCES
[1] Velazquez-Dimas, J., Ehsani, M. R., and Saadatmanesh, H., (1998). "Cyclic Behavior of Retrofitted URM Wall," Proceedings of the 2nd int. conf. on composite in infrastructure, Tucson, Arizona.

[2] Triantafillou, T. C., (1998). "Strengthening of Masonry Structures Using Epoxybonded FRP Laminates," Journal of composites for construction, Vol. 2, No. 2, May.

[3] Ehsani, M. R., Saadatmanesh, H., and Al-saidy, A., (1997). "Shear Behavior of URM Retrofitted with FRP Overlays," Journal of Composites for Construction, Vol.1, No.1, February.

[4] Fattal, S. G., Gattaneo, L. E., (1976). "Structure Performance of Masonry Walls Under Compression and Flexure," National Bureau of Standards, Washington, D.C.

[5]   Haroun, M., A. (1998). "Seismic Retrofit of Reinforced Concrete Columns and Unreinforced Masonry-Infilled Walls by Advanced Composite Materials," Proceedings, State-of-the-Art Techniques for Seismic Repair and Rehabilitation of Structures Conference, Edt. A. Mosallam, November 9, pp. 91-101.

Session 6 : **Analysis and design**

6.1   Analytical study of FRP deck on a bridge

M Chiewanichakorn[1], A J Aref[1] and S Alampalli[2]
[1]*SUNY at Buffalo, NY, U.S.A.*
[2]*NYSDOT, NY, U.S.A.*

ABSTRACT
Many of the old truss bridges in New York State are restricted to legal loads due to deterioration, increased dead loads due to the addition of overlays, and increased live loads, which are larger than that allowed in the past. By replacing the heavy decks of these bridges with lighter ones, live load carrying capacities can be improved cost effectively. Hence, the New York State used Fiber reinforced polymer (FRP) composite systems, in few such cases, to remove allowable live load restrictions. Recently, a lighter FRP deck was used to replace a concrete bridge deck of an old truss bridge, on an experimental basis. Load tests were conducted to evaluate the deck behavior at serviceability limit state level. Validated finite element models were then developed to predict the failure modes of this bridge when subjected to overload and thermal effects. This paper will summarize that study.

BACKGROUND
The New York State Department of Transportation (NYSDOT) had identified 7585 (or 38.9%) bridges as structurally or functionally deficient. Almost 2000 of these 7585 bridges were classified deficient due to poor deck conditions or weight restrictions [1]. As a result, fiber reinforced polymer (FRP) composite systems were used as an alternative material to replace a concrete bridge deck of an old truss bridge in New York State. These materials have high strength-to-weight ratios and excellent durability. Consequently, the allowable live load capacity would increase as a result of dead load reduction.

The bridge, carrying State Route 367 over Bentley Creek in New York, was erected in 1940. It was a simply supported, single span truss bridge with a concrete deck. The bridge is 42.7 m long, 7.3 m wide from curb-to-curb and had a skew of 27 degrees. The floor system consists of steel wide-flange floor beams and stringers. The existing concrete deck of 180 mm thick was replaced by the fiber reinforced concrete (FRP) deck for rehabilitation purposes. The FRP deck was a cell core structure made of E-glass stitched fiber fabric wrapped around 150 mm x 300 mm x 350 mm isocycrinate foam blocks used as stay-in-place forms. The deck panels were designed to span between the floor beams. The steel stringers were left in place to provide bracing to the structure. Each panel was 3.8 m wide and the length varied from 12.8 m to 16.2 m as shown in Figure 1. For the design and load rating of Bentley Creek bridge, composite action was not assumed. The panels were connected to one another by the field splice plates and the gaps were filled with two-part epoxy adhesive. These field splices were constructed in both longitudinal and transverse directions.

FINITE ELEMENT ANALYSIS
Modelling Method
Geometry and Boundary Conditions

Bentley Creek bridge structural system was composed of a FRP deck on floor beams, stringers and two main steel trusses which were simply-supported across the two abutments. Only one portion of the bridge was modeled in MSC PATRAN [2] to simplify the finite-element analysis procedure. The selected modeling area was a region between the first and the sixth floor beam as shown in Figure 1, which were adjacent to the south abutment. Floor beams and stringers were constructed by wide-flange structural steel. Webs of the floor beams were modeled using the four-node shell elements, where the ratio of the width-to-length ranged between 1.8 to 3.5. Flanges were modeled by the eight-node solid elements. Similarly, neoprene shims, beam-to-shim interface and deck-to-shim interface were modeled using the eight-node solid elements.

FRP deck panels were composed of two main parts, i.e. core material inserted between the two faceskin layers as a sandwich type construction. Top and bottom faceskins were compositely fabricated from two types of materials named "QM6408" and "Q9100", which are shown in Table 1. Then, thin FRP plies were laminated with different fiber orientations up to the thickness of 15 mm. Consequently, both faceskins were modeled using the solid composite elements. Similarly, the core material was composed of web-type FRP layers, QM6408, oriented in a grillage configuration of 150 mm by 300 mm. They were modeled using the shell elements (see Figure 1).

Longitudinal and transverse panel joints were constructed using composite plates as field-splice connections. Acrylic adhesive that joined the splice plates to the top faceskins was modeled by solid elements. Composite plates were assumed to have the thickness of 5 mm. These connections were modeled to evaluate the stress transfer mechanism between the FRP panels.

As the floor beams were simply-supported on the main trusses, the appropriate boundary conditions were chosen. At one support (East direction), all movements in translation degree of freedom (x-, y- and z-axes) and two rotation degree of freedom (about y- and z-axes) were restrained along the edge of the bottom flanges. At the other end, the supports were restrained in two translation directions (x- and z-axes) and about two rotation axis (y- and z-axes).

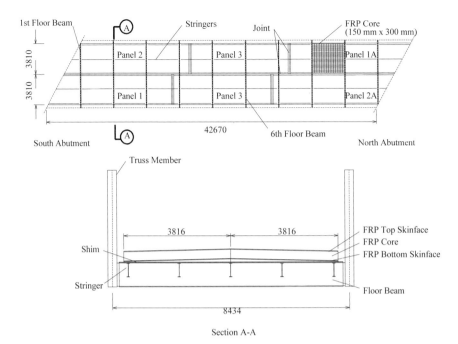

Figure 1. Plan and Cross-Section of Bentley Creek Bridge

Materials
A typical stress-strain relationship of structural steel Grade 36 (AASHTO M270) was used for the floor beams. Modulus of elasticity is 200 GPa (29,000 ksi). Materials named "QM6408" and "Q9100" were used for the modeling of the FRP deck. The mechanical properties of these two FRP materials are shown in Table 1.

From the experimental results, composite action did not exist under the applied truck loads. Hence, the interface elements were modeled with the orthotropic properties by introducing a very low in-plane stiffness and high transverse stiffness to take into account the noncomposite action.

Table 1. FRP Material Properties

Material	Elastic Modulus MPa (ksi)	Shear Modulus MPa (ksi)	Thermal Expansion °C^{-1}
QM6408	18,479 (2,680)	5,860 (850)	14.4 x 10^{-6}
Q9100	29,724 (4,311)	6,206 (900)	14.4 x 10^{-6}

Static Overload and Thermal Stress Analyses
Static Analysis
M-18 (H-20) AASHTO live loads were imposed on the FRP deck at the specified test locations. Concentrated loads were used instead of distributed tire loads due to the lack of the tire contact area information. Nonlinear geometric properties were not

included in the static analyses. The results were used to verify the finite element model prior to any further analyses.

Overload static analysis was conducted to evaluate different possible failure modes of the bridge–for example FRP decks, floor beams, stringers, and panel joints. Configurations of the applied load were chosen based on the maximum stress results from the model verification stage. The loads were applied with an increment of one truck until the bridge reached failure. Additionally, a linear buckling analysis was performed to evaluate other possible failure modes, i.e. local buckling failure.

Thermal-Stress Analysis
In thermal-stress analysis, the linear, steady state temperature gradient between top and bottom surfaces of the FRP top faceskin was assumed. The appropriate values for thermal conductivity of all materials were chosen [3]. The analyses were processed in two stages [4]. Firstly, the steady state heat transfer analysis was performed which yielded the nodal temperatures. Secondly, the static stress analysis was conducted by applying the given nodal temperature.

Two worst case scenarios were considered in the thermal-stress analysis, (a) uniformly distributed high ambient temperature and (b) concentrated extremely high temperature in an event of fire. The temperature gradient for case (a) was $60^{\circ}C$, which represented a hot day condition. The temperature gradient for case (b) was $500^{\circ}C$, which represented a burning truck. It must be noted that the total temperature of the burning truck and the ambient temperature was less than the glass transition temperature of vinylester resins, i.e. $504^{\circ}C$ ($939^{\circ}F$).

RESULTS AND DISCUSSIONS
Comparisons of Analytical and Experimental Results
Results from the finite element analysis were compared with the field test results to verify the finite element model. Strains of the floor beams and FRP deck, which were obtained from finite element analysis, matched well with the measured strains from the experiment (see Table 2). However, the lack of the actual interface material properties resulted in the differences in the strain results. Strain profiles obtained from the finite element analysis showed that the location of neutral axis was approximately coincided with the geometric centroid of steel beam. This was an indication that the bridge behaved noncompositely when subjected to the truck loads. Because of the unknown interface properties, the analytical strain values appeared to be greater than the measured strains in the floor beam. However, the measured strains on bottom faceskin of the FRP deck were relatively close to the analytical results. The case of both trucks on the bridge (one truck per lane) was identified to be the critical case that induced the maximum longitudinal strain on the FRP deck.

Table 2. Comparisons of Analytical and Test Results (Alampalli & Kunin 2001)

Gauge No.	Strain ($\mu\varepsilon$)		Gauge No	Strain ($\mu\varepsilon$)		Gauge No	Strain ($\mu\varepsilon$)	
	FEM	EXP.		FEM	EXP.		FEM	EXP.
1	-93	-89	6	-26	-18	11	91	126
2	-14	0	7	24	25	12	12	12
3	-14	4	8	-39	-26	13	3	20
4	97	90	9	-37	-16	14	46	33
5	97	95	10	92	111	15	142	148

Failure Mode Predictions
In order to predict the failure modes of the bridge, overload static, and buckling analyses were conducted separately. As mentioned earlier, having both trucks on the bridge would maximize the longitudinal strains. Therefore, the same truck configurations were used in the overload static and buckling analyses.

FRP Deck Failure
In the overload static analysis, all truck axle loads were applied on the FRP deck with an increment of one truck load. For instance, all axle loads of the test trucks would be multiplied by a factor of 2.0 at the second load level. At each increment, Tsai-Hill failure criterion [5] was applied to the composite deck at different ply levels. As the FRP deck was made up of two parts, faceskins and core materials, every single ply was monitored closely to identify the occurrence of the first ply failure. As soon as the first ply failed, the entire laminate was defined as a failure condition. This definition leads to a conservative assessment of capacity. It was found that as the applied axle loads increased, Tsai-Hill failure index approached unity, which defined as the "first-ply failure".

Tsai-Hill failure index for FRP deck core increased as the applied load increased. Obviously, Bentley Creek Bridge would reach its failure in the core material of the FRP deck (transverse direction) at the loading magnitude of seven times truck loads as shown in Figure 2. Both of the FRP deck faceskins were modeled using the solid composite elements. Hence, Tsai-Hill failure equation [5] was employed to calculate their failure indices. The results showed that there would be no ply failure taken place within both faceskins at the load level of seven times truck loads. It must be noted that this failure mode was entirely based on the FRP deck only.

Longitudinal and Transverse Panel Joints
Joint failure was another mechanism that must be considered for this type of deck system, which had two or more FRP deck panels joined together by the splice plates and the epoxy adhesive. Theoretically, the joints were expected to transfer forces across from one panel to the others. The maximum stresses induced in the splice plate at the load level of seven times truck loads were small. Shear stress along the glue material was not significantly high.

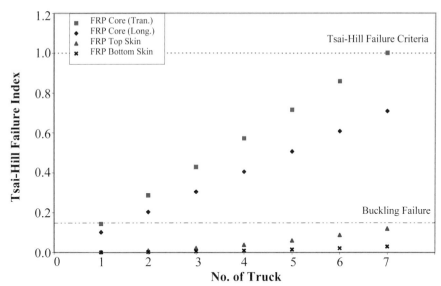

Figure 2. Tsai-Hill Failure Index of FRP Deck in Static Overload Analysis

Steel Floor Beams

There are two types of failures which would occur in the floor beams–local buckling failure, and member failure due to flexure and shear stresses. The latter type of failure will be addressed first as the result of static overload. When considered the state of stress in the isotropic material, Von Mises failure criterion was employed. At the loading magnitude of seven truck loads, Von Mises stresses in the floor beam were smaller than the yield strength of 248 MPa (36 ksi). Hence, the floor beams were not expected to fail before the FRP deck reached its failure.

Finally, local buckling failure can be predicted by performing a buckling analysis. The same truck axle configurations were imposed on the FRP deck. Five buckling modes of the bridge are depicted in Table 3. The results implied that the buckling failure would govern the overall structural failure of this bridge. The first buckling mode would occur when the bridge is subjected to the loading magnitude of 1.46 times truck loads, which is much smaller than seven times truck loads obtained in conjunction with FRP material failure. Local buckling failure was predicted that would take place in the top flange of the second floor beam near the south abutment.

Table 3. Buckling Analysis Results

Mode	1	2	3	4	5
Eigenvalues	1.46	1.54	2.27	2.46	2.53

Thermal-Stress Analysis
At a high ambient temperature gradient, Tsai-Hill failure indices for the top and bottom skinfaces were relative small. However, the maximum Tsai-Hill failure index occurred in the FRP core (transverse direction) was 0.113, which implied a large amount of reserved capacity.

As the temperature gradient was increased at a certain location in the event of a fire, Tsai–Hill failure index for the top faceskin rised rapidly more than the FRP cores and bottom faceskin. At the stage near the glass transition temperature of vinylester resins, i.e. 504°C (939°F), Tsai-Hill failure index for the FRP top skinface reached 0.619. The rapid rise in the failure index was resulted by an extremely high temperature gradient, that was constrained in a limited area, i.e. a burning truck. The applied nodal temperature contributed the same effects as the applied concentrated loads. The results of the different applied temperature gradients can be shown in Figure 3. Figure 4 shows a contour plot of the vertical displacement induced by the temperature gradient of 500°C. The Tsai-Hill failure index of 0.619 combined with a truck at high temperature may lead to a failure of the FRP in that specific location.

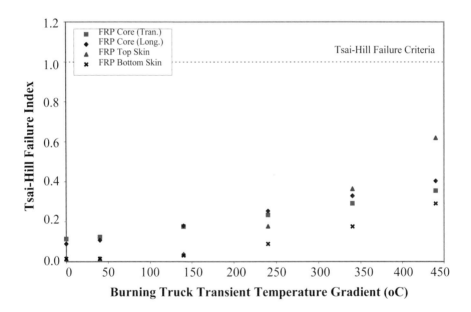

Figure 3. Tsai-Hill Failure Index of FRP Deck in Thermal–Stress Analysis

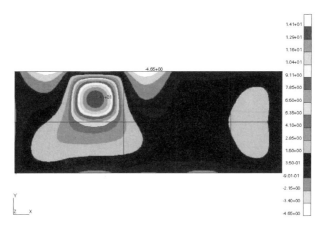

Figure 4. Displacement of FRP Deck at Temperature Gradient = 500°C (Unit = mm)

CONCLUSION

The failure mode of the bridge was verified by means of overload static and buckling finite element analyses. Local buckling at the top flange of the floor beam would be triggered at the loading magnitude of 1.46 times test truck loads, which would govern the failure mode of Bentley Creek bridge. This prediction was based on the assumption that the actual shear strength of the glue at the field splice was adequate.

Thermal-stress analysis indicated that the FRP deck would have a sufficient reserve capacity during a high ambient temperature gradient, i.e. 60°C. In an event of fire, Tsai-Hill failure index for the top skinface would increase significantly and it will be more than the FRP cores and bottom faceskin due to the high concentrated applied nodal temperature, i.e. 500°C temperature gradient.

REFERENCES

[1] Alampalli, S. and J.Kunin (2001), "Load Testing of an FRP Bridge Deck on a Truss Bridge", Special Report 137, Transportation Research and Development Bureau, NYSDOT, Albany, New York

[2] "MSC. Patran Version 9 - User's Guide" (1999), MSC. Software Corporation.

[3] Hyer, M.W. (1997), "Stress Analysis of Fiber-Reinforced Composite Materials", McGraw-Hill, 627 p.

[4] Hibbitt, Karlsson and Sorensen (2000), "ABAQUS/Standard User's Manual (Version 6.1)", Habbitt, Karlsson & Sorensen, Inc.

[5] Jones, R.M. (1999), "Mechanics of Composite Materials", 2nd Edition, Taylor & Francis, 519 pp.

6.2 Calculation of the critical local buckling load in PFRP shapes

J T Mottram,
University of Warwick, Coventry, UK

ABSTRACT
An assessment is presented on closed form equations to calculate the critical compressive load for local flange buckling in pultruded wide flange profiles. Since there are 17 physical test results the study considers the standard 203x203x9.53mm profile. It is shown that Equation (4), taken from the American Society of Civil Engineers 1984 Manual No.63, is suitable for adoption in the future preparation of simple design guidance having universal recognition.

INTRODUCTION
Pultruded Fibre Reinforced Plastic (PFRP) shapes are used as structural members in frames. Columns have Wide Flange (WF) shape, which is characterised by the same dimension of breadth ($2b_f$) and depth (D). They also have equal flange and web wall thicknesses (i.e. $t_f = t_w$). If isolated and concentrically loaded, column design can be by a manual prepared by the pultruders [1]. If column length is 'short' the ultimate mode of failure is by local flange instability [2], at load P_L. This paper considers the calculation of P_L for standard 203x203x9.53mm profiles by way of closed form equations (which may require software to generate buckling coefficients). Since its flange outstand-to-thickness ratio is high ($b_f / t_f = 10.7$) this WF profile is susceptible to local buckling when column length is < 3000mm.

Figure 1 shows the distinct local flange buckling mode shape in the upper flanges of a WF profile. The left-sided sketch gives a side view and shows four half-wavelengths, of length a, along a concentrically loaded column of length L. The right-hand sketch is an end view and this shows the mode shape has a wave pattern that is antisymmetric about the minor axis. The onset of this distinct instability happens when the uniform stress over the cross-sectional area A is its minimum critical value, given by $\sigma_{cr,L}$.

Figure 1 shows a rectangular plate element under uniform compression with the *x*-direction aligned to the applied load (see Figure 1). It is convenient when defining elastic constant notation for PFRP material to use subscript 'L' for the Longitudinal (pultrusion) *x*-direction, and subscript 'T' for the Transverse *y*-direction. Elastic constants are therefore E_L for the Longitudinal Young's modulus, E_T for the

Transverse Young's modulus, G_{LT} for the in-plane shear modulus, and v_{LT} and v_{TL} $(= (v_{LT}E_T)/E_L)$ for the major- and minor-Poisson's ratios, respectively.

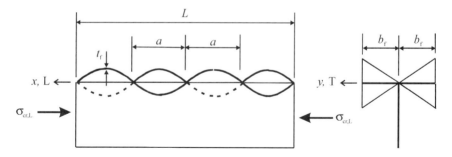

Figure 1. Local flange buckling in WF profiles.

There is recognition that mechanical properties of PFRPs have an inherent variability which is often neglected when equations predicting instability loads are assessed. The load predictions given in Table 1 should not be taken as precise; they are expected to be, if independent of G_{LT}, within ±10% of the actual buckling load.

CRITICAL LOCAL FLANGE BUCKLING LOAD: PHYSICAL TESTING
Mottram [2] has considered the accuracy of 17 physical tests for P_L. A number of these tests were performed with column length too low. Mottram observed that the minimum length in a 'short' column test must be four half-wavelengths (as in Figure 1). It is also essential that the elastics constants are known if a meaningful comparison between theory and practice is to be made. For these reasons data from Lane and Mottram [3] is used herein. Geometry for the standard 203x203x9.53mm profile is given by A = 5630 mm^2, D = 203.2mm, b_f = 101.6mm, and t_f = 9.53mm. The flange outstand is assumed to be half the flange width. Other data required is: P_L = 360kN, a = 365 mm (i.e. a/b_f = 3.6), E_L = 23GPa, E_T = 10GPa, G_{LT} = 3.0 or 4.5GPa, and v_{LT} = 0.3. It is assumed that the inplane E_L and E_T are the same as the equivalent flexural modulus associated with the instability. The lower G_{LT} is that given in the pultruders' design manuals, while the value 50% higher is obtained by resin burn-off with micromechanical modelling. The higher G_{LT} is used when calculating the P_Ls in Table 1.

CRITICAL LOCAL FLANGE BUCKLING LOAD: THEORY
Theory of Elasticity
When a thin plate is loaded in compression within its own plane, it is subject to sudden instability, or lateral deflection at a stress ($\sigma_{cr,L}$) that depends upon its stiffnesses, edge lengths and edge restraints. Figure 2 shows such a plate buckling for the outstand of a WF profile with half-wavelength a. For this length of plate the buckling stress will be a minimum and $\sigma_{cr,L}$ will be substantially less than the compressive strength of the material (which is > 200MPa [1]. To calculate P_L it is assumed that the uniform compressive buckling stress acts over the cross-sectional

area A (i.e. $P_L = A\sigma_{cr, L}$). While the initial buckling causes rippling in the flanges, it does not immediately result in catastrophic failure.

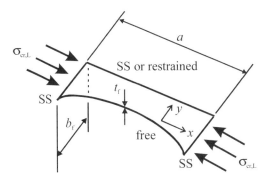

Figure 2. Local buckling of a rectangular plate.

The American Society of Civil Engineers (ASCE) present equations for isotropic and specially orthotropic rectangular plates with various edge restraints [4]. Consider an isotropic flange with longer sides of length L. The general equation for the compressive critical buckling stress is

$$\sigma_{cr,iso,L} = k_{iso} \frac{\pi^2 E}{12(1-v^2)}\left(\frac{t_f}{b_f}\right)^2. \tag{1}$$

Here k_{iso} is a buckling coefficient that depends on edge restraint conditions, the plate proportions (given by L/b_f), and the variation of stress over the plate width. Here the stress is uniform over the width. The buckling coefficient has minimum values for those L/b_f ratios that result in buckling of the plate in an integral number of half-wavelengths. The plate model in Figure 2 is for the situation when $L = a$, the two shorter sides are Simply Supported (SS) and one of the longer side is free. The other longer side is where the two flanges and web connect and this edge will have a rotational stiffness. To predict P_L using Equation (1), the supported long edge conditions of Simply Supported (SS) and CLamped (CL) give 'minimum' k_{iso} values of 0.45 and 1.33, respectively. For the SS situation a more accurate value to k_{iso}, is $0.45+(b_f / a)^2$. With $a/b_f = 3.6$ coefficient k_{iso} increases to 0.53.

Equation (2), also from ASCE Manual No, 63, gives the approximate critical buckling stress when the plate material is specially orthotropic and the supported long edge is SS (other three edge boundary conditions as before).

$$\sigma_{cr,ortho,L} = \left[\frac{\pi^2 E_L}{12(1-v_{LT}v_{TL})}\left(\frac{b_f}{a}\right)^2 + G_{LT}\right]\left(\frac{t_f}{b_f}\right)^2 \tag{2}$$

If the half-wavelength is very large and b_f / a is very small, Equation (2) simplifies to

$$\sigma_{cr,ortho,L} = G_{LT} \left(\frac{t_f}{b_f} \right)^2 . \tag{3}$$

Equation (3) gives the lowest critical buckling stress prediction. Equations (1) to (3) take no account of shear deformation. Buckling relations become very complex when shear is not neglected. It has been found that shear deformation will be significant if $E_L t_f / G_{LT} b_f > 2$. Since this ratio is less than one for all WF profiles, it is deemed acceptable to neglect the small reduction to $\sigma_{cr,ortho,L}$ due to shear deformation.

ASCE also provides an approach when information about the in-plane shear modulus (G_{LT}) is not available, and this is useful. Shear modulus of PFRP is a difficult elastic constant to determine accurately. An estimate of the effect of the differing stiffnesses in the L and T directions can be obtained by modifying the isotropic Equation (1) to

$$\sigma_{cr,ortho,L} = k_{iso} \frac{\pi^2 \sqrt{E_L E_T}}{12(1-v^2)} \left(\frac{t_f}{b_f} \right)^2 , \text{ or} \tag{4a}$$

$$\sigma_{cr,ortho,L} = k_{iso} \sqrt{\frac{E_T}{E_L}} \frac{\pi^2 E_L}{12(1-v^2)} \left(\frac{t_f}{b_f} \right)^2 \tag{4b}$$

Poisson's ratio is left undefined; v^2 can be assumed to be $v_{LT} v_{TL}$.

Design Manuals
In Section 10 of its design manual Strongwell [1] give Equation (C-2), which is

$$P_L = \frac{0.5E_L A}{\left((2b_f)/t_f \right)^{1.5}} . \tag{5}$$

For a 203x203x9.53mm profile Equation (5) predicts $P_L = 657$kN ($E_L = 23$GPa). This is 1.8 times higher than measured. Equation (5) was derived by Strongwell from 'in-house' testing. To calculate an allowable loads for tabulation in the design manual buckling loads from Equation (5) are divided by a safety factor of 3. Thus the short column allowable load would be 219kN, which is 60% of 360kN. The overall factor of safety is closer to 1.6, just over half that implied by the manual.

Creative Pultrusions Inc. have released three volumes of their design manual and Volumes 1 and 2 were revised. Chapter 5 "Load Tables for Compression Members" in Vol. 2 Rev. 2 (1999) and Vol. 3 Rev. 1 (2001) are the same [5, 6]. To determine the critical stress for flange buckling of short Pultex® WF columns, R. Yuan (University of Texas) has modified Equation (1) or Equation (4b), to

$$\sigma_{cr,ortho,L} = \Phi k_{Yuan} \frac{\pi^2 E_L}{12(1-v^2)} \left(\frac{t_f}{b_f}\right)^2 . \tag{6}$$

The coefficient Φ is 0.8 (to account for the orthotropic material) and the buckling coefficient k_{Yuan} is recommended to be 0.5 for SS restraint along the supported longer edge. v is taken to be v_{LT}. Equation (6) has the form of Equation (4b), if Φk_{Yaun} replaces $k_{iso}\sqrt{E_T/E_L}$. For a 203x203x9.53mm profile $k_{iso}\sqrt{E_T/E_L} = 0.35$, which is 14% lower than $\Phi k_{Yaun} = 0.4$. P_L by Equation (6) is higher than by Equation (4).

To calculate allowable compressive loads Creative Pultrusions use a factor of safety of 3 (same as Strongwell), with $E_L = 21$GPa and $v = 0.36$. The unfactored allowable P_L is 393kN; being 9% higher than 360kN. Applying the factor of three gives an acceptable overall factor of safety of > 2.5 (with $E_L = 21$GPa).
In Vol. 2 Rev. 2 (1999), Chapter 4 "Load Tables for Flexural Members and Connections" gives Equation (2) to determine the critical stress for local buckling failure in beam members. This guidance is changed in Vol. 3 Rev. 1 (2001) and Equation (7), from the theoretical development by Qiao [7] is now recommended.

Numerical Simulation
The closed form equations above ought to give a low P_L, even if the precise elastic constants are known, because of the edge stiffness that exists at the web-flange connection. Several advanced analytical studies have been made for orthotropic plates where the supported edge restraint is modelled with rotational stiffness [7-9]. Zureick and Shih [10] analysed the stability of PFRP columns by adopting a structural sections' approach, whereby the whole profile is modelled and thin-walled orthotropic panel interaction is accounted for in a particular way. Column instability can be solved by eigenvalue finite element analysis, and Barbero and DeVivo [11], Pecce and Cosenza [12], and Qiao et al [9] have adopted this simulation method.

Zureick and Shih [10], Bank and Yin [8], and Davalos and Qiao [13] produced charts and additional information for general application. The three groups solved the instability problem using a transcendental equation. It is difficult to use their charts to predict P_L with good accuracy, either because specific profile parameters are unavailable or because the elastic constants used to generate a chart are different to those of real profiles (e.g., G_{LT} is closer to 4.5GPa than 3.0GPa).

Using the computer programme of Zureick and Shih [10] the author obtained a minimum $k_f = 0.577$, with the specific WF profile properties given earlier. Buckling coefficient k_f replaces k_{iso} in Equation (4a). k_{iso} is 0.53 if the supported longer edge is SS, material is isotropic and $a/b_f = 3.6$. The presence of the edge restraint and orthotropic material is seen to increase P_L by only 8.8%. P_L is 371kN which is only 3% higher than 360kN. To further support Zureick and Shih's analysis the minimum k_f was for $a/b_f = 3.6$, the measured half-wavelength.

The solution by Bank and Yin [8] gives parametric information on buckling for specific anisotropy ratios in terms of L/b_f and R, a normalised 'coefficient of rotational restraint'. $R = 0$ corresponds to the SS and $R = \infty$ to the CL edge condition. Bank and Yin recognised that R is difficult to determine and therefore presented charts to determine R. One chart is for PFRP material based on the Strongwell elastic constants (from Strongwell 1988), and on taking $L/b_f = 10$.

Qiao [7] developed a comprehensive discrete laminated plate approach. By considering the case of a long, elastically restrained plate Equation (7) was developed, and it is adopted in Chapter 4 of the current Creative Pultrusions design manual (Volume 3 Revision 1 2001) for local flange buckling in beam members.

$$\sigma_{cr,ortho,L} = \frac{\pi^2}{12}\left[2\sqrt{q\,E_L\,E_T}\right) + p\left(v_{LT}\,E_T + 2G_{LT}\right)\right]\left(\frac{t_f}{b_f}\right)^2. \tag{7}$$

To account for the rotational stiffness at the web-flange connection ζ is the coefficient of restraint (= $1/R$). It is zero for CL and infinite for SS edge conditions. p and q depend on ζ and the plate's properties and they are obtained by analysis. Using data for the standard 304x304x12.7mm WF profile Davalos et al [13] present $p = 0.3 + 0.004/(\zeta - 0.5)$ and $q = 0.025 + 0.065/(\zeta + 0.5)$. Coefficient of restraint ζ is given by $(2\,D\,(E_T)_f)/(b_f\,(E_L)_w)$, where $(E_T)_f$ and $(E_T)_w$ are the transverse elastic modulus of the flange and web, respectively. Letting $(E_T)_f = (E_T)_w$, Equation (7) predicts P_L to be 394kN, which is 10% higher than 360kN.

Qiao et al [9] have proposed a generalised buckling load equation, in terms of resultant force pre unit width, $(N_L)_{cr}$. P_L is obtained from $(N_L)_{cr}\,A/t_f$. It is

$$\frac{(N_L)_{cr} - (N_L)_{cr}^{\infty}}{(N_L)_{cr}^{0} - (N_L)_{cr}^{\infty}} = \frac{1}{p\zeta^q + 1}. \tag{8}$$

p and q in Equation (8) are not same as in Equation (7). $(N_L)_{cr}^{\infty}$ and $(N_L)_{cr}^{0}$ are the lower (SS) and upper (CL) bounds to the local buckling load. No closed form equations to these bounds are given. Using SuperStructural WF profile elastic constants, which have higher values than for standard WF profiles, Qiao et al. obtained $P_L = 411$kN by Equation (8). This compared favourably with an experimental load of 394kN using a 737mm column of standard profile. One explanation for the good correlation is that the test load was too high because the number of half-wavelengths was below four.

Pecce and Cosenza [12] developed a local buckling curve from an extensive FE eigenvalue parametric study. These workers recognised that elastic constants vary from source to source and that suitable experimental procedures are needed to account for this. By varying profile geometry and elastic constants (ratio G_{LT} / E_L was assumed invariant) Pecce and Cosenza established a lower bound to P_L for b_f / t_f ratios to 30. Their proposed design equation for the critical stress is

$$\sigma_{cr,ortho,L} = \alpha \left(\frac{E_T}{E_L} \right)^{0.85} \frac{\pi^2 E_L}{12 \left(1 - v_{LT} v_{TL}\right)} \left(\frac{t_f}{b_f} \right)^2. \tag{9}$$

α is the coefficient which involves the restraint at the web-flange connection. A linear relationship for α against $b_f t_w / D t_f$ $(0.3 > b_f t_w / D t_f < 1.15)$ was derived from the FE results. This gives $\alpha = 0.525$ for WF profiles with $b_f t_w / D t_f = 0.5$. Equation (9) takes the from of Equation (4a), when $\alpha \left(E_T / E_L \right)^{0.85}$ replaces the buckling coefficient k_{iso}. The author has confirmed that a typographical mistake in Pecce and Cosenza [12] gives the equation with $\left(E_L / E_T \right)^{0.85}$. Using Equation (9) P_L is 300kN, which is well below 360kN. Sensibly, Pecce and Cosenza have intended that their design equation always gives safe buckling load predictions.

Table 1. Comparison between theory and practice for P_L.

	Source	P_L kN	Note
	Lane and Mottram (2002)	360	Physical testing, $a / b_f = 3.6$
Equation (3)	ASCE (1984)	222	Lower bound, a / b_f very large
Equation (2)	ASCE (1984)	298	Orthotropic plate, SS long edge
Equation (9)	Pecce and Cosenza (2000)	300	FEA parametric study for α and β
Equations 4a) & (4b)	ASCE (1984)	341	Derived from isotropic plate equation, SS long edge
Equation (4a)	Zureick and Shih (1994)	371	Numerically determines k_f
Equation (7)	Qiao (1997)	394	Numerically determines p, q and ζ
Equation (8)	Qiao et al. (2001)	411	Numerically determines p, q and ζ
Equation (6)	Creative Pultrusions Vol.3 Rev. 1 (2001)	411	Physical testing with modified Equation (4b)
Equation (5)	Strongwell (1998)	657	Physical testing and curve fitting

COMPARISON AND CONCLUSIONS

Table 1 presents a comparison of buckling loads ($P_L = A\sigma_{cr,L}$) predicted by Equations (2) to (9) with the measured load of 360kN. Note that the compressive load to crush the profile would be > 1100kN. The analytical predictions of P_L have been entered in ascending order, starting with Equation (3) and ending with Equation (5).

Ideally the equation used to predict P_L for the purpose of design should be exact. Even if theory was able to model reality (e.g. inherent imperfections and the stiffness restraint at the web-flange connection) accurate knowledge of elastic constants and other profile properties will often be lacking.

It is therefore judgement and experience that must guide us to decide which equation should be used. Simplicity is a virtue, which engineers desire, and this helps to discard those predictive approaches which require software (that only their writers might have access to). The two design Equations (5) and (6), from pultruders, can also be eliminated because P_L is on the high side. This leaves the three equations from the ASCE Manual No. 63. Equation (2) is discarded because it requires knowledge of G_{LT} and a/b_f, and both these parameters are not always known with accuracy. Equation (4) is the most accurate equation remaining with P_L = 341kN being within 20kN of the test load. Equation (4) has the virtues of being developed from the well-known isotropic Equation (1) and not requiring knowledge of G_{LT}. To be acceptable Equation (4) does, however, require knowing a/b_f. Given reliable and relevant physical test results for a range of WF profiles and lengths, the generality of Equation (4), with $k_{iso} = 0.45 + (b_f/a)^2$ can be assessed. The contribution by Pecce and Cosenza [12] will be useful in making this assessment since Equation (9) is for general application. In the event that Equation (4) is found to be deficient, it is recommended that a coefficient, similar to Φ in Equation (6), be introduced to cope with the specific local buckling response of PFRP shapes.

REFERENCES
[1]   "Strongwell design manual." (1998). Strongwell, Bristol, Va.

[2]   Mottram, J. T. (2000). Review of beam-to-column design equations for wide-flange pultruded structural shapes, Civil Engineering Research Group, Report CE65, School of Engineering, Univ. of Warwick, pp. 50.

[3]   Lane, A. and Mottram, J. T. (2002). "The influence of modal coupling upon the buckling of concentrically PFRP columns," under review Proc. Inst. of Mech. Engineers Part L: J. Mater. - Design and Applications.

[4]   "Structural plastics design manual." (1984). American Society of Civil Engineers manuals and reports on engineering practice, No. 63, ASCE, New York.

[5]   Vol. 2 Rev. 2 (1999). "The new and improved Pultex® pultrusion design manual of pultex standard and custom fiber reinforced polymer structural profiles." Creative Pultrusions Inc., Alum Bank, Pa.

[6]   Vol. 3 Rev. 1 (2001). "The new and improved Pultex® pultrusion Global design manual of pultex standard and custom fiber reinforced polymer structural profiles." Creative Pultrusions Inc., Alum Bank, Pa.

[7]   Qiao, P. (1997). Analysis and design optimization of fiber-reinforced plastic (FRP) structural beams, PhD thesis, West Virginia Univ., Morgantown, WV.

[8]   Bank, L. C. and Yin, J. (1996). "Buckling of orthotropic plates with free and rotationally restrained unloaded edges," Thin-Walled Structures, 24 (1), 83-96.

[9]   Qiao, P. Z., Davalos, J. F. and Wang, J. L. (2001). "Local buckling of composite FRP shapes by discrete plate analysis," J. Struct. Engrg., ASCE, 127 (3), 245-255.

[10] Zureick, A. and Shih, B. (1994). 'Local buckling of fiber-reinforced polymeric structural members under linearly-varying edge loading,' Structural Engineering and Mechanics Research Report No. SEM 94-1, Georgia Institute of Technology, Atlanta, pp. 106.

[11] Barbero, E. J. and DeVivo, L. (1999). "Beam-column design equations for wide-flange pultruded structural shapes," J. Compos. for Constr., ASCE, 3 (4), 185-191.

[12] Pecce, M. and Cosenza, E. (2000). "Local buckling curves for the design of FRP profiles," Thin-walled Structures, 37 (3), 207-222.

[13] Davalos, J. F., Barbero, E. J. and Qiao, P. Z. (1999). Step-by-step design equations for fiber-reinforced plastic beams for transportation structures, Contract RP#147 West Virginia Division of Highways, WV.

6.3 Transient analysis of composite sandwich plates using assumed strain plate bending elements based on Reddy's higher order theory

A K Nayak, R A Shenoi, S S J Moy, and J R Blake
University of Southampton, Highfield, Southampton, UK

ABSTRACT

Two new C° assumed strain finite element formulations of Reddy's higher order theory are used to determine the transient response of layered anisotropic composite and sandwich plates. The material properties that are typical of glass fiber polyester resins for skin and HEREX C70.130 PVC foam material for the core are used to show the parametric effects of boundary conditions, length to thickness ratio on the dynamic responses. A consistent mass matrix is adopted in the present formulation. The Newmark direct integration scheme is used to predict the transient response of layered composites. The results presented in this investigation should be useful in better understanding the transient response of composite sandwich plates and potentially beneficial for designers of sandwich structures.

Key Words: Transient response; Composite sandwich plates; Assumed strain approach; Higher order theory

INTRODUCTION

Laminated composites are increasingly being used now days in aerospace, marine, civil and other weight related sensitive applications. The study of the transient response of laminated composite panels has received widespread attention in recent years. The use of composites in construction industry has been gaining attention and the sandwich panels are more suitable for use to resist the dynamic response due to various loadings. Hence to meet the economic and performance requirements of the construction industry, it is important to study the behaviour of sandwich panels under various loadings.

Transient analysis of laminated composite plates has been investigated by several investigators [1]. The same does not hold true for transient analysis of sandwich panels, which is in sharp contrast to the analyses of laminated composite plates. Different theories over the years have been proposed which differ mainly in the inclusion of the effects of the shear deformation and rotary inertia in the formulation. The classical lamination plate theory (CPT) [2] is primarily based on the Kirchoff assumption and is found to be inadequate for the analysis of thick laminated plates since it under predicts the dynamic response. . The first order shear deformation theory (FSDT) [2] gives a state of constant shear strain through the thickness of the plate. Higher order shear deformation plate theories (HSDT) [2] are those in which the displacements are expanded up to quadratic or higher powers of the thickness coordinate. Of all the higher order theories, that proposed by Reddy [3] was the first to obtain the equilibrium equations in a consistent manner using the principle of virtual displacements. This give rise to self adjoint equations, as a result of which the type and form of the boundary conditions were defined uniquely. Rohwer [4] showed that the higher order theory of Reddy [3] is one of the best considering various parameters.

A review of the literature therefore reveals that: (1) Both classical plate theory (CPT) [2] and first order shear deformation (FSDT) theory [2] are inadequate for the analysis of composite sandwich plates. The reason being attributed to the use of constant shear correction factors in FSDT and lack of shear formulation in case of CPT. Hence there is a need to either go for a higher order theory or to use a first order theory with the addition of shear correction factors as discussed in the work of Vlachoutsis [5]. The later work may be cumbersome with various lamination orientations, constitutive properties, boundary conditions. (2) Possibly the best higher order theory with minimum possible displacement parameters is Reddy's higher order theory (see Rohwer [4]). However this theory requires C^1 continuity in the formulation of finite elements so that it performs well in most of the situations. As seen over the years (see Yang et al [6]) it is very difficult to get generalised finite elements based on C^1 continuity. (3) Spurious zero energy modes are detected in Mindlin plate bending elements due to the uniform or selective reduced integration [6] of the troublesome shear energy terms. In order to eliminate these spurious zero energy modes, the assumed strain approach was followed in the work of Hinton and Huang [7] which gave rise to a family of new generation of plate bending elements, popularly known as assumed strain plate bending elements. However very few attempts have been made to apply these elements to transient analysis of composite sandwich plates.

Recently Nayak, Shenoi and Moy [8] developed two new C^o assumed strain plate bending elements based on Reddy's higher order theory and analysed damping in composite sandwich plates. The above mentioned points and the formulation criteria of these elements are described in detail in Reference [8]. In this paper, we extend further the applicability of these elements to analyse the transient response of composite sandwich plates, a subject that is ill covered in the literature

PRESENT FORMULATION OF REDDY'S HIGHER ORDER THEORY
Kinematics
Hence displacement field is chosen according to Reddy's higher order theory [3] (see Figure 1)

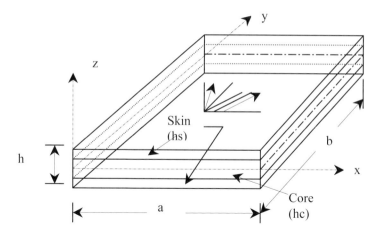

Figure 1: Sandwich plate geometry with laminate reference axes, and fibre orientation

$$u_1 = u_o + z\left[\psi_x - \frac{4z^2}{3h^2}\left(\frac{dw_o}{dx} + \psi_x\right)\right] \quad u_2 = v_o + z\left[\psi_y - \frac{4z^2}{3h^2}\left(\frac{dw_o}{dy} + \psi_y\right)\right]$$

$$u_3 = w_o \tag{1}$$

where u_o, v_o and w_o denote the displacements of a point (x,y) on the mid plane, ψ_x and ψ_y are the rotations of normals to the mid plane about the y and x axes respectively. The equation (1) is the displacement field of Reddy's celebrated

$$u_1 = u_o + z\left[\psi_x - \frac{4z^2}{3h^2}(\phi_x)\right] \quad u_2 = v_o + z\left[\psi_y - \frac{4z^2}{3h^2}(\phi_y)\right]$$

higher order theory [3]. However, the finite element based on equation (1) requires C^1 continuity because of the presence of $\dfrac{dw_o}{dx}$ and $\dfrac{dw_o}{dy}$ terms. For convenience, it is assumed that $\dfrac{dw_o}{dx} + \psi_x = \phi_x$ and $\dfrac{dw_o}{dy} + \psi_y = \phi_y$, Where ϕ_x and ϕ_y are warping functions. After substituting the above mentioned values in equation (1); we get

$$u_3 = w_o \tag{2}$$

Equation (2) forms the displacement field which requires C^o continuity in the finite element formulation. We use this displacement field to derive the governing equations of motion in the present paper.

Equilibrium Equations

Here Hamilton's principle is used to derive the equations of motion appropriate for the displacement field in equation (2), The principle can be stated in analytical form as

$$\int_{o}^{t} \int_{\Omega} (\sigma_1 d\varepsilon_1 + \sigma_2 d\varepsilon_2 + \sigma_6 d\varepsilon_6 + \sigma_4 d\varepsilon_4 + \sigma_5 d\varepsilon_5) dVdt + \int_{o}^{t}$$

$$\int_{A} (qdwdxdydt) = \int_{o}^{t} \int_{\Omega} \rho(\ddot{u}_1 du_1 + \ddot{u}_2 du_2 + \ddot{u}_3 du_3) dVdt \qquad (3)$$

where Ω is the volume, A is the cross sectional area, $\rho(x,y,z)$ is the density of the plate, $\varepsilon_i (i = 1,2,4,5,6)$, $\sigma_i (i = 1,2,4,5,6)$ are the strain and stress components respectively. We get the principle of virtual work equation from equation (3) in the following form

$$\int_{o}^{t}\int_{A} (\delta\varepsilon^{o^T}[A]\varepsilon^o + \delta\varepsilon^{o^T}[B]\kappa^o + \delta\varepsilon^{o^T}[E]\kappa^2 + \delta\kappa^{o^T}[B]\varepsilon^o + \delta\kappa^{o^T}[D]\kappa^o + \delta\kappa^{o^T}[F]\kappa^2 + \delta\kappa^{o^T}[E]\varepsilon^o +$$

$$\delta\kappa^{2^T}[F]\kappa^o + \delta\kappa^{2^T}[H]\kappa^2 + \delta\varepsilon^{s^T}[A^s]\varepsilon^s + \delta\varepsilon^{s^T}[D^s]\kappa^s + \delta\kappa^{s^T}[D^s]\varepsilon^s + \delta\kappa^{s^T}[F^s]\kappa^s)$$

$$dA\ dt\ + \int_{o}^{t}\int_{A} qdwdAdt =$$

$$\int_{o}^{t}\int_{A} [I_1(\ddot{u}_o \delta u_o + \ddot{v}_o \delta v_o + \ddot{w}_o \delta w_o) + I_2(\ddot{\psi}_x \delta u_o + \ddot{\psi}_y \delta v_o + \ddot{u}_o \delta\psi_x + \ddot{v}_o \delta\psi_y)$$

$$+\left(\frac{-4I_4}{3h^2}\right)(\ddot{\phi}_x \delta u_o + \ddot{\phi}_y \delta v_o + \ddot{u}_o \delta\phi_x + \ddot{v}_o \delta\phi_y) + \left(\frac{-4I_5}{3h^2}\right)(\ddot{\psi}_x \delta\phi_x + \ddot{\psi}_y \delta\phi_y + \ddot{\phi}_x \delta\psi_x + \ddot{\phi}_y \delta\psi_y) +$$

$$I_3(\ddot{\psi}_x \delta\psi_x + \ddot{\psi}_y \delta\psi_y) + \left(\frac{16I_7}{9h^4}\right)(\ddot{\phi}_x \delta\phi_x + \ddot{\phi}_y \delta\phi_y)]dxdydt \qquad (4)$$

where

$$(A_{ij}, B_{ij}, D_{ij}, E_{ij}, F_{ij}, H_{ij}) = \int_{\frac{-h}{2}}^{\frac{h}{2}} Q_{ij}(1, z, z^2, z^3, z^4, z^6) dz (i, j = 1,2,6)$$

$$\left(A_{ij}^s, D_{ij}^s, F_{ij}^s\right) = \int_{-\frac{h}{2}}^{\frac{h}{2}} Q_{ij}\left(1, z^2, z^4\right) dz (i, j = 5,4) \tag{5}$$

The inertias $I_i(i = 1,2,3,4,5,7)$ are defined by

$$\left(I_1, I_2, I_3, I_4, I_5, I_7\right) = \int_{-\frac{h}{2}}^{\frac{h}{2}} \rho\left(1, z, z^2, z^3, z^4, z^6\right) dz \tag{6}$$

where Q_{ij} is the constitutive matrix [3,8]. $\varepsilon^o, \kappa^o, \kappa^2, \varepsilon^s, \kappa^s$ are the strain components [3,8]. Note that ε^s in equation (4) is to be replaced by $\bar{\varepsilon}^{(s)}$ (substitute shear strain) [8,6] in order to eliminate parasitic spurious zero energy modes.

Finite Element Formulation
The general procedure for obtaining finite element formulations from the principle of virtual work is well known and hence a brief overview of the same is given. The finite element equations are obtained by discretizing the plane region R into a number of isoparametric elements. Each element 'e' has 'n' nodes, where each node $I(I=1,...,n)$ is identified with seven degrees of freedom $U_{(e)}^i = \left(u_o^i, v_o^i, w_o^i, \psi_x^i, \psi_y^i, \phi_x^i, \phi_y^i\right)_{(e)}$. For simplicity, we assume over each element the same interpolation for all seven variables, i.e

$$u_o = \sum_{i=1}^{n} N_i u_o^i \quad v_o = \sum_{i=1}^{n} N_i v_o^i \quad w_o = \sum_{i=1}^{n} N_i w_o^i \quad \psi_x = \sum_{i=1}^{n} N_i \psi_x^i$$

$$\psi_y = \sum_{i=1}^{n} N_i \psi_y^i \quad \phi_x = \sum_{i=1}^{n} N_i \phi_x^i \quad \phi_y = \sum_{i=1}^{n} N_i \phi_y^i \tag{7}$$

Where N_i, i= 1,...,n are the interpolation functions. Knowing the generalised displacement vector $\left(U_{(e)} = [N]^{(e)}\{\delta\}_{(e)}\right)$ at all points within the element 'e', the generalised mid-surface strains at any point given can be expressed in terms of nodal displacements as follows:

$$\varepsilon^{o(e)} = \left[B_\varepsilon^o\right]^{(e)}\{\delta\}_{(e)} \quad \kappa^{o(e)} = \left[B_\kappa^o\right]^{(e)}\{\delta\}_{(e)} \quad \kappa^{2(e)} = \left[B_\kappa^2\right]^{(e)}\{\delta\}_{(e)} \quad \bar{\varepsilon}^{s(e)} = \left[\bar{B}_\varepsilon^s\right]^{(e)}\{\delta\}_{(e)}$$

$$\kappa^{s(e)} = \left[B_\kappa^s\right]^{(e)}\{\delta\}_{(e)}$$

where $\left[B_\varepsilon^o\right]$, $\left[B_\kappa^o\right]$, $\left[B_\kappa^2\right]$, $\left[\bar{B}_\varepsilon^s\right]$, $\left[B_\kappa^s\right]$ are generated strain-displacement matrices. For arbitrary value of virtual displacements, equation (4) finally leads to the following assembled equations for forced vibration analysis:

$$[M]\{\ddot{\Delta}\} + [K]\{\Delta\} = \{F\} \tag{8}$$

Here the unknown vector $\{\Delta\}$ is generated by the assemblage of element degrees of freedom $\{d\}_e^T$, e=1,...., total degrees of freedom in the region R. [K], [M] and {F} are the assembled stiffness matrix, mass matrix and force vector respectively.

Details of The Newmark Integration Scheme
To complete the discretization, we must approximate the time derivatives appearing in equation (8). Here we use the Newmark direct integration method in which the vectors $\{\dot{\Delta}\}$ and $\{\ddot{\Delta}\}$ at the end of a time step
Δt are expressed in the form

$$\{\dot{\Delta}\}_{n+1} = \{\dot{\Delta}\}_n + \left[(1-\alpha)\{\ddot{\Delta}\}_n + \alpha\{\ddot{\Delta}\}_{n+1}\right]\Delta t$$

$$\{\Delta\}_{n+1} = \{\Delta\}_n + \{\dot{\Delta}\}_n \Delta t + \left[\left(\frac{1}{2} - \beta\right)\{\ddot{\Delta}\}_n + \beta\{\ddot{\Delta}\}_{n+1}\right](\Delta t)^2 \qquad (9)$$

where α and β are parameters that control the accuracy and stability of the scheme, and the subscript 'n' indicates the solution is evaluated at nth time step (i.e at time t= nΔt). We choose $\alpha = \frac{1}{2}$ and $\beta = \frac{1}{4}$ (which corresponds to the constant average acceleration method) in our analysis. By rearranging equation (8) and equation (9), we arrive at

$$\left[\hat{K}\right]\{\Delta\}_{n+1} = \{\hat{F}\}_{n,n+1} \qquad (10)$$

$$\left[\hat{K}\right] = [K] + a_o[M]$$

$$\{\hat{F}\} = \{F\}_{n+1} + [M]\left(a_o\{\Delta\}_n + a_1\{\dot{\Delta}\}_n + a_2\{\ddot{\Delta}\}_n\right), \ a_1 = a_o\Delta t \quad a_2 = \frac{1}{2\beta} - 1 \qquad (11)$$

$$a_o = \frac{1}{(\beta\Delta t^2)}$$

Once the solution $\{\Delta\}$ is known at $t_{n+1} = (n+1)\Delta t$, the first and second derivatives (velocity and accelerations) of $\{\Delta\}$ at t_{n+1} can be computed from, rearranging the expressions in equation (9),

$$\{\ddot{\Delta}\}_{n+1} = a_o\left(\{\Delta\}_{n+1} - \{\Delta\}_n\right) - a_1\{\dot{\Delta}\}_n - a_2\{\ddot{\Delta}\}_n, \{\dot{\Delta}\}_{n+1} = \{\dot{\Delta}\}_n + a_3\{\ddot{\Delta}\}_n + a_4\{\ddot{\Delta}\}_{n+1} \ (12)$$

where $a_3 = (1-\alpha)\Delta t$, and $a_4 = \alpha\Delta t$. All of the operations indicated above, except for equation (12), can be carried at element level, and the assembled form of equation (10) can be obtained for the whole problem. The equation is then solved for the global solution vector at time $t = t_{n+1}$ (after imposing the specified boundary conditions of the problem).

Results and Discussions

Example 1: In order to validate the new Reddy type elements, the problem for which the analytical [11]   solution exists is solved. The problem consists of a simply supported rectangular plate (with a=1.414, b=1, h=0.2, ρ=1, $v = 0.3$, E=1) subjected to a uniform pulse loading q=1 on a square (side 0.4a) area at the center of the plate. A non uniform 4×4 mesh as shown in Reference [9] (with $\Delta t = 0.10$ seconds) was employed. The problem was also solved analytically by Reissman and Lee [11]. A comparison of the non dimensional deflection $\left(wEah \middle/ qb^2 \right)$ obtained by present 9, present 4 and analytical solution is shown in Figure 2. The present finite element solutions for the central deflection are in excellent agreement with the thick plate solution. However from our numerical experiments, present 9 gives better result than present 4 element for a given mesh size. Hence for the remaining examples, present 9 will be used to predict the transient response of laminated composite and sandwich plates.

Example 2: Next, results for plates of layered composite plates are discussed. The 2×2 mesh in quarter plate was used because of biaxial symmetry of the problem. The example of composite plates consists of a simply supported cross ply (0/90/0) square plate subjected to various loads:

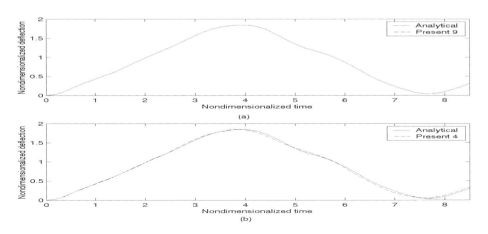

Figure 2: Nondimensionalized central deflection versus time for simply supported rectangular isotropic plate subjected to suddenly applied patch loading at the center for 9 node and 4 node assumed strain higher order elements

$$q(x, y, t) = q_o \sin\left(\frac{\pi x}{a}\right) \sin\left(\frac{\pi y}{b}\right) F(t),$$ where the time dependent loading F(t)

can be any of the following types: for step loading; 1 for $0 \le t \le t_1$ and 0 for t_1, for triangular loading ; $1 - \dfrac{t}{t_1}$ for $0 \le t \le t_1$ and 0 for t_1, for sine loading;

$\sin\left(\dfrac{\pi t}{t_1}\right)$ for $0 \le t \le t_1$ and 0 for t_1, for exponetial blast loading ; $e^{(-\gamma t)}$. The

time period t_1 is 0.006 seconds and q_o is 68.9476 Mpa. For blast loading, γ is set

to 330 s^{-1}.   The geometrical and material properties used in this example are

a=b=5h, h=0.1524m, E_1= 172.369 GPa, E_2= 6.895 GPa, $G_{12} = G_{13} = 3.448$

GPa, $G_{23} = 1.379$ GPa, $V_{12} = 0.25$, $\rho = 1603.03$

$kg\Big/ m^3$. A comparison of the higher order and exact solution [2] is presented in

Figure 3. The present formulation is in excellent agreement with the exact solution.

Here the time step Δt is taken as $1.0e^{-04}$ second.

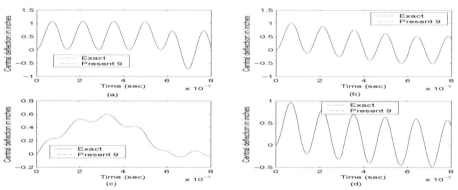

Figure 3: Central deflection versus time for simply supported square laminated
         plate subjected to a) step b) Triangular c) Sine pulse d) Blast loading for
         Present 9 and analytical solution.

Example 3: In order to investigate the effect of thickness on the transient response
of a layered sandwich plate (0/90/0/core/0/90/0) made with PVC foam core, the
following material properties of unidirectional glass fiber in a polyester resin matrix
for skin are taken [14]; E_1= 27.86 GPa, E_2= 8.07 GPa, G_{12}= G_{13} =3.17 GPa,

$G_{23} = 2.07$ GPa, ρ_s= 1650 $kg\Big/ m^3$, V_s= 0.32. A 6×6 mesh in full plate is

used. Note that the modeling in full plate rather than quarter plate plays a significant
role in the transient analysis [10]. The time step used here is according to the
conditions from finite difference scheme [12,13]. From our numerical convergence
studies for a given mesh, the solution at a given time increases with decreasing
values of the time step. However, the difference gets smaller with decreasing time

step. The solution obtained by using the 6×6 mesh with $\Delta t = 1.0e^{-04}$ second is acceptable for most purposes.   The loading on the plate is due to sine load as mentioned in example 2. The length of the plate is taken as unity. The hc/h ratio is taken as 0.94. The results from this analysis are shown in Figure 4. As the thickness decreases, there is increase in the response.

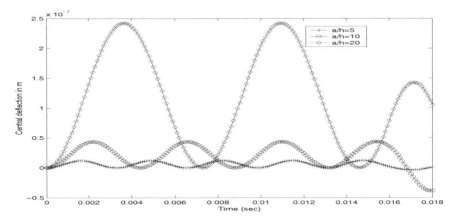

Figure 4: Central deflection versus time for simply supported square composite plate subjected to sinusoidal load.

Example 4: In the fourth example, a parametric study has been undertaken to study the effect of boundary conditions on the transient response of composite sandwich plates. The material and geometric property of the sandwich plate is same as that of example 3. The notations S, C and F correspond to simply supported, clamped and free boundary condition respectively. As seen from the results, the response is higher for simply supported plates followed by CFCF and CCCC plates.

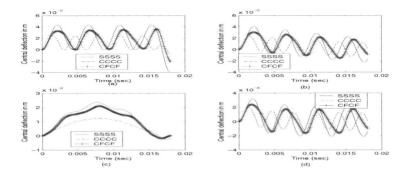

Figure 5: Central deflection versus time for different boundary conditions for a square composite sandwich plate subjected to different loadings.

CONCLUDING REMARKS

In this paper, two new C^o Reddy type elements with the assumed strain approach have been developed.

The following conclusions can be made

1) Based on Reddy's higher order theory, these elements have been implemented with a very simple and understandable mathematical framework and are easily programmed.

2) To control spurious zero energy modes, the assumed strain method as suggested by earlier investigators has been implemented in the present formulation.

3) In terms of variational principle, the resulting equations are similar to the FSDT except that the higher order terms are added to replace the shear correction factors.

4) The applicability of Reddy type elements covers a wide range of dynamic problems with varying material combinations, geometric features and boundary conditions.

5) As seen from the behaviour of these elements, the results are reliable. Several parametric studies depending on the needs of the designers can be undertaken to suit the requirement of a particular need of the composite structures.

REFERENCES

[1]   Noor AK, Burton WS and Bert CW. " Computational models for sandwich panels and shells," Applied Mechanics Reviews, Vol 49, No 3, 1996, pp.155-199.

[2]   Khdeir AA and Reddy JN. " Exact solutions for the transient response of symmetric cross ply laminates using a higher order plate theory," Composite Science and Technology, Vol 34, 1989, pp.205-224.

[3]   Reddy JN. " A simple higher order theory for laminated composite plates," Journal of Applied Mechanics, Transactions of ASME, Vol 51, 1984, pp. 745-752.

[4]   Rohwer K. " Application of higher order theories to the bending analysis of layered composite plates," International Journal of Solids and Structures, Vol 29, 1992, pp. 105-119.

[5]   Vlachoutsis S. " Shear correction factors for plates and shells," International Journal for Numerical Methods in Engineering, Vol 33, 1992, pp.1537-1552.

[6]   Yang HTY, Saigal S, Masud A and Kapania RK. " A survey of recent shell finite elements," International Journal for Numerical Methods in Engineering, Vol 47, 2000, pp. 101-127.

[7]   Hinton E and Huang HC. " A family of quadrilateral Mindlin plate elements with substitute shear strain fields," Computers and Structures, Vol 53, 1986, pp. 51-54.

[8]   Nayak AK, Shenoi RA and Moy SSJ. " Damping prediction of composite sandwich plates using assumed strain plate bending elements based on Reddy's higher order theory," 43rd AIAA/ ASME/ ASCE/ AHS conference, Colorado, 25-27 April, 2002. (accepted for publication).

[9]   Reddy JN. " Dynamic (transient) analysis of layered anisotropic composite materials," International Journal for Numerical Methods in Engineering, Vol 19, 1983, pp. 237-255.

[10] Reddy JN. " A note on symmetry considerations in the transient response of unsymmetrically laminated anisotropic plates," International Journal for numerical Methods in Engineering, Vol 20, 1984, pp. 175-194.

[11] Reismann H and Lee Y. " Forced motions of rectangular plates," in Developments in Theoretical and Applied Mechanics, Vol 4(Ed D. Frederick), Pergamon Press, New York, 1969, pp. 3-18.

[12] Leech JW. " Stability of finite difference equations for the transient response of a flat plate," Journal of AIAA, Vol 3, No 9, 1965, pp. 1772-1773.

[13] Tsui TY and Tong P. " Stability of transient solution of moderately thick plate by finite difference method," Journal of AIAA, Vol 9, 1971, pp. 2062-2063.

[14] Meunier M. " Dynamic analysis of composite sandwich plates using higher order theory," PhD thesis, university of Southampton, April 2001.

6.4 An Australian industry code of practice for the structural design of fibre composites

G M Van Erp and S R Ayers
Fibre Composites Design and Development, University of Southern Queensland, Toowoomba, Australia

ABSTRACT

In 2000, the Composites Institute of Australia (CIA) initiated a program to develop an Industry Code of Practice for the structural design of fibre composites. This paper discusses two of the initial topics addressed by this program. These are the development of a national system of grading and certification for composite constituents, and the characterisation of unidirectional lamina behaviour.

INTRODUCTION

For several years now it has been apparent that fibre composites have significant potential for application in civil engineering structures. However the uptake of these materials continues to be slow, and while a number of large scale projects have now been constructed around the world, this type of application remains the exception rather than the rule. One of the issues which has been identified as a major impediment to the broader utilisation of fibre composites in civil engineering is the lack of relevant design standards. Even though most building codes do allow for the use of new materials which are not covered by design codes, there is little incentive for civil engineers to embark on such a development. The costs and liabilities associated with the use of new and "unproven" technology can generally not be justified. Recognising this significant barrier to the utilisation of fibre composites in civil engineering, the CIA has embarked on a major program to develop a Code of Practice for the structural design of fibre composites. This paper outlines two of the initial projects undertaken as part of this development program. The first concerns the implementation of a characterisation and grading system for all composite constituent materials used in Australian civil and structural engineering projects. The second involves a new system of lamina characterisation that has the potential to significantly simplify the design and specification of fibre composites.

CERTIFICATION AND GRADING OF CONSTITUENTS

At present, publicly available information on constituent behaviour normally takes the form of "technical data sheets" supplied by material manufacturers. While these sources provide an initial introduction to the material, the information contained therein is typically insufficient for engineering purposes. The reliability and relevance of the data presented is also open to debate. It has been found that the information presented in most data sheets is limited and generally obtained under idealised laboratory conditions and in many cases not representative of "real world" material performance. To highlight this situation, a study was made of vinyl ester resin systems currently available in the Australian market. Samples of five different vinyl ester resins were obtained from the relevant suppliers along with the "technical data sheet" for each material. Specimens of each material were prepared to determine the maximum tensile stress and associated strain at failure. All testing was performed in accordance with testing standard ISO 527.

Table 1 shows the mean result and standard deviation obtained through this testing program. Values reported in corresponding manufacturer literature are also shown. It can be seen that few of the values reported in the data sheets agree with those determined experimentally. Of most concern are the strain-to-failure figures where all of the manufacturer values exceed experimental results by a significant margin. Manufacturers do seem to be somewhat aware of this situation, with most putting disclaimers on their literature, stating that the values given should not be taken as design values. However, it has generally been found that when information on the performance of materials has been requested, this type of data sheet is all that is offered to the engineer, which is clearly unacceptable for civil engineering projects.

Table 1. Comparison of reported resin tensile properties with experimentally determined values for a range of vinylester resin systems

Property	Published Value	As Tested	Standard Deviation
Maximum Stress (Mpa)			
Resin A	86	63	4
Resin B	76 – 83	75	4
Resin C	86	81	1
Resin D	69 - 76	71	0.7
Resin E	83	78	3
Strain at Failure (%)			
Resin A	6.5	2.45	0.28
Resin B	7 - 8	3.20	0.39
Resin C	5 – 6	4.95	0.93
Resin D	10 – 12	6.46	1.20
Resin E	4.2	2.80	0.29

To address these problems a new approach to material certification has been proposed. Under this program, materials would be assessed as to their ability to meet a prescribed set of performance criteria and compliant materials would attain a certified grading. Only materials possessing an official certification would be permitted for use in civil engineering projects. The program is to be administered by the CIA.

The proposed certification program will require manufacturers of constituent materials to submit their material for assessment. Each material would be run through a complete characterisation program at a certified testing institute. Under the four primary categories of Resins, Reinforcements, Core Materials and Adhesives, individual constituents would be classified into a series of performance grades. For example, in the Reinforcement category, the "Grade A Reinforcement" rating would be given to high grade structural fibres and will cover reinforcement fibres suitable for the primary load carrying function of a structure. Similarly the "Grade A Resin" rating would cover structural resins suitable for primary load carrying elements, while Grade B and Grade C resins would only be allowed for non-structural applications. In this way the engineer only has to specify the required performance grade and it is then up to the manufacturer to make a final selection of a suitably graded material.

It is believed that the characterisation programs for constituent materials in civil engineering should be extremely thorough. For example, the proposed resin certification program covers areas such as; basic chemical composition, physical characteristics of uncured and cured resins, cure and processing characteristics, mechanical performance, environmental exposure characteristics and fire performance.

It is recognised that some of the information sought by the certification program, particularly in regard to material composition, has traditionally been regarded by manufacturers as confidential. In order to safeguard the proprietary and confidential nature of this information, it is proposed that the final certification document issued by the CIA only shows the obtained grading. All test results and material information will be treated as confidential. The ability of the certifying body to obtain this information is seen as vitally important to ongoing industry development. Such information will enable the development of a comprehensive materials database. This information is seen as essential for long-term evaluation of materials. There are currently many unknowns in respect to long-term performance of composite materials and there is a reasonable probability that some materials may exhibit problems over the longer term. The development of a high quality database of materials information would assist in identifying and rectifying key material problems should they arise.

Current Progress
The first area to be targeted for development is that of polymer resin materials. A twelve-month project to develop an appropriate system of assessment and grading for polymer resins used in fibre reinforced laminates has recently been initiated. A number of major, international resin companies are providing financial support for this project and are involved in the formulation of policies. The project aims to present an industry backed proposal for assessment and grading of polymer resins, including details of testing and assessment criteria and recommendations for the implementation of the system by August 2002.

CHARACTERISATION OF FIBRE COMPOSITE LAMINA BEHAVIOUR

The second issue being addressed as part of the development program is the structural characterisation of unidirectional lamina. Composite structures are generally in the form of laminates consisting of multiple laminae, or plies oriented in different directions and bonded together in a structural unit. The virtually limitless combinations of lamina materials, lamina orientations and stacking sequences can be used to generate a wide range of physical and mechanical properties. The structural behaviour of composite laminae has traditionally been characterised using the same stress based approach as for homogeneous material. Load carrying capacity is characterised by ultimate stress and stiffness by Modulus of Elasticity. For homogeneous materials, these are excellent parameters to use, but for materials such as fibre reinforced laminae they are far less suitable. The main reason for this is that stress values depend on lamina thickness, which for a given amount of reinforcement is essentially a processing variable. As a result, wildly varying stress data are obtained for similar laminae [1], which is highly undesirable from a design perspective.

The authors have previously demonstrated how a significantly improved system of characterisation for unidirectional fibre laminae can be created by replacing the thickness parameter in the traditional stress characterisation with the mass of reinforcement [2]. In the case of load carrying capacity, this results in the traditional ultimate tensile stress (ultimate force per unit area) parameter being replaced by a new quantity of Normalised Unit Strength (ultimate force per unit width per mass of reinforcement). The SI units for this parameter are: N/mm_{width} per kg/m^2 of reinforcement. Similarly, the traditional Modulus of Elasticity (N/mm^2) can be replaced by Normalised Unit Modulus (N/mm per kg/m^2 of reinforcement). This alternative characterisation approach assumes that the primary structural behaviour of unidirectional laminae is fibre dominated and that the contribution of the resin to this behaviour is negligible. This results in a linear relationship between the amount of lamina reinforcement and the resulting load carrying capacity and stiffness along the fibre axis. The Unit Modulus and Unit Strength of a specific lamina (in N/mm width of lamina), can thus be calculated by linear interpolation of the normalised property and the areal reinforcement weight of the lamina, ie:

Unit Modulus = Normalised Unit Modulus x Areal Reinforcement Weight   (1)
Unit Strength = Normalised Unit Strength x Areal Reinforcement Weight   (2)

This approach results in significantly improved consistency for material data and also simplifies the specification of composites [1].

Before adopting this approach for the Australian design standard, a survey of the literature was undertaken to determine whether this type of characterisation system had been utilised previously. It was found that the British Tank Standard BS4994 [3] uses a similar approach. Also, a recent report from the American Concrete Institute Committee 440 [4] has suggested a related system of characterisation. The ACI approach allows the characterisation of uni-directional laminae based on the thickness of dry fibre reinforcement. While the survey did highlight the use of a unidirectional characterisation system based on the Normalised Unit Strength and Modulus approach, no experimental studies examining this approach were found.

It is the authors opinion that this alternative characterisation approach has significant merit and that it offers simplicity and clarity over the traditional stress approach. However, without sound experimental backing, it is unlikely that the system would be adopted by the civil engineering community. To address this problem, an experimental study was initiated to examine and verify the foundational tenets of this alternative characterisation method.

Experimental Program
The new method of characterisation is based on the following assumptions:
- Existence of a linear relationship between the amount of reinforcement in a lamina and its resulting load carrying capacity and stiffness along the fibre axis.
- Independence of unidirectional strength and stiffness characteristics from the resin or reinforcement fabric form used.

The experimental program involved the testing of a large range of reinforcement fibres, weights and resins in order to examine these key assumptions. Due to space restrictions, detailed results for the different combinations of resins and reinforcements cannot be provided.

Summary of Experimental Results
Strength and Stiffness Variation with Areal Reinforcement Weight
One of the foundational assumptions of the alternative characterisation system is that there is a direct linear relationship between the stiffness and strength of a lamina and the amount of reinforcement fibre contained therein. To investigate this assumption, ten different E-glass laminae were constructed with increasing areal reinforcement weights. Figure 1 shows the variation of Unit Modulus found with these laminae, and Figure 2 shows an example of the variation of Unit Strength for two different glass fabrics. From Figure 1 it can be clearly seen that there exists a direct linear relationship between Unit Modulus and the weight of reinforcement in a lamina, in accordance with Equation 1. There appears to be no significant difference in the Normalised Unit Modulus values obtained with each of the fabrics tested. Figure 2 shows that there is again a definite linear relationship between the load capacity of a lamina and the amount of reinforcement contained therein. This is consistent with Equation 2. However, unlike the Normalised Unit Modulus, there appears to be a noticeable difference in the linear trends yielded by different fabrics. As will be discussed later, this is related to the different failure strains of each material.

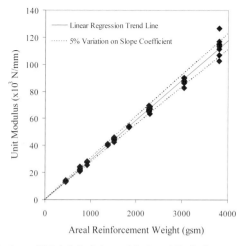

Figure 1. Variation of Unit Modulus with Areal Reinforcement Weight

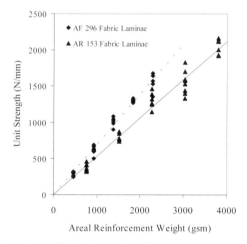

Figure 2. Variation of Unit Strength with Areal Reinforcement Weight

The Influence of Resin Type on Strength and Stiffness
The second assumption of the new approach is that the influence of resin type on unidirectional lamina stiffness and strength can be ignored for most engineering calculations. To investigate this assumption, a series of tests were undertaken using five different polymer resins. All test panels utilised the same reinforcement in an effort to isolate resin influences. The Normalised Unit Modulus values obtained with each resin were relatively consistent. An examination of the individual specimen values indicates that 94% of the values fall within 5% of the overall mean Normalised Unit Modulus value. On this basis it would appear that for most engineering purposes, the influence of the resin on unidirectional lamina stiffness

can be neglected. There were however distinct differences in the Normalised Unit Strength figures yielded by each resin. The polyester resin lamina gave significantly lower strength figures than those found with the other resins, while the higher quality epoxy resins gave the best strength figures. These strength results seem not to be in agreement with the assumption that ultimate load capacity is independent of the resin used in the lamina. This phenomena will be discussed in more detail shortly.

The Influence of Reinforcement Format on Strength and Stiffness
In addition to an independence from resin type influences, it is also assumed that the tensile characteristics of a unidirectional lamina will be largely independent of the reinforcement format used. Strength and stiffness characteristics are expected to only be influenced by the type of reinforcement fibre, and not by the actual fabric used. To examine this assumption, a series of laminae were fabricated with different reinforcement fabrics. Again the Normalised Unit Modulus values obtained in testing are highly consistent between the different fabrics used. However, significant variation is found between the strength results for each type of fabric. These figures indicate that the reinforcement form plays a significant role in the ultimate load carrying capacity of unidirectional laminae.

Discussion and Conclusions from Experimental Program
Based on limited experimental data obtained to date it would appear that from the perspective of stiffness characteristics, the foundational premises of the new characterization approach are consistent with experimental findings. The Normalised Unit Modulus approach appears to be a much more consistent method of characterizing lamina stiffness. It does not appear to be significantly influenced by either the type of resin used or the format of reinforcement. Further the experimental data appears to support the existence of a linear relationship between the stiffness of a specific lamina and the amount of reinforcement contained therein. In a similar manner, the test data appears to support the existence of a linear relationship between the load capacity of a lamina and the amount of reinforcement therein. However, the Normalised Unit Strength values appear to be influenced by both the format of reinforcement used and the type of resin.

In assessing the experimental results in detail, it became apparent that the scatter found in the strength data is primarily due to large variations in the failure strain of the laminae. Strain to failure values ranged from as low as 1.5%, up to nearly 3%. This high level of strain variation displayed by unidirectional laminae is not ideal from a civil engineering design perspective. There are many factors which have an influence on the failure strain level of a unidirectional lamina including, fibre waviness and out-of-plane distortion, low strain to failure in matrix materials etc. It is the authors opinion that the most reasonable course of action to adopt at this point, is to account for the high levels of strain variation observed and to determine a statistically based characteristic failure strain for the different combinations of resin and reinforcement grades. For each combination of resin and reinforcement grade the Normalised Unit Strength can then be determined by the relationship:

Normalised Unit Strength = Normalised Unit Modulus x Char Failure Strain   (3)

Table 2 shows an example of the proposed format for a table with laminae properties for unidirectional A Grade glass reinforcement for resin grades A, B and C. Similar tables will be provided for other grades of glass and carbon reinforcement.

Table 2. Characteristic Material Properties for Fibre Reinforced Laminae Using A Grade Unidirectional Glass Fibre Reinforcements

Direction	Action	Normalised Unit Modulus	Characteristic Failure Strain (%)		
		N/mm per kg/m^2 reo.	Grade A Matrix	Grade B Matrix	Grade C Matrix
Grade UGA Fibre Reinforcements					
Parallel to fibres	Tension				
	Compression				
Perpendicular to fibres	Tension				
	Compression				
In-Plane	Shear				

CONCLUSIONS

Due to the nature of the civil engineering design process, significant attention has to be paid to lamina characterisation and specification. This paper has discussed two proposals currently under consideration for the Australian Code of Practice for fibre composites. The proposals are at present being discussed with industry stakeholders.

REFERENCES

[1]   Van Erp G. M., Ayers S. R. & Simpson P. E. 2000, Design of Fibre Composite Structures in a Civil Engineering Environment. In: S. Bandyopadhyay et al. (eds), Proceedings of the ACUN-2 International Composites Conference: 16 - 22. Sydney: University of New South Wales

[2]   Van Erp G. M. & Ayers S. R. 2001, An Alternative System for the Structural Characterisation of Unidirectional Fibre Reinforced Laminae in Civil and Structural Engineering. In: E. Pereloma & K. Raviprasad (eds), Proceedings of Engineering Materials 2001 Conference: 149-154. North Melbourne: Institute of Materials Engineering Australasia Ltd.

[3]   BS 4994. 1987, British Standard Specifications for Design and Construction of Vessels and Tanks in Reinforced Plastics. London: British Standards Institute.

[4]   ACI Committee 440-F. 1999, Guidelines for the Selection, Design and Installation of Fibre Reinforced Polymer (FRP) Systems for External Strengthening of Concrete Structures. Farmington Hills: American Concrete Institute International.

6.5 Numerical Investigation on the Structural Performance of PFRP Frame Connections

A Mosallam,
California State University, Fullerton, California, USA

ABSTRACT

This paper presents the results of a comprehensive numerical investigation program on structural behavior of semi-rigid composite connections. Several finite element codes were used in this program to confirm the validity of the different models developed in this study. Both linear and non-linear analyses were performed. The FE model included the effect of the flexibility of beam-to-column connections as well as the partial restrain of the web/flange junction of open-web pultruded fiber reinforced polymer (PFRP) profiles. All models were compared to full-scale experimental results from a prior investigation. This paper presents a part of the study using NASTRAN FE code.

INTRODUCTION:

Understanding the behavior of frame connections for pultruded fiber reinforced polymer (PFRP) structures is an essential key to satisfying both the safety and efficiency requirements of such structures. This issue is particularly important when designing load-bearing pultruded composite structures such as bridges and building skeletons.

The efficient connection design must produce a joint that is safe, economical and practical (i.e., easily built at both shop and site). For other construction materials such as steel, varieties of structurally sound connection details are available. Until recently, most available connection details were duplications of steel details [1]. The two major obstacles limiting the use of pultruded structural composites in the construction industry are: i) Lack of information on both short- and long-term performance of PFRP structures, and ii) Lack of design standards and acceptance by building codes. Lately, two major initiatives have been accomplished in Europe and United States. The development of the EUROCOMP provided a great design tool for the structural engineers. In this code, a chapter was devoted to design and analysis of composite joints (Chapter 5). Recently, a design guide for FRP composite connections has been developed by the author under a contract with the American Society of Civil Engineers (ASCE), and will be available to public in 2002. The ten-chapter manual presents a comprehensive coverage of all important aspects related to the design and analysis of composite joints and frame connections.

ACIC 2002, Thomas Telford, London, 2002

Chapter 9 of the ASCE manual is devoted to numerical techniques for designing composite joints and framing connections.

MODELING OF PFRP COMPOSITES

In order to build a finite element model for any composite element, material properties of the fibers and matrix material as well as lay-up must be well defined. For commercially produced PFRP materials, the first two pieces of information are defined. However, it is very difficult to precisely identify the material layup due to the irregularity of the fiber positions along the beam. This may be attributed to two factors; the nature of the pultrusion process and the quality control of the available commercially produced PFRP sections. For this reason, the available mechanical properties of the materials are only provided as effective moduli which are determined experimentally. The apparent modulus of this material is given as an average value. The variation of the layup along the cross section of the PFRP test specimens can be detected by the naked eye. This observation indicates the possibility of deviation between the results calculated from the FE model and those obtained from the test. The effect of the irregularity in the fiber positioning is minimum in the case of in-plane loading condition [1]. This is due to the fact that the axial stress distribution is uniform throughout the section; consequently, positioning of the fibers in the section will have little impact on the overall stiffness characteristics of the PFRP section. However, under combined loading conditions, for example, tensile/flexural (out-of-plane), and/or torsional loads, the stress distribution is non-uniform and the inaccurate positioning of the fibers will greatly affect the stress distribution along the section. Consequently, the stiffness will be altered.

As an example, Figure 1 shows an approximation for the material lay-up and fiber architecture of a typical open-web unidirectional PFRP H-profile. The materials data is listed in Table 1. The material information presented in Table 1 was supplied by the manufacturer (Bedford Reinforced Plastics, U.S.A.), while the ply stiffness in terms of engineering constants of laminae (e.g. E_{11}, E_{22}, G_{12}, $\nu_{12.}$) was approximated according to Figure 1.

LINEAR ANALYSIS USING NASTRAN FE CODE
Description of the Finite Element Models:

Typical analysis of thin-walled frame structures assumes a rigid connection between the beam and columns. In addition, the junction between the flanges and the web of an open-web PFRP sections is commonly assumed rigid. Although these assumptions may be applicable to steel connections and open-web sections made of steel, it is not satisfactory for unidirectional PFRP open-web profiles. In order to illustrate the impact of ignoring these inherent characteristics of PFRP connections and thin-walled open-web profiles, on predicting the overall structural behavior of PFRP frame structures, the following models were developed.

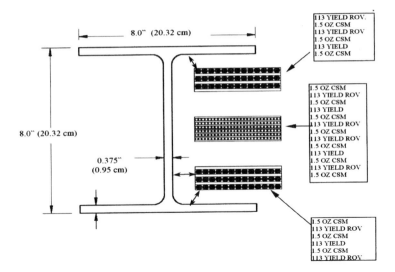

Figure 1. Cross-sectional Dimensions and Fiber Architecture of an 8" X 8" X 3/8" (20.32 cm X 20.32 cm X 0.95 cm) PFRP Profile

Table 1. Engineering Constants for Pultruded H-Profile *[Bedford Reinforced Plastics, Inc. (1999)]*

Mechanical Properties	1.5 oz CSM, psi (GPa)	113 YIELD ROVING, psi (GPa)
E_1	1,716,000 (11.83)	4,320,000(29.79)
E_2	1,716,000(11.83)	959,000(6.61)
E_3*	10(0.00007)	10(0.00007)
G_{12}	605,000(4.17)	371,000(25.58)
G_{23}*	10(0.00007)	10(0.00007)
G_{31}*	10(0.00007)	10(0.00007)
v_{12} (dimensionless)	0.419	0.293
v_{21} (dimensionless)**	0.419	0.065

*   These were assumed engineering constants for the orthotropic materials needed for FEA input. (note that these values are negligible as compared properties at other directions)

** These values were calculated per the reciprocal relation $v_{12}/E_1 = v_{21}/E_2$

Model I -- Rigid Frame Assumptions:
This model was developed using the rigid frame assumption. The NASTRAN composite shell element (LAMINATE) was used in building this model. The flanges and web of the PFRP beam were connected directly through common nodes with 6 degrees of freedom (DOF). This regime was also used to connect both ends of the PFRP to the frame columns. In the experimental program, 2" (5.08 cm) diameter steel rods were used as rollers to apply the loads to the top flange of the frame

girder [2]. To ensure a uniform load distribution and to prevent the possible web crippling of the thin-walled section under the load points, two 6" x 8" x ½" PFRP (15.24 cm x 20.32 cm x 1.27 cm) bearing plates were placed (not bonded) under the steel rollers. This was also incorporated in the finite element model for accurate comparison with the experimental results. In the experimental program, the column was connected to the base via two equal leg PFRP angles 6" x 6" x ½" (15.24 cm x 15.24 cm x 1.27 cm) using FRP threaded rods and nuts (refer to Figure 2).

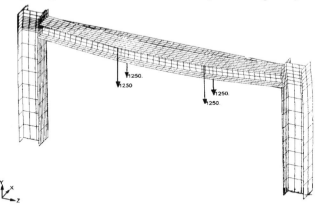

Figure 2.   An Isometric View of FEA Model I (with deformed shape shown)

Model II -- Flexibility of Web/Flange Junction and Semi-Rigid Beam to Column Connection:

In this model, the flexibility of the web/flange junction and the semi-rigid characteristics of the beam-to-column connections of the PFRP frame are considered.

Web/Flange Junction Modeling:
The connection between the flanges and the web of the H-beam was modeled with NASTRAN rigid elements in T_x, T_z, R_x, and R_y degrees of freedom (refer to Figure 3). The flexibility of the web/flange junction was modeled using two spring elements, as follows:
i)   a single DOF spring element with rotational stiffness in the degree of freedom of R_z. The bi-linear behavior of the M/θ relationship of the connections was considered in this model, and
ii) a single DOF extensional spring element in the T_y direction.

Both the axial and rotational stiffness were obtained from full-scale test results (refer to Figures 4 and 5).

Beam-to-Column Connection Modeling:
The beam-to-column connections were modeled using NASTRAN Rigid Elements connected in the degree of freedom of T_x, T_y, R_z, R_y. The rotational flexibility of the connection in the degree of freedom of R_x was introduced indirectly using

equivalent extensional spring stiffness of fourteen spring elements located at the upper and lower flanges of the beam ends (7 on the upper flange and 7 on the lower flange).

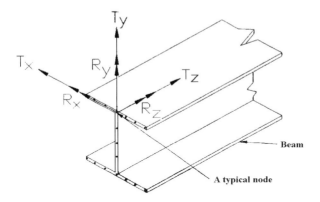

Figure 3. A Sketch Showing the Six DOF used in the FEA Model II
Relative to the Global Coordinate System

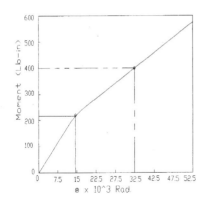

Figure 4. Moment vs. Relative Rotation Curve of TSW Frame Connection

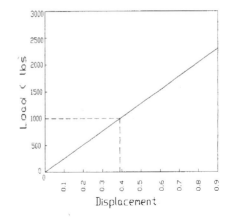

Figure 5. Pulling Load vs. Displacement for Flange /Web Junction.

The axial stiffness of these springs was indirectly obtained from the actual connection rotational stiffness obtained experimentally from a beam-to-column connection full-scale test [2] shown in Figure 6. Loading and frame support conditions were identical to those of Model I.

Model III: Semi-Rigid Beam to Column Connection:
As a part of sensitivity analysis, a third model was developed to measure the effect of the web/flange flexibility of the thin-walled beams on the linear behavior of the

frame structure. This model is identical to Model II, except that a rigid connection was assumed between the web and flanges.

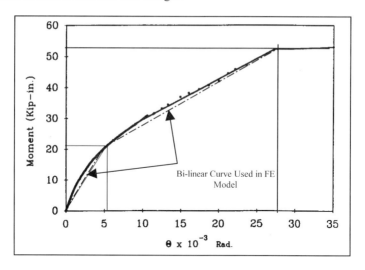

Figure 6. Moment-Rotation Curve for Beam-To-Column Connection Type TSW (Mosallam [2])

COMPARISONS BETWEEN EXPERIMENTAL & NUMERICAL RESULTS
In order to verify the validity of the numerical models, a comparison between full-scale experimental test results [2] with the closed-form solution results was performed.

Figure 7 shows the experimental, theoretical and numerical deflection values at the mid-span of the PFRP frame girder tested by Mosallam [2]. Because the numerical analysis was conducted in the linear range, without including the effects of both geometrical and buckling non-linearities, an expected deviation from the experimental buckling load is noticed in this plot. However, good agreement between the experimental and the numerical deflection results using Model II, up to a total load of 17,500 lb (77.78 kN). Both Model I and III resulted in a relatively stiffer behavior as compared to both Model II and the experimental results in the linear range. The strain distribution along the depth of the mid-span section of the PFRP frame girder is shown in Figure 8. Figure 9 shows that the compression strain on the top surface of middle span on all three models was stiffer than the analytical (both rigid and semi-rigid assumptions) and the experimental results. Because of the semi-rigid behavior of base-column connections, a realistic modeling of this support can be achieved by introducing a rotational spring with rotational stiffness obtained from a full-scale connection M/θ diagram. Experimental data on the rotational stiffness of the base-column connections is unavailable. Therefore, the best approach in modeling these supports is by connecting only the column's webs directly to the PFRP base plate.

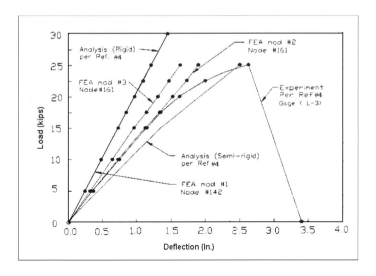

Figure 7. Comparison between Theoretical and experimental Mid-span Frame
Deflection

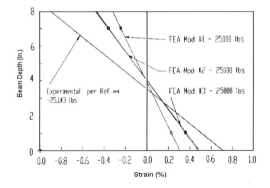

Figure 8. Strain Distribution along the Web Depth at Frame Girder's Mid-Span

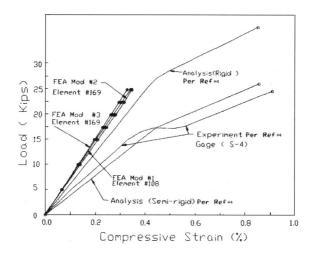

Figure 9. Load versus Girder's Compression Strain   (Mid-span Top Flange)

REFERENCES

[1]   Mosallam, A.S., Abdel Hamid, M. K., and Conway, J. (1993) "Performance of Pultruded FRP Connections under Static and Dynamic Loads," Journal of Reinforced Plastics and Composites, Vol. 13-May, pp. 386-407.

[2]   Mosallam, A.S. "Short and Long-term Behavior of Pultruded Fiber-Reinforced Plastic Frame," Ph.D. Thesis, The Catholic University of America, Washington, D.C. USA.

6.6 Swedish design guidelines for CFRP strengthening of concrete structures

B Täljsten,

Department of Civil and Mining Engineering, Division of Structural Engineering, Luleå University of Technology, Luleå, Sweden and Skanska Teknik AB, SE-16983 Solna, Sweden

INTRODUCTION

All over the world there are structures intended for living and transportation. The structures are of varying quality and function, but they are all ageing and deteriorating over time. Of the structures needed in 20 years from now about 85-90 % of these are already built. Since a concrete structure usually has a very long life, there are quite common that the demands on the structure changes with time. The structures may have to carry larger loads later or fulfil new standards. In extreme cases, a structure may need to be repaired due to an accident. Another reason can be that errors have been made during the design or construction phase so that the structure needs to be strengthened before it can be used. It has also to be remembered that rehabilitation and strengthening of existing concrete structures has become increasingly in focus during the last decade. Some of these structures will need to be replaced since they are in such a bad condition. Nevertheless, there exist many different ways to strengthen an existing concrete structure, such as sprayed concrete, different types of concrete overlays, pre-tensioned cables placed on the outside of the structure, just to mention a few. A strengthen method that was used quite extensively during the mid 70-ties is steel Plate Bonding, this method has gained renaissance the last decade, but now as FRP (Fibre Reinforced Polymers) Plate Bonding. This technique may be defined as one which a composite plate or sheet of relatively small thickness is bonded with an epoxy adhesive to in most cases a concrete structure to improve its structural behaviour and strength. The sheets or plates do not require much space and give a composite action between the adherents. Extensive investigations shows that the method is a very effective and considerably strengthening effect can be achieved. Nevertheless, if the method shall be successfully used it is of utmost importance that a proper design forms the base for the strengthening work to be carried out. Therefore, design guidelines are of utmost importance. In Sweden this was first made for steel Plate Bonding in the end of the 80-ties and during the end of the 90-ties design guidelines for FRP Plate Bonding was written and incorporated in the Swedish Bridge Code: BRO 94. This paper presents a short summary of the existing Swedish guideline for FRP Plate Bonding.

ACIC 2002, Thomas Telford, London, 2002

STRENGTHENING PHILOSOPHY
Our opinion is that a structure shall only be strengthened if it is absolutely necessary. The cheapest and easiest way to do this can be an administrative upgrading where refined calculation methods are used in connection with exact material parameters to show that the existing structure has a higher load-carrying capacity than what has earlier been assumed. However, if it is found that a structure needs to be strengthened and that FRP can be the solution then a number of steps to verify this shall be taken. First of all shall the existing structure be closely investigated. Not only shall existing documentation be studied also material properties and loading history shall be investigated. If then, it is shown that FRP Plate Bonding can be used an accurate design methodology shall be followed. In this paper a summary how to calculate for strengthening is purposed.

SAFETY FACTORS
In normal design of a structure exists unsafeness of different nature. These can be divided in unsafeness due to, the size and frequency of the loads, the properties of the materials used, deviation in structure dimensions and form in relation to the nominal figures and tolerances that is given on the drawings and used design models. To compensate for these uncertainties factors of safety are used. It is important that these factors give so correct description of the reality that is possible. It must be possible to compare different materials so that competition can be formed from the same base [1]. The dimensioning material values can be decided with the partial coefficient method:

$$f_d = \frac{f_k}{\eta \gamma_m \gamma_n} \tag{1}$$

where f_k is the characteristic value used on the material, lower 5 % fractile, γ_m is the partial factor for material properties in which the random distribution is considered. η consider the systematic difference between a test specimens properties and the structures and finally γ_n is related to the class of safety. In Sweden we have three safety classes, class 1 to 3. In design for FRP Plate Bonding the partial factor γ_m is different compared to steel of concrete. In the Swedish guideline this factor is used to compensate for uncertainties due to durability issues, manufacturing process of the FRP material, method to carry out the strengthening work. Also uncertainties due to short or long time is considered. In addition, structures that is considered to be strengthened with FRPs are divided into four different environmental classes.

DESIGN FOR STRENGTHENING IN BENDING
Strengthening for bending has been the commonest way to strengthen concrete structures with FRP´s. CFRP (Carbond Fibre Composite Polymer) has dominated as well as the use of laminate in comparison to fabrics. The design process is quite straightforward and is based to great extent on reinforced concrete design with special considerations to FRP Plate Bonding.

In design with FRP Plate Bonding in bending, [2] the following assumptions are made; Bernoulli's hypothesis is valid – plane cross sections remain plain after deformation, which means that the strain is linearly distributed over the cross section and implies complete composite action between the materials, cracked concrete has no tensile strength and the FRP used is linear elastic up to failure. In the design, seven primary failure modes are considered and these are controlled in the design. These failure modes are presented in Figure 1. However, yielding in the tensile reinforcement in the ultimate limit state is assumed. Yielding of the reinforcement in the service ability state is not permitted. Also other failure modes are possible, for example laminate peeling off at a shear crack or peeling off at a section with extensive yielding at the bending reinforcement. However, these failures are considered as secondary failures. In the next section brief presentations of the design equation for strengthen a concrete structure in bending are given.

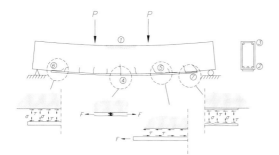

1. Concrete crushing
2. Yielding of reinforcement
3. Yielding of reinforcement
4. Laminate failure
5. Anchorage failure
6. Peeling in concrete
7. Delamination laminate

Figure 1   Failure modes for a concrete beams strengthen for bending

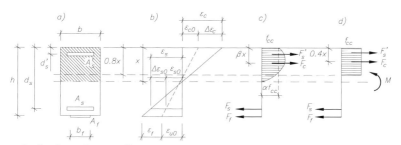

Figure 2   Strain and stress figures

In the design first of all the actual strain distribution over the beam is calculated, the initial calculated strain due to e.g. dead load is the used in design. In Figure 2 the initial strain is denoted ε_{u0}.

When the initial strain distribution is known necessary material for strengthening can be calculated. Here different equations are used with considerations to calculated failure mode, i.e. if the failure will arise in the laminate with or without yielding the compressive reinforcement or if concrete crushing with or without yielding in the compressive reinforcement.

The failure criterion for concrete is $\varepsilon_c < \varepsilon_{cu} = 3.5$ ‰ respectively $\varepsilon_f < 0.6\,\varepsilon_{fu}$ for the composite. It is assumed in the ultimate limit state that the tension reinforcement yields. The design equations is based on equilibrium conditions. In this paper only the simplest form, the case when we have laminate failure with yielding in the compressive reinforcement, is presented, see also reference [2]:

$$M_d = A_s'\sigma_s'\left(\beta x - d_s'\right) + A_s f_y\left(d_s - \beta x\right) + \varepsilon_f E_f A_f\left(h - \beta x\right) \tag{2}$$

A horizontal equilibrium equation gives, see Figure 3:

$$x = \frac{A_s f_y + \varepsilon_f E_f A_f - A_s'\sigma_s'}{\alpha f_{cc} b} \tag{3}$$

Anchorage of the laminate is essential to ensure force transfer between the concrete structure and laminate. The design equation for anchorage is quite simple to use, Equation 4 is based on fracture mechanics and laboratory testing, [3, 4]. It has been shown that there exist a critical anchor length above which longer anchor length not will carry any extra load, but contribute to the safety. The maximum load that can be transferred into the concrete in the anchorage zone is approximately 20 % of the ultimate tensile capacity of the laminate. This give then an anchor length. If the calculation give longer anchor lengths than the critical one, 250 mm, then mechanically anchorage is necessary.

$$\ell_a = \ell_{cr}\frac{0.2 f_{fu}}{\sqrt{f_{ct}E_f w/t_f}} \tag{4}$$

In the design guidelines the anchorage is calculated from the section where the laminate stress is below 20 % of the ultimate tensile stress for the laminate. Special considerations has to be taken to the peeling stresses at the end of the laminate. This can be complicated, [3, 5} but in the guidelines a simplified approach on the safe side is used.

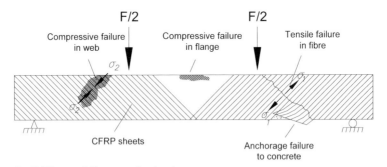

Figure 3   Different failure modes in shear

DESIGN FOR STRENGTHENING IN SHEAR
Strengthen a concrete member for shear is from a theoretical standpoint in many cases more complicated than design for strengthening in bending. The reason for this

is that the shear failure is not as well understood as the bending failure. However, in the guideline a simplified approach is used. For shear strengthening it is quite common that uni- or bidirectional fabrics are used, however, laminates can also be used for this purpose. The base in the design is taken in the truss model and the superposition principle with special consideration of the orthotropic CFRP material used. The shear capacity of a concrete beam with steel stirrups and CFRP is expressed as in Equation 5.

$$V_d = V_c + V_s + V_f \tag{5}$$

Where V_c, V_s and V_f, is the contribution form the concrete, steel and FRP respectively. The following assumptions have been made; Bernoulli's hypothesis is valid – plane cross sections remain plain after deformation, the shear crack arises in main principal stress direction, the FRP is linear elastic up to failure and the load is presupposed to act so that the shear force can be considered even distributed on the structures width perpendicular against the beam span. In figure 3 failure modes typical for shear strengthening are shown. If the structure is over strengthen concrete crushing can arise. Fibre failure in the most stressed fibre is another possible failure mode. Anchorage failure can arise if the sheets or laminates are not anchorage properly. In this paper a shorten version of a comprehensive derivation for concrete beams strengthen with CFRP in shear is presented. For the full derivation see references [6-8]. The design equation for shear can be written:

$$V_f = 2t_f \varepsilon_f E_f 0.9d(1 + \cot \beta) \sin^2 \beta \cos^2 \theta \tag{6}$$

where, σ_f is the stress and ε_f is the strain in the fibre direction. E_f is the Young's modulus in the fibre direction, $s = b_f/\sin\beta$, is the distance between adjacent sheets if the whole side is covered and t_f is the thickness of the fibre or laminate by definition. There exist a considerable difference between the contribution from steel stirrups and CFRP sheets or laminates to the shear capacity. The reason for this is that in ultimate limit design the steel stirrups is assumed to yield. However, CFRP, does not have a yield limit and is linear elastic up to failure. If a rectangular beam is considered, the maximum shear stress is at the centre of the beam. The most stressed steel stirrup is in the mid section of the beam. When a crack is formed this stirrup yields at a certain strain level, if the load increases the neighbouring stirrup start to yield and so on. At a defined load level all stirrups have yielded and the condition stipulated in V_s is fulfilled. However, this is not possible for CFRP materials and we will have an increasing strain the CFRP strip up to failure, see Figure 4.

It can be noticed in Figure 4 that all stirrups reach the yield limit, S1 to S5 in the curve, but for the carbon fibre sheet a uneven strain distribution can be noticed where C3 is the sheet that has the highest strain (stress). However, it can also be noticed that the contribution from the CFRP sheets can be considerable if fibre failure is reached. In design this can be difficult to taken care of and a simplified approach is necessary where a assumed strain distribution is considered. It is important to anchorage the sheets or laminates in the compressive zone. This is often taken care of mechanically, however, if possible it is suggested that a closed "stirrup" is formed by the composite.

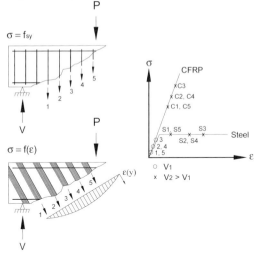

Figure 4 Principal for shear capacity contribution for steel and CFRP

DESIGN FOR STRENGTHENING IN TORSION

Strengthen a concrete structure for torsion is quite unusual. However, tests [9] have shown that it is fully possible to strengthen concrete members for torsion with FRP. The torsion cracks are formed by the same mechanism as for shear. However, when it comes to torsion it is important to enclose the structure with the strengthening system used. It should also be mentioned that the theory for torsion is not as well founded as the theory for bending and shear. The same assumptions as for shear is considered also the failure modes are more or less the same as for shear failure of a FRP strengthen concrete member. In Figure 5 the a figure is shown that forms the base for the derivation of design equations for shear, which then can be written as:

$$\frac{T}{bh} = t_f \varepsilon_f E_f \left(1 + \cot\beta\right) \sin^2\beta \cos^2\theta \tag{7}$$

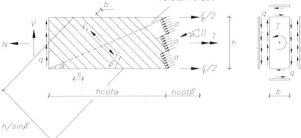

Figure 5 Principal for torsion capacity for FRP

DESIGN WITH CONSIDERATION TO FATIGUE
In the design guidelines is it suggested that the capacity against fatigue failure in the composite is sufficient if:

$$\varepsilon_f \leq 0.7\varepsilon_{fu} \tag{8}$$

However, in general the CFRP has a very good fatigue resistance and it is most likely that other parts of the concrete structure will fail in fatigue before the FRP used.

EXECUTION OF THE STRENGTHENING WORK
If the investigation shows that Plate Bonding is the most suitable method to solve the problem mainly four steps has to be carried out. First the aggregates in the concrete must be uncovered, the surface must be dust free before the strengthening moment starts. The commonest way to uncover the aggregates is by sand blasting, but also water blasting or grinding can be used. Pressurised air or a simple vacuum cleaner removes the dust. The next moment in the strengthening process is to apply a primer. The function of the primer is to enhance the bonding for the adhesive. After the primer has hardened the first layer of resin is applied, thereafter can FRP-laminates or sheets be bonded to the surface of the structure. If sheets are used the process with resin and carbon fibre wrap can be repeated until enough numbers of layers have been applied. Finally a finishing layer of plaster, polymer concrete, paint, shotcrete can be applied upon the reinforcing material for aesthetic look or for fire protection It is of utmost important that the surfaces that shall be bonded are dust free and free from contamination such as grease or oil. Different adhesive suppliers can have different demands on the bonding environment but in general shall the temperature on the concrete surface exceed 10 °C and it is recommendable that the working temperature is 3 °C over the existing dew point. The relative humidity in the air be shall be below 80 %. The tensile strength of the concrete surface shall be over 1.5 MPa. Allowed unevenness on the concrete surface depend on strengthening system used and should be given by the manufacturer. It is also important that the strengthening work is carried out with skilled workers with experience from these type of works.

SUMMARY AND CONCLUSIONS
In this paper a brief summary of the Swedish design guidelines for concrete structures strengthen with FRP is presented. The design guidelines has now been in use for almost three years and an increasing use of FRP´s for strengthening due to the guidelines has been noticed. It is important to compile guidelines since without their existents the consultants will fall back to old proven strengthen methods and FRP´s will mostly been used in special cases. The summary of the guidelines presented here covers strengthening for bending, shear as well as for torsion.

ACKNOWLEDGEMENT
The author acknowledge the financial support to carry out this work provided by Skanska AB, The Swedish Road Authorities, The Swedish Railroad Authorities and

SBUF (The Development Fund of the Swedish Construction Industry) and BPE®
Systems AB.

REFERENCES

[1] Nilsson M., 1998, 30 ton på Malmbanan, Teknisk Rapport, Luleå Tekniska Universitet, Avdelningen för Konstruktionsteknik, Institutionen för Väg- och Vattenbyggnad, 1998, ISSN 1402-1528 (In Swedish)

[2] Täljsten B., 2000, Förstärkning av befintliga betongkonstruktioner med kolfiberväv eller kolfiberlaminat, Dimensionering, material och utförande, Teknisk Rapport, Institutionen för Väg- och vattenbyggnad, Avdelningen för Konstruktionsteknik, 2000:16, ISSN 1402 – 1536, 2000, p 133 (In Swedish).

[3] Täljsten B., 1994, Plate Bonding, Strengthening of Existing Concrete Structurers with Epoxy Bonded Plates of Steel or Fibre reinforced Plastics, Doctoral Thesis 1994:152D, Div. of Structural Engineering, Luleå University of Technology, ISSN 0348 – 8373, p 308.

[4] Täljsten B., 1996, Strengthening of concrete prisms using the plate-bonding technique, International Journal of Fracture 82: 253-266, 1996, 1996 Kluwer Academic Publishers, Printed in the Netherlands.

[5] Täljsten B., 1997, Defining anchor lengths of steel and CFRP plates bonded to concrete, Int. J. Adhesion and Adhesives Volume 17, Number 4, 1997, pp 319-327.

[6] Täljsten B., 1997, Shear Strengthening with CFRP sheets, Unpublished material Luleå University of Technology.

[7] Täljsten B., 2002, CFRP strengthening of concrete beams in shear – a test and theory, paper to be published.

[8] Täljsten, B. (2002): "Strengthening of existing concrete structures with externally bonded Fibre Reinforced Polymers – design and execution". Technical report. Luleå University of Technology, Division of structural engineering. Under printing.

[9] Täljsten B., 1998, Förstärkning av betongkonstruktioner med stålplåt och avancerade kompositmaterial utsatta för vridning, Forskningsrapport, Luleå Tekniska Universitet, Avdelningen för konstruktionsteknik, Institutionen för väg och vattenbyggnad, 1998:01, ISSN 1402-1528 p. 56 (In Swedish).

6.7 Characterisation and assessment of manufacturing distortions in buckling-sensitive composite cylinders

M K Chryssanthopoulos,
School of Engineering, University of Surrey, UK

INTRODUCTION

Glass fibre-reinforced plastic (GFRP) shells could be an attractive structural form for a number of applications in construction due to their good strength-to-weight ratio, inherent corrosion resistance, and relatively low cost compared to other composite systems. However, their efficient use is restricted by the limited availability of design criteria and by fairly scant test data on which any such criteria can be validated.

Research on shell buckling has been carried out over many years, with emphasis on its manifestation in isotropic materials. In parallel with theoretical research on stability [1], early design investigations provided estimates of buckling strength using formulae for linear critical loads together with test-based 'knock down' factors, e.g. [2]. Physical tests have always played a central role in shell buckling but in the last thirty years, and in response to the theoretical results, they are specifically geared towards validation of analytical and numerical procedures. This entails accurate measurement of shell specimen characteristics prior to load application, particularly initial geometric imperfections [3], which have long been singled out as the primary cause of discrepancy between actual and predicted strength values.

In the case of composite cylinders, some buckling studies have been carried on very thin carbon fibre-epoxy systems used in the aerospace industry [4-6]. In comparison, fewer GFRP shells have been tested [7,8] and reported results are not sufficiently detailed for analytical or numerical comparisons to be undertaken with confidence.

This paper gives a brief account of a series of tests on GFRP cylinders, with radius-to-thickness ratio (R/t) between 70 and 110, under concentric and eccentric compression. Two types of laminae, consisting of E-glass woven rovings within polyester resin, and four laminate configurations were examined. Specimen dimensions were selected so that failure occurs by local shell (rather than column) buckling, and so that theoretically (i.e. for perfect geometry and ideal boundary conditions) the initiation of buckling precedes material failure. Detailed records of thickness variations and initial geometric imperfections were taken, in order for the data to be utilised within numerical simulation studies. The tests and the associated imperfection measurements have been described in detail elsewhere [9-11].

ACIC 2002, Thomas Telford, London, 2002

Simulations of shell buckling tests may be pursued through formulations which model the non-linear effects arising from geometric distortions and other irregularities [12], or through general finite element (FE) programs combined with specialised software to account for specific composite material failure phenomena [13]. However, there are constraints to the patterns and forms of distortions that can be modelled analytically, and the development and validation of specialised software is not always practical, especially in the context of civil engineering structures. Thus, if composite structures are to gain wider acceptance in structural applications, it is desirable to assess the possibility of undertaking analysis and design tasks using general purpose FE programs. This was attempted in this study through linear (eigenvalue) and geometrically non-linear analyses [14,15], the results from which are summarised herein. Good correlation between numerical and experimental results was found, except where local buckling was influenced by premature material failure. Relevant design and assessment ratios, including 'knock down' factors, were determined in order to contribute to the development of design guidelines for composite shells.

SUMMARY OF BUCKLING TESTS

A total of 22 cylinders were tested under concentric or eccentric compression within a purpose-built displacement control rig. In addition to load, end shortening and strain measurements, an automated laser scanning system, operating inside the cylindrical specimens, was used for recording the initial geometric imperfections, as well as the progressive change in deformations of the inner wall during loading [16].

Two GRP systems were employed, namely 'Rovimat 1200' and 'DF 1400'. In both types, the fibre mat is E-glass woven roving and the resin is isophthalic polyester. In the case of the Rovimat 1200 material, the roving is sprinkled on one surface with chopped glass fibres to enhance bonding of plies. This chopped mat is not present in the 'DF 1400' roving. Mechanical properties of flat two-ply $0°/0°$ characterisation specimens are given in Table 1, where directions '1' and '2' refer to the weft and the warp directions of the roving [17]. The nominal thickness of the two-ply specimens was 2.8mm for Rovimat 1200 and 3.5mm for DF 1400.

Table 1. Mechanical properties obtained from flat two-ply $(0°/0°)$ specimens

Property	Direction	Rovimat 1200	DF 1400
Elastic Modulus (GPa)	E_1	14.7	16.4
	E_2	13.2	12.7
Poisson's Ratio	ν_{12}	0.26	0.20
Shear Modulus (GPa)	G_{12}	3.5	3.1
Tensile Strength (MPa)	σ_1	199.6	192.4
	σ_2	178.5	163.2

The shell specimens were produced using hand lay-up around a cylindrical mandrel. Nominal dimensions are shown in Figure 1(a). Figure 2 depicts the lay-up process during which overlaps of 50 mm nominal width were introduced in a staggered pattern.

Two types of distortion were primarily present in the cylinders as a result of the lay-up process, namely thickness variations and geometric imperfections. The former are caused by overlaps or are due to excess resin being placed. On the other hand, geometric imperfections are related to wrapping around the mandrel. Both forms of distortion may have a significant influence on buckling behaviour, hence accurate mapping of their amplitude was necessary. In addition, discrepancies from nominal lay-up angles are expected, though random checking did not reveal any significant deviations in the outer layers.

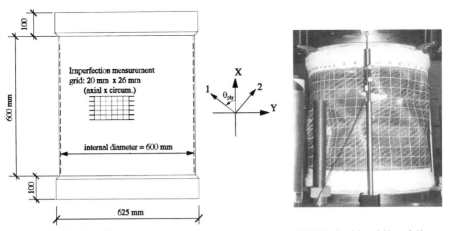

Figure 1: (a) Specimen geometry   (b) Typical buckling failure

The typical coefficient of variation of the thickness within a normal or an overlap zone varied between 5% and 10%. Measured thicknesses in normal zones were found to be up to 20% higher than the nominal values [9-11]. The average measured thickness (excluding overlaps) for a number of specimens is given in Table 3.

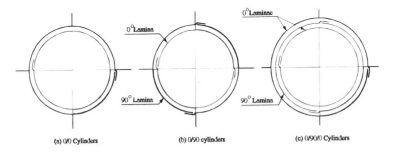

Figure 2: Specimen cross-section indicating overlaps for different laminates
Measurement of geometric imperfections was carried out using the laser scanning system operating on a grid consisting of 31 axial and 72 circumferential stations.

The 'raw' imperfection data were processed using a 'best fit' procedure to account for possible misalignments between the axis of the specimen and the measuring frame. Following the 'best fit' procedure, the resulting imperfections were analysed using 2D harmonic analysis [3] to produce a set of Fourier coefficients

$$w_0(x,\theta) = \sum_{m=1}^{M}\sum_{n=0}^{N}\xi_{mn}\sin\frac{m\pi x}{L}\sin(n\theta + \phi_{mn})$$

where $w_0(x,\theta)$ is the initial imperfection function of the cylinder at any point ($0 \le x \le L$ and $0 \le \theta \le 2\pi$), m is the number of axial half waves, n is the number of circumferential waves, ξ_{mn} and θ_{mn} are the imperfection amplitude and phase angle associated with mode (m,n) respectively. This technique allows the measured imperfections to be readily fed into analytical and numerical models. Furthermore, it facilitates comparative studies on the effect of the manufacturing process on the magnitude and spatial distribution of initial imperfections [18].

Table 2. Dominant imperfection modes for Rovimat 1200 models

Specimen Reference	Fourier coefficients Amplitudes and corresponding modes ξ_{mn} (mm)			(m,n)		
RVM01	1.13	0.52	0.37	(1,2)	(2,2)	(3,2)
RVM02	0.47	0.26	0.20	(1,2)	(2,2)	(3,2)
RVM03	0.56	0.38	0.27	(1,2)	(2,2)	(1,4)
RVM04	0.47	0.34	0.22	(2,2)	(1,2)	(1,4)
RVM05	0.60	0.44	0.31	(1,2)	(2,2)	(1,3)
RVM05A	0.58	0.27	0.25	(1,2)	(1,3)	(2,2)
RVM05B	0.55	0.41	0.22	(1,2)	(2,2)	(3,2)
RVM06	0.44	0.22	0.19	(1,2)	(2,2)	(1,4)
RVM07	0.29	0.23	0.12	(2,2)	(1,2)	(1,3)
RVM08	0.62	0.31	0.22	(1,2)	(2,2)	(3,2)
RVM09	0.35	0.14	0.13	(1,2)	(3,2)	(1,4)
RVM10	0.44	0.37	0.36	(1,2)	(2,2)	(1,3)
RVM11	0.63	0.27	0.24	(1,2)	(3,2)	(1,4)
RVM12	1.63	0.56	0.18	(1,2)	(3,2)	(2,3)
RVM13	0.25	0.25	0.25	(1,3)	(1,5)	(2,2)

For the two-ply models, the maximum wall imperfection normalised by the nominal thickness was found to have a mean of 0.55 with a coefficient of variation of 0.25. As these imperfections are largely related to the mandrel, the average imperfection-to-thickness value reduces in the three-ply models to about 0.27. The results of the Fourier analysis indicated that dominant imperfection wavelengths along the axial and circumferential direction were barrelling ($m=1$) and ovalisation ($n=2$) respectively (Table 2). However, as the experiments revealed non-axisymmetric buckling modes with shorter wavelengths (e.g. $n=4$ or $n=6$) and the full matrix of Fourier coefficients (MxN) indicated the presence of higher order modes (albeit with decreasing amplitude), the full set of coefficients was used (rather than just the dominant ones) to generate the imperfection surface for the FE models.

The buckling tests were carried out under displacement control. Typical load-displacement plots for a set of concentrically loaded Rovimat 1200 models are shown in Figure 3. All two-ply models exhibited a load-end shortening response similar to that shown in Figure 3, with a degree of post-buckling stiffness being evident. On the other hand, failure of the three-ply models resulted in total loss of axial capacity and stiffness. Moreover, whereas failure of the two-ply models was dominated by elastic buckling, see Figure 1(b), failure of the three-ply models was associated with through thickness fibre cracking, uniformly spread around the circumference. As a qualitative assessment of residual strength, it should be noted that further buckling cycles were possible for the two-ply specimens, albeit with lower peaks being reached, whereas the three-ply models could not be reloaded beyond the first cycle.

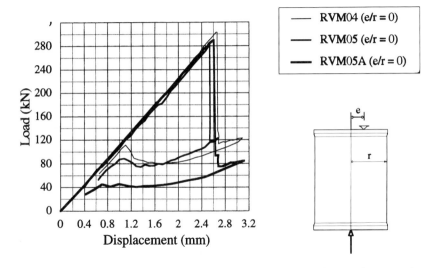

Figure 3: Load-end shortening relationships for concentric 0/90 Rovimat specimens

Significant differences were observed between the behaviour of Rovimat and DF cylinders. In relative terms, the buckling strength of the two-ply Rovimat cylinders was higher than their DF counterparts, despite the higher elastic moduli and thickness of the latter. The response of the DF specimens was comparatively brittle with some resin cracking (prior to buckling) and fibre fracture from the peak load onwards. The presence of a chopped fibre mat in the Rovimat laminates may have enhanced the overall ductility of the cylinders, hence leading to a higher buckling capacity.

NUMERICAL SIMULATION

Numerical simulations were carried out using the general purpose FE program ABAQUS [20]. The 9-noded doubly curved shell element, which is suitable for laminated composite shells and accounts for the orthotropic properties of each ply, was selected for modelling the shell wall. This element offers a reasonably efficient alternative shell buckling analysis provided that transverse shear effects are small, which is the case here since the radius-to-thickness ratio of the cylinders is greater than 70. The element formulation accounts for geometric non-linearity but material behaviour is confined to the linear elastic domain, in line with CLT (Classical Laminate Theory) applied to laminated composite structures.

For all analyses, the FE model consisted of the complete circumference of the cylinder. This is because the spatially varying thickness and the measured geometric imperfections were introduced in the models. Clearly, neither of the two effects complies with axisymmetry. Variation in top boundary conditions, including in some cases the need to provide for eccentric loading, also pointed towards modelling of the complete circumference. Lengthwise, the central 600 mm length of the cylinder was considered after confirming (via eigenvalue analysis) that the significantly thicker end parts (100mm at either end) do not participate in the buckling response. Based on convergence studies, a mesh consisting of 24 and 72 elements in the axial and circumferential directions, respectively, was adopted. This choice was influenced by the wavelength of the linear buckling mode [19], and took into account the localised behaviour that might be expected under eccentric compression or in the presence of thickness variations.

An upper bound to the buckling strength was first investigated through linear eigenvalue analysis, before any geometric imperfections were introduced. The numerical results summarised below are for the case of a zoned thickness approach in which the cylinder is divided into normal and overlap sectors, and a uniform thickness representing the average measured value is adopted for each of these sectors. It is worth pointing out that the influence of various thickness modelling approaches - including uniform, zoned (uniform for normal or overlap zones) and spatially varying thickness (based on nodal variations) - on the linear buckling strength was found to be within 10% [14,15]. Inclusion of detailed thickness variations (as opposed to average zone values) resulted in lower buckling strength in most cases, the reason being that buckling may be triggered in relative thinner zones, providing these are large enough to accommodate the relevant buckling wavelengths.

The buckling modes for the concentrically loaded model RVM01 are shown in Figure 4(a) for the uniform thickness approach and in Fig 4(b) using the zoned approach. Note that, although the wavelengths do not change, the mode becomes localised. The effect of eccentric loading was also to produce mode localisation, but this time confined in the region of higher compression. In the 45/-45 specimens, for which the mode assuming uniform thickness is axi-symmetric, the effect of thickness variations is to confine the mode within a zone (thus changing the circumferential wavelength). This implies that, whereas an ideal cylinder would be most sensitive to axi-symmetric imperfections, the real shell is also adversely affected by imperfections with a circumferential half wavelength equal to the zone width. This further justifies the use of the full imperfection surface in validating FE models, rather using than a sub-set of dominant or structurally-relevant modes.

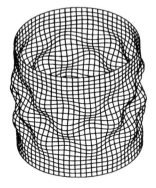 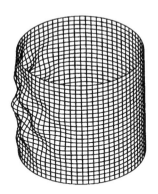

Figure 4: Eigenmode for RVM01
 (a) uniform thickness   (b) zoned thickness variation

The ratio obtained by dividing the experimental buckling load (P_{exp}) with the critical load (P_{cr}) from linear analysis gives the traditional 'knock-down' factor. Using the test results and the FE eigenvalues from a zoned thickness model (i.e. uniform thickness within each normal or overlap zone), the range of 'knock-down' factors estimated for all twenty two specimens in the two test series is presented in Table 3. Note that direction 0° indicates that the ply is laid with the weft parallel to the axis of the cylinder. Other ply orientations may be identified through the axis system shown in Figure 1. Several remarks can be made with respect to the calculated values:

(i)   The Rovimat 1200 cylinders are associated with higher 'knock-down' factors than the corresponding DF 1400 cylinders.
(ii)   As might be expected from theoretical considerations, 'knock-down' factors under eccentric compression are higher than for concentric compression.
(iii)   'Knock-down' factors for cross-ply laminates show less variation than those for angle-ply laminates.

On the basis of these results, it is proposed that, in a traditional design procedure, a knock-down factor of 0.5-0.6 may be used for these GFRP shells which exhibit R/t ratios between 70 and 120, and L/R ratio in the intermediate range (i.e. where local buckling is dominant). These factors are generally higher (i.e. lead to more economical design) than those derived for isotropic shells. This indicates that GFRP cylinders are less sensitive to initial imperfections than their isotropic counterparts, especially since the manufacturing process used (hand lay-up) has produced relatively severe imperfection amplitudes as well as thickness (hence, stiffness) variations.

Table 3. Summary of numerical (FE) study

Specimen	Laminate	R/t (nom)	e/R	$t_{measured}$ (mm)	P_{exp}/P_{nl}	P_{nl}/P_{cr}	P_{exp}/P_{cr}
RVM01	0/0	108	0.0	3.35	1.0	0.68	0.68
RVM02	0/0	108	0.10	3.46	0.90	0.76	0.68
RVM03	0/0	108	0.25	3.42	0.99	0.76	0.75
RVM05A	0/90	108	0.0	3.20	0.86	0.78	0.67
RVM05B	0/90	108	0.25	3.26	0.98	0.78	0.77
RVM07	0/90/0	72	0.0	4.78	0.90	0.81	0.72
RVM09	45/-45 VO	108	0.0	3.26	0.74	0.89	0.66
RVM11	45/-45 VO	108	0.25	3.26	0.90	0.88	0.79
RVM12	45/-45 IO	108	0.0	3.18	0.72	0.77	0.56
DF01	0/0	86	0.0	3.65	0.62	0.87	0.54
DF02	0/0	86	0.0	3.72	0.70	0.76	0.53
DF03	0/90	86	0.0	3.94	0.63	0.86	0.54
DF04	0/90	86	0.0	4.00	0.62	0.87	0.54

Geometrically non-linear analyses were also undertaken in order to assess the capability of general purpose FE programs in simulating the buckling tests. In addition to thickness variations, experimentally measured geometric imperfections were introduced, using the procedure outlined in the preceding section. From these runs, the ratio (P_{exp}/P_{nl}) is obtained which is relevant to model validation, since a value of unity represents perfect correlation between tests and FE modelling. The relatively lower ratios for the DF1400 models are attributed to the relative brittleness of this material system, which as mentioned earlier may, in turn, be related to the absence of the random chopped mat. Inclusion of pre-buckling material degradation effects in the analysis of the DF 1400 cylinders would necessitate employing micro-mechanical models in conjunction with appropriate material failure criteria, all integrated within a geometrically non-linear finite element analysis.

Finally, the ratio (P_{nl}/P_{cr}) demonstrates that FE-based reduction factors are generally higher than the traditional test-based 'knock-down' factors. This is important for design, since it implies that if a numerical procedure were to be used in deriving 'knock-down' factors, it would be prudent to assume higher levels of geometric imperfections than those actually encountered or expected in practice, in order to account for manufacturing effects which are not explicitly covered in a geometrically non-linear FE analysis. The imperfections introduced in a numerical model should thus be treated as 'effective' rather than actual, and an appropriate magnification factor could be developed for different composite material systems and manufacturing methods. In other words, a numerically estimated reduction factor related to actual geometric imperfection levels should be multiplied by a second factor (also lower than unity, and accounting for distortions/imperfections) before it can be considered as equivalent to a traditional test-based 'knock-down' factor.

CONCLUSIONS

An overview of an experimental and numerical investigation on the buckling behaviour of laminated composite cylinders was presented. As with shells made from isotropic materials, it was shown that linear eigenvalue analysis significantly overestimates real buckling loads. The importance of incorporating the actual wall thickness in the numerical analysis was highlighted even if the effect of thickness variations in linear buckling loads is small compared to the effect of non-linearities arising from boundary conditions and imperfections. The composite GFP cylindrical shells examined herein appear to be less sensitive to the effects of initial geometric imperfections than their isotropic counterparts. Nevertheless, including geometric imperfections in a geometrically non-linear analysis did improve the comparison between tests and FE results, and is considered essential for the derivation of numerical reduction factors, which in turn can lead to simple 'knock down' factors.

For the cylindrical shells considered in this study (with $R/t > 70$), design against local buckling can also be achieved through the direct use of geometrically non-linear analysis with due consideration of initial imperfections, thickness irregularities and load eccentricities, provided that the material has sufficient local ductility to avoid brittle failure modes. Otherwise, for accurate prediction of the buckling strength, geometrically non-linear FE analysis would need to be coupled with non-linear material and local damage models.

ACKNOWLEDGEMENTS

The author would like to thank the partners in the Brite-Euram Project 7550 for fruitful discussions but in particular his collaborators Dr A Elghazouli, Dr I Esong, Mr R Millward, Dr A Spagnoli and Prof C Poggi for their contributions to the work summarised herein and Mr H Wickens for his invaluable help with presentation aspects.

REFERENCES

[1]   Budiansky B. (Ed), Buckling of Structures, Springer-Verlag, 1976.

[2]   Baker E.H., Cappelli A.P., Kovalevsky L., Rish F.L. & Verette R.M., Shell Analysis Manual, NASA Contractor Report, NASA-CR-912, April 1968.

[3]   Arbocz J. & Babcock C.D., Prediction of Buckling Loads Based on Experimentally Measured Imperfections, in Buckling of Structures, B. Budiansky (ed), Springer-Verlag, 1976, 291-311.

[4]   Tennyson R.C., Buckling of laminated composite cylinders: a review, Composites, January 1975, 17-24.

[5]   Simitses G.J., Shaw D. & Sheinman I., Stability of imperfect laminated cylinders: a comparison between theory and experiment, AIAA Journal, Vol. 23, 1985, 1086-1092.

[6]   Fuchs H.P., Starnes Jr. J.H. & Hyer M.W., Pre-buckling and collapse response of thin-walled composite cylinders subjected to bending loads, in Proc. 9th Int. Conf. Comp. Mat. (ICCM-9), A. Miravete (ed), Woodhead Publishing, 1993, Vol. 1, 410-417.

[7]   Abu-Farsakh G.A.F.R. & Lusher J.K., Buckling of Glass-reinforced plastic cylindrical shells under combined axial compression and external pressure, AIAA Journal, Vol. 23, 1985, 1946-1951.

[8]   Wooton A.J. & Scowen G.D., Axial buckling of reinforced plastics cylinders with surface imperfection, National Engineering Laboratory, NEL Report 687, October 1983.

[9]   Elghazouli A.Y., Chryssanthopoulos M.K. & Esong I.E., Buckling of Woven GFRP Cylinders under Concentric and Eccentric Compression, Composite Structures, 45, 1999, 13-27.

[10] Chryssanthopoulos M.K., Elghazouli A.Y. & Esong I.E., Compression Tests on Anti-symmetric Two-Ply GFRP Cylinders, Compression Tests on Anti-Symmetric Two-Ply GFRP Cylinders, Composites, Part B: Engineering, 30, 1999, 335-350.

[11] Elghazouli, A.Y., Chryssanthopoulos, M.K. & Spagnoli, A., Experimental Response of Glass-Reinforced Plastic Cylinders under Axial Compression, Marine Structures, 11(9), 1998, 347-371.

[12] Li Y.W., Elishakoff I. & Starnes Jr. J.H., Axial Buckling of Composite Cylindrical Shells with Periodic Thickness Variation, Computers and Structures, 56, 1995, 65-75.

[13] Hilburger M.W. & Starnes Jr. J.H., Effects of imperfections on the buckling response of compression-loaded composite shells, AIAA-2000-1387, 2000.

[14] Chryssanthopoulos M.K., Elghazouli A.Y. & Esong I.E., Validation of FE Models for Buckling Analysis of Woven GFRP Shells, Composite Structures, 49, 2000, 355-367.

[15] Spagnoli A., Elghazouli A.Y. and Chryssanthopoulos M.K., Numerical Simulation of Glass-Reinforced Plastic Cylinders under Axial Compression, Marine Structures, 14, 2001, 353-374.

[16] Esong, I.E., Elghazouli A.Y. & Chryssanthopoulos M.K., Measurements Techniques for Buckling-Sensitive Composite Cylinders, 'Strain', 1998, 11-18.

[17] Poggi, C. Characterisation of materials, Brite-Euram Project: DEVILS, Report No. WP04.DR/PM(1), Politecnico di Milano, Italy, 1996.

[18] Chryssanthopoulos M.K. & Poggi C., Stochastic Imperfection Modelling in Shell Buckling Studies, Thin Walled Structures, 23, 1995, 179-200.

[19] Vinson J.R. & Sierakowski R.L., The Behaviour of Structures Composed of Composite Materials, Kluwer Academic Publishers, 1987.

[20] ABAQUS, Theory and User's Manual, Version 5.8 Hibbit, Karlsson and Sorensen, Providence, Rhode Island, 1998.

Session 7 : **All-composite structures and materials issues**

7.1 Experimental and analytical evaluation of all-composite highway bridge deck

M Haroun[1], A Mosallam[2] and F Abdi[3]
[1]University of California, Irvine, California, USA
[2]California State University, Fullerton, California, USA
[3]Alpha Star Corporation, Long Beach, California, USA.

ABSTRACT

This paper presents a summary of the results of both the experimental and the numerical studies conducted on evaluating the structural behavior of all-composite hybrid bridge deck. The carbon/fiberglass reinforced-polymeric deck was designed to replace existing low-profile welded steel gratings that is deteriorating due to high fatigue loadings on the Schuyler Heim Bridge in Long Beach, California. In this study, a total of nine tests, both small- and full-scale, were conducted. The experimental test results indicated that the composite bridge deck has exceeded both the predicted design and ultimate capacities. The span-to-deflection ratio at the mid-span was L/738 based on a span length of 48" (1.22 m). This deflection range is satisfactory for the majority of highway bridges decking systems with other materials. The average safety factor (SF) of the composite deck design was 6. In all tests, the ultimate failure was initiated either by a punching shear under the loading steel plate, or/and by the delamination of the curved portion of the drop sandwich panel. The latter local failure mode is attributed to the high radial stress components generated at the tension side of the deck. In modeling the performance of the composite deck, the GENEA progressive failure analysis numerical code was used to perform virtual testing of the composite decks under both quasi-static and fatigue loading conditions.

INTRODUCTION

For the past few decades, the aerospace industry was the major user of composites in worldwide, and in the U.S.A. Recently, civil engineers realized the potential of these materials in different applications where conventional construction materials had poor performance records. Namely, at seismic and corrosive environments, where the major problem of concrete damages due to the lack of ductility and corrosion of the steel reinforcements is a great challenge to the engineer.

OBJECTIVE AND MOTIVIATION

The majority of composite bridge deck replacements reported earlier were focused on solving the chronic corrosion problem of corroded reinforced concrete bridge decks by taking the advantages of high corrosion performance of polymeric decks. However, the application described herein is one of the first reported cases where corrosion was not the motive behind the use of composites. The problem with the 55-year old Schuyler Heim Steel Bridge (Figure 1), is the continuous local damage of the welded steel gratings of the lift span due to high fatigue and impact loadings resulting from the large truck traffic in and out the terminal island of the Long Beach Harbor. The California Department of Transportation (Caltrans), being the leading governmental organization in the U.S. since early 1990's in promoting the use of composite and high performance materials and systems in both seismic repair and bridge applications, decided to take the advantages of the known high fatigue characteristics of polymer composites to establish an effective remedy to this problem.

Figure 1. The Schuyler Heim Steel Bridge, Long Beach, California, U.S.A.

EXPERIMENTAL PROGRAM
Loading and Instrumentation
The composite deck specimens were instrumented with electronic strain gauges, and some with fiber optical sensors, at several critical locations and at different directions. The vertical deflections were captured using several calibrated Linear Variable Differential Transducers (LVDT's) positioned at several locations along the span(s) and the width of each specimen. In these tests, the pre-load was 25% of HS25 maximum wheel load (5,500 lb/ 24,464 N). At this pre-load level, the load was held constant for few minutes, to monitor any sign of localized damages. The load was then released to zero with the same rate as for the loading ramp (1 kip/min or 4.448 kN/min.). All tests were operated using a load-control regime. The loading magnitudes include 25% (pre-load), 50% (11,000 lb/ 48,928 N), 75% (16,500 lb/ 73,392 N), and up to the failure load corresponding to each specimen.

Composite Deck Construction

The composite deck specimens were constructed as sandwich panels with unidirectional E-glass/polyester 4" X 4" X ¼" (101.6 mm X 101.6 mm X 6.35 mm) pultruded square beams as the core. The face sheets were constructed using 4 plies of unidirectional carbon fibers ($0°/90°/0°/90°$), with two layers of quasi-isotropic ($45°/0°/-45°/90°$) E-glass resin-saturated stitched fabrics as outer skins. The resin used for fabricating the top and bottom face sheets was Derakane 450-311. Due to the severe deflection requirements, a drop sandwich panel at the middle third of the span was used to increase the section modulus, and consequently reduced the deflection at the maximum moment zone. The paper honeycomb core of the drop panels was injected by high-density foam materials to enhance the shear properties. The edges were sealed with a compatible resin to protect the core material from water intrusion and other environmental agents. The top surface of some of the specimens had a ½" (12.7 mm)-thick layer of polymer pavement that contains ¼" (6.35 mm) rock chips.

Small-Scale/Simply-Supported Tests

Test Matrix: Four tests were performed on 1' X 5' X 5 ¼"/ 0.3 m X 1.52 m X 0.13 m (width X length X depth (outside the drop panel zone)) simply supported specimens with and without polymeric pavement layer (refer to Figure 2). Three test protocols were performed on each small-scale specimen:

- Test I: Service load test up 20.53 kips (91.32 kN),
- Test II: 50% of average ultimate load (53.56 kips/238.2 kN), and
- Test III: up to ultimate failure.

No local failures were observed during Test I of each specimen. The first common local failure was observed during Test II and Test III for each small-scale specimen. This local failure occurred at about 40 kips (178 kN) in the form of delamination of the bottom face sheet near the drop panel (refer to Figure 3). This delamination is attributed to the development of high radial stresses at the curved portion of the drop panels. As the load increased, the delamination propagated towards the supports as shown in Figure 3. It should be noted that this local delamination failure did not only occur at the edges, but also was extended along the entire width of the specimen. Test II results indicated that the local delamination failure resulted in a sudden loss in the section modulus (about 30% of the flexural stiffness). Figure 4 shows the load-deflection curves of the small-scale specimens with no pavement polymeric layer. Experimental results for Tests II and I can be found in Mosallam *et al* [1].

Large-Scale/Two-Span Tests

In this phase, three specimens were tested. The specimens had exact materials and lay-up as for the small-scale specimens described earlier. The first specimen, designated herein as Sample (3), had no wearing polymeric pavement layer. The same specimen was retested after it was repaired (designated herein as Sample (3R)). The third specimen tested in this program (Sample (4)) was identical to

Sample (3) except that a polymeric-wearing layer was casted on the deck top surface.   The dimensions and the test setup are shown in Figure 5.   Both conventional electronic strain gauges and fibre optical gauges were used in these tests.

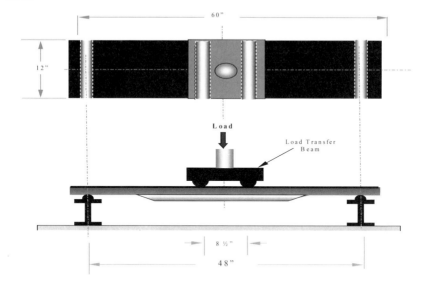

Figure 2. Details and Dimension of the Simply Supported Tests *(1"=25.4 mm)*

Connection Details:
A flexible connection detail between the composite deck and the supporting steel girders, as shown in Figure 6, was selected to accommodate for the thermo-mechanical mismatches of the two materials. In this flexible connection detail, no polymer concrete was used to fill up the hollow cavity around the Nelson™ studs. In addition, the holes were enlarged in the two perpendicular directions to allow for the expected relative movements during the service environment of the bridge. In addition, each connection had oversized steel washers, rubber pads under the steel washers and between the deck and the steel girder top flanges, and the sharp corners of the steel girder top flanges were rounded to minimize the stress concentration at these locations.

Figure 7 presents the load vs. the loaded mid-span deflection for Sample (3). As shown in this figure, the ultimate load was 162.5 kips (722.8 kN) with a maximum mid-span deflection of the loaded span of 1.06" (27 mm).   One important observation is that, at the design load of 25 kips; the mid-span deflection was 0.073" (1.85 mm) which can be translated to L/657 (L = 48" or 1.22 m).

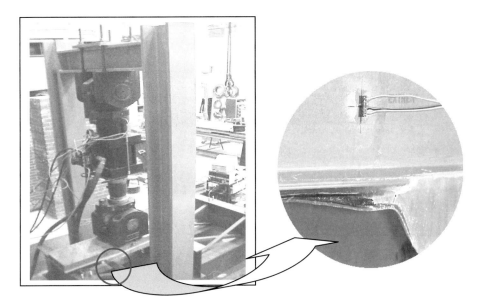

Figure 3. Local Delamination Failure initiated at the Curved Portion of the Drop Panel

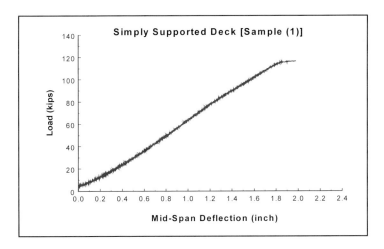

Figure 4. Loading/Unloading Curves for Simply Supported Specimen NPL1

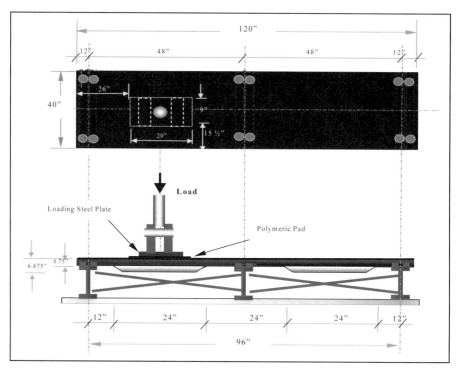

Figure 5. Dimensions and Test Setup of 2-Span Bridge Deck Specimens

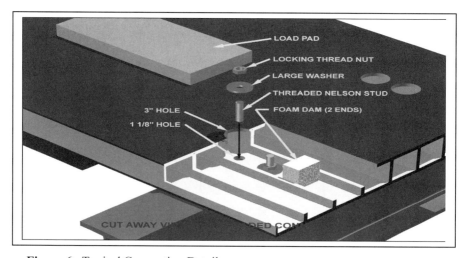

Figure 6. Typical Connection Details

The ultimate failure mode of this specimen was a combination of local failure of the pultruded core beams under the loading area followed by a punching shear due to the stress concentration at the sharp edges of the loading steel plate. As mentioned earlier, the same specimen was repaired and retested. Test results confirmed the success of the repair system not only in restoring the original capacity of the undamaged deck but also in adding about 13% strength gain to the repaired deck as compared to the original undamaged specimen. However, the initial stiffness of the repaired specimen was far less than the undamaged stiffness (about 52% of the linearized initial stiffness of Sample (3) at 25-kip/111-kN load level).   Similar results were obtained for Sample (4) with polymeric pavement layer.   Table 1 summarizes the results of the three large-scale tests.

Table 1. Results Summary of Large Scale Tests

Property	Sample (3)	Sample (3R)	Sample (4)
Span-to-deflection ratio *(at 111 kN- load)*	657	345	738
Deflection at *111 kN, mm*	1.85	3.53	1.65
Deflection at Ultimate Load, *mm*	26.9	44.45	49
Ultimate Load (P_{ult}), *kN*	723	819	644

PROGRESSIVE FAILURE ANALYSIS USING GENOA CODE

The use of progressive failure analysis approach, the uncertainty and variability, that are typically associated with advanced composites, are scrutinized for their contribution to potential premature failures. Unlike conventional finite element analyses, the GENOA life prediction approach allows for optimizing the deck design and allows for a quick assessment of composite material candidates, by performing "virtual tests" that minimize or eliminate the need for expensive and time consuming large-scale tests, and, more importantly simulating a realistic micro-mechanical behavior of the composite deck. This study using GENEO is the first reported investigation using this state-of-the-art code. In this study, virtual tests of the large-scale deck specimens were conducted simultaneously with the large-scale tests. In this scenario, no prior knowledge on the ultimate capacity nor the mode of failures were provided to the operator of the "virtual" test that was conduced in the same time as the "real" laboratory test was performed. The virtual test predicted the ultimate failure load to be 140 kips (622.72 kN), as compared to "real" full-scale laboratory ultimate load capacity test result, for Sample (4), of 144.85 kips (644 kN) as shown in Table (1) with less than 3.5% deviation. The deflected shape of the model (prior to failure) matches the imprint observed during testing (refer to Figure 7). Figure 8 shows one of the actual failure mode observed during the test as compared to the GENEO predicted failure mode. In addition to the static test simulation GENOA was used to simulate a fatigue test. According to GENOA prediction, the first node damage occurred after 1.342×10^8 cycles of loading, while the first node failure occurred only at the 2.684×10^8-cycle. The ultimate fatigue failure of the bridge deck occurred after 6.221×10^8 cycles of loading according to GENOA prediction.

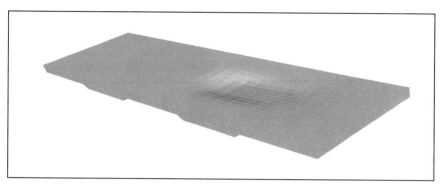

Figure 7. The Model as Deflected Prior to Failure Matches the Imprint Observed
During Testing

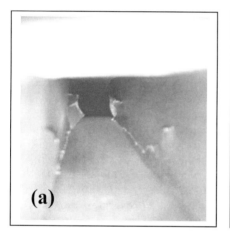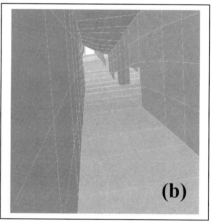

Figure 8. Ultimate Failure Mode of Pultruded Sandwich Core Beams:
(a) "Real" Test   (b) "Virtual" Test.

REFERENCES

[1]   Mosallam, A., Haroun, M., Abdi, F., and C. Dumlao (2001). "Structural
Evaluation of the All-Composite Deck for the Schuler Heim Bridge," To
appear in the Proceedings of SAMP'2002 Conference, May 12-16, Long
Beach, California, U.S.A.

7.2 The analysis, design and optimisation of an advanced composite bridge deck

L Canning, J Hodgson, A Jarman, R Karuna and S Luke
Mouchel Consulting Limited

INTRODUCTION

Throughout the world, a significant percentage of RC bridges are in a deteriorated state. The main reasons for the poor state of these bridges are inadequate cover, cracking of the concrete leading to corrosion of the steel reinforcement and spalling of the surface concrete (worsened by the use of de-icing salts), inadequate initial detailing and the increase in traffic loads over time. In the United States of America, approximately 20% (100,000) of all bridges are structurally deficient or obsolete, with 35% of these exhibiting poor deck conditions. In Europe, a significant number of bridge decks are also structurally deficient. The cost of reinstating or reconstruction of these bridge decks is economically prohibitive using conventional means, and would be equally prone to deterioration.

Fibre reinforced polymer (FRP) bridge decks provide the possibility of a more durable, lightweight and easily installed alternative. To this end, project ASSET (a four-year EC-funded research programme) was initiated to develop an optimised FRP bridge deck for deck replacement and new-build, culminating with the replacement of the West Mill bridge in Oxfordshire, UK. The development of the ASSET FRP bridge deck consisted of the preliminary analysis, optimisation, large-scale testing of the ASSET profile and design of West Mill bridge. The partners in the ASSET project are Mouchel, UK (Lead Partner), Fiberline, Denmark (FRP Pultruder), KTH, Sweden (Academic), Skanska, Sweden & UK (Contractor), HIM, Netherlands (Surfacing Supplier) and Oxfordshire County Council, UK (Client/Owner).

The detailed analysis and optimisation of the ASSET profile and the design and practical issues for West Mill Bridge are now described.

PRELIMINARY ANALYSIS OF ASSET PROFILE

The preliminary analysis of the ASSET FRP profile was carried out using analytical techniques (finite element analysis and laminate analysis theory). To ascertain the geometric form, size, fibre type and fibre architecture of the FRP profile a number of criteria were first determined:

- Structural form

ACIC 2002, Thomas Telford, London, 2002

- Span of ASSET profile
- Load conditions
- Compatibility with deck replacement
- Manufacturing considerations.

For the structural type, the ASSET profile would span transversely across the main longitudinal beams of a bridge deck. Following a review of typical bridge decks a suitable span length of 2 metres was determined. To ascertain the likely loads on the ASSET profile, a review of design codes throughout Europe was undertaken, and it was proposed that the British Standards were generally the most onerous with regard to loads. Therefore, the ASSET profile was developed based on a 2m span and 40 tonne load, based on HA and HB loading.

In addition to the span and load conditions, the depth and size were governed by the suitability for the pultrusion manufacturing method and also compatibility with existing bridges. Therefore, the general maximum depth of the ASSET profile for bridge deck replacement was chosen as approximately 250mm, with a maximum width in the region of 500mm, including overlaps (See Figure 1).

Based on economical considerations and discussions with the pultrusion specialist in the consortium, the fibre type was chosen as E-glass in an isopthalic polyester matrix.

To find the optimum form for the FRP profile, a number of shapes were investigated based on multi-cellular construction. In addition, a review of the research on FRP bridge deck forms [1-3] was undertaken to determine an appropriate form. These forms consisted essentially of box or trapezoidal or arch sections with internal webs either vertical or at an angle. The preliminary form that was chosen for further analysis is shown in Figure 1. This section was then used as the basis for the optimisation of profile.

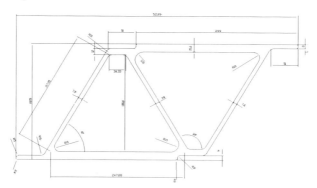

Figure 1   Preliminary Form of ASSET Profile For Optimisation Procedure

OPTIMISATION OF ASSET PROFILE

To perform the optimisation of the ASSET profile under the determined boundary conditions, the FE package ANSYS was used to create a 3D layered shell model. To carry out the optimisation process, an initial 'Basic Feasible Solution' is required. The variable items to be optimised (fibre orientation, ply thicknesses, total web thickness, total flange thickness) can then be used to optimise the 'Objective Function', in this case the volume of the ASSET profile. To ensure that the optimised solution was economically viable, the fibre orientations were restricted to $0°$, $90°$ and $+/-45°$. The optimum solution also had to satisfy the 'State Variables', which were the strength and deflection criteria of the ASSET profile. A deflection limit of span/300 (based on a span of 2m) was imposed; the strength criterion was dependent on the laminate lay-up and was based on various failure criteria (Tsai-Wu, Maximum Stress, Maximum Strain, etc).

DETAILED ANALYSIS OF ASSET PROFILE

The next stage in the analysis of the ASSET profile, and to ensure its adequacy at the serviceability and ultimate limit states, was to model a full representative bridge deck. The FE model for the full deck is shown in Figure 2 and used first-order shell elements. Two such FE models were created; one using shell elements with the average properties of the full laminate calculated from classical laminate analysis, the other using layered shell elements with individual layer properties. The results shown in this paper are based on the 'average properties' shell model. The results from the more complex layered shell model will be reported upon at a later date.

The material properties of the profile and longitudinal beams were defined as linear elastic. The primary structural characteristics of the full deck model were:

- Longitudinal span of 10 metres (with a deck consisting of 34 bonded ASSET profiles).
- Spacing of longitudinal beams of 2 metres, i.e. transverse span of ASSET profile was 2 metres.
- Deck width was assumed as 4 metres, i.e. the ASSET profiles consisted of two continuous spans.
- End fixity of longitudinal beams was assumed as simply supported.
- Longitudinal beams were generically designed to provide longitudinal stiffness to satisfy a deflection limit of span/300.

The load effects applied from one profile to another were determined from the application of HA, HB and KEL traffic loads to the full deck model. The inter-profile load effects essentially consisted of forces applied at the top and bottom flanges. This then allowed a sub-model consisting of three ASSET profiles with two continuous spans to be created for detailed stress analysis (see Figure 3).

The sub-model consisting of three ASSET profiles used the same first-order shell elements as in the full representative deck model, but a much finer mesh was possible. In general, the aspect ratio of the elements was less than 1.2. Using a linear elastic analysis, the deflection, strain and stresses imposed on the profile were determined from the HA and HB traffic loads described in BD37/88 [4]. A number of load cases were used to determine the worst hogging, sagging, shear and deflection effects.

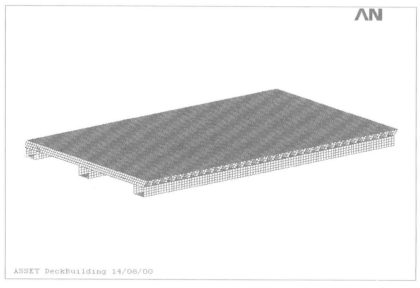

Figure 2 Finite Element Model of Full Representative Deck

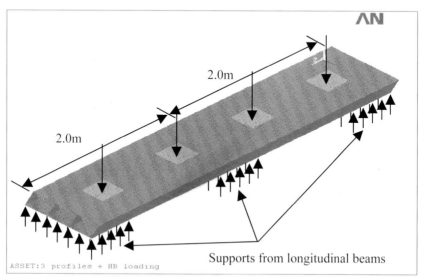

Figure 3 Sub-model Consisting of Three Profiles Over Two Continuous Spans (Hogging Load Case With HB Loads)

FE RESULTS FROM DETAILED SUB-MODEL

A typical deflection plot under HB loading is shown in Figure 4. In general, the deflection limit of span/300 was the critical design factor. The dependence of the design on the deflection criteria has been reported in a number of investigations where advanced composite components are used as primary structural members [5]. Due to the geometry of the FE model and loading conditions, local stress concentrations were particularly noticeable near the loading points and at the web/flange interfaces. The shell FE model did not model the fibre architecture and geometry at the web/flange interface and therefore it was important that relatively high factors of safety were present for strength. In general the factor of safety on strength at the ultimate limit state was over 3. The actual stress field present at the web/flange interfaces was determined from the large scale testing programme on the ASSET profile.

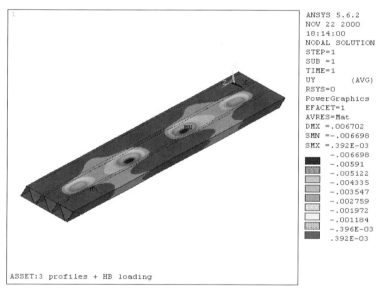

Figure 4 Typical Deflection Plot For HB Load Case

It is worth noting that for the ASSET profile, and for primary structural advanced composite members in general, the shear deflection of the member can be a significant proportion of the total deflection. For structural members using conventional materials the shear deflection is typically negligible, furthermore, the stiffness of the member is most often not the critical design factor.

PULTRUSION TECHNOLOGY

The development of the pultrusion technology used to manufacture the ASSET profile was important due to the relatively complex nature of the geometry and fibre architecture. The size/surface area of the profile was such that very large traction forces were required, beyond that typically used. Based on the optimised geometry and fibre architecture of the profile, the pultruder partners (Fiberline A/S) further enhanced their pultrusion technology with the following aspects:

- Semi-automatic control systems for surveillance of roving, speed control, continuity of all glass fibre mats and automatic error reporting systems.
- Precise control of impregnation pressure of the resin matrix into the fibre architecture and also temperature.
- Traction units capable of achieving the large pulling force required for the ASSET profile.
- Position and design of injection chamber to ensure effective production and proper impregnation of the resin matrix into the fibre reinforcement pack.
- Positioning and fixing of mandrels in the die.

Figure 5 shows the pultrusion of the ASSET profile, from guidance and entry of the fibre reinforcement into the injection chamber through to the final pultruded profile. The optimisation of the ASSET profile was undertaken using not only an analytical procedure but also to ensure that the profile could be efficiently manufactured using state-of-the-art pultrusion techniques.

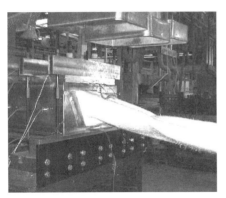 

Figure 5 Pultrusion of the ASSET Profile:
Pultrusion Test (No Resin) and Finished Pultrusion

SMALL AND LARGE-SCALE TESTING
To complement and validate the numerical and theoretical design of the ASSET profile, small and large-scale testing was undertaken. To reduce the number of test variables in terms of environmental conditions, creep and fatigue, a thorough literature review was conducted to ascertain the most severe and relevant load and environmental conditions.

The small-scale testing provided information on the mechanical properties of various ply lay-ups for the theoretical and FE analyses. The tested properties consisted of tensile, compressive and shear strength and modulus, ILSS (inter-laminar shear strength) and Poisson's ratios of the GFRP composite [6].

A summary of the large-scale testing programme is shown in Table 1. These tests were used to characterise the flexural strength and rigidity and the shear strength. Tests were undertaken on single and triple-ASSET profiles to ascertain the load sharing capability and using single and double spans to determine both sagging and hogging capacity. Impact tests were also used to obtain the resistance to impact of the top surface of the ASSET profile. Table 1 also shows the tests that are being undertaken on ASSET profiles manufactured using an isopthalic polyester or vinylester matrix. In certain cases the tests included UD CFRP fibres in the flanges and a 30mm thick polymer concrete slab on the top flange of the profile.

Table 1 – Large Scale Testing Programme

Test Type	Description
Static Flexure	Sagging and hogging tests on single and multiple profiles over one or two spans.
Static Shear	Shear tests on single and triple profiles over one span.
Fatigue (10 million cycles)	Fatigue tests on single and multiple profiles over one or two spans.
Creep	Creep tests on triple profile over two spans.
Impact	Standard impact test on single profile.
Additional tests	Flexural tests on profiles with CFRP/GFRP hybrid flanges.
Additional tests	Static, fatigue and creep tests on single and multiple profiles over one or two spans, using a vinylester matrix.

Finite element models of the ASSET profile large-scale tests were also used to validate the structural behaviour. The load configuration was chosen such that HB wheel loads were represented.

In general, the numerical and experimental results were within 5% of one another. A number of important observations were made based on the data obtained from the instrumentation:

- For the single ASSET profile tests the longitudinal flange strains at the edges of the flange were affected by torsionally induced strains, due to the unsymmetric geometry of the ASSET profile and also the eccentricity of the applied load.
- The strains in the bottom flange were generally lower than those measured in the top flange due to local strains caused by the concentrated loads; the FE model generally provided a better prediction of these strains and stresses.
- The shear strain in the outer and inner webs were evaluated from the use of strain rosettes; in general the outer webs provided a greater proportion of the shear resistance than the inner web.

- The failure mode for the bare ASSET profiles with no polymer concrete slab was observed as local buckling and delamination of the top flange/web interfaces at midspan.

DESIGN AND BUILDABILITY ISSUES FOR PROTOTYPE BRIDGE DECK

To prove the advanced composite bridge deck concept, a prototype bridge deck will be constructed in April 2002. The bridge owner partner within the ASSET project, Oxfordshire County Council identified the opportunity of not only replacing a bridge deck but a complete bridge including the substructure and superstructure. Therefore, the ASSET profile was used as an integral part of an advanced composite bridge deck to replace West Mill Bridge in Oxfordshire. The existing West Mill Bridge dates from the 1870's and carries a single lane of traffic over a span of 10m over a river. The replacement of the bridge will improve the crossing to a 7m wide highway bridge and uses the ASSET profile spanning transversely as the decking system. The deck is supported upon four longitudinal GFRP/CFRP hybrid composite box beams that provide the longitudinal flexural rigidity. A preliminary cross-section at midspan of the bridge is shown in Figure 6. It is also possible to support the ASSET profiles on steel or concrete longitudinal beams.

Buildability of West Mill Bridge

The structural performance of the transverse ASSET profile and longitudinal composite box beams was of importance, however, the buildability and practicality of the advanced composite bridge deck system and joining mechanisms were also crucial to the success of the project. Therefore, a catalogue of possible joining techniques for various bridge components was developed by the contractor partner (Skanska):

- ASSET profile end-to-end joint and edge-edge joint.
- Edge details; supporting barriers and light poles, verge details.
- ASSET profile/longitudinal beam connection.
- Abutment/ASSET profile and longitudinal beam connection detail.

Significant client input from Oxford County Council was given for a number of the aesthetic and buildability issues, which ensured that the various connection options were acceptable. In particular, two types of profile edge-edge joint were investigated; an adhesive bonded and a bolted joint. The bolted joint was found to be considerably more complicated than the adhesive bonded joint.

Design of West Mill Bridge

The preliminary design of West Mill bridge, with regard to the required flexural rigidity and strength, was determined using theoretical techniques and with knowledge of the results from the previous FE analysis on the generic ASSET profile and bridge deck. Once this preliminary design was complete, a finite element model was developed to model West Mill bridge (Figure 7), with the following level of detail:

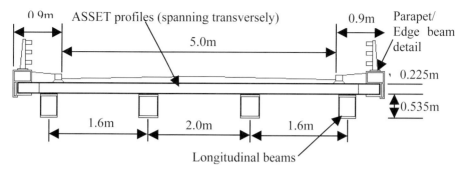

0 9m   ASSET profiles (spanning transversely)   0.9m   Parapet/
                                                            Edge beam
                        5.0m   /detail

                                                           0.225m

                                                           0.535m

        1.6m   2.0m   1.6m

                    Longitudinal beams

Figure 6 Midspan Cross-section Of West Mill Bridge

- Shell elements with anisotropic material properties were used for the ASSET profile and longitudinal beams. The connection between the transverse ASSET profiles and longitudinal beams was effected by using springs.
- The concrete parapet edge beam and concrete diaphragms were modelled using solid elements with isotropic properties. The end bearings for the longitudinal beams were represented with spring elements.
- The verge section and surfacing material of the bridge deck were note modelled, although the associated dead load was applied.

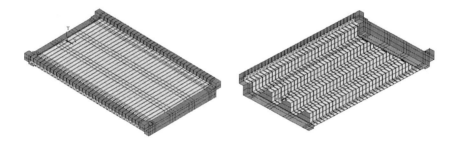

Figure 7   Finite Element Model of West Mill Bridge (Top of Deck and Underside)

A linear elastic analysis was used to determine the deflections, strains and stresses within the various components of the bridge under the full range of load combinations. These results were then checked against the serviceability and ultimate limit states, essentially consisting of deflection and stress limits (which were based on experimental results and laminate analysis). In particular, when using advanced composite materials, it is well known that the stress conditions should be restricted so that creep rupture or fatigue failure does not occur, especially in the case of GFRP composites.

One aspect of the design of the bridge of particular interest was the degree of composite action present between three components; the ASSET profile, the longitudinal beams and the concrete edge beams. The effect of these sections all acting compositely to some degree has a significant beneficial effect on the flexural rigidity and strength of the bridge superstructure. However, in the absence of extensive data on this structural aspect, it was deemed suitable to design the bridge assuming no or very little composite action. However, both the 'no composite action' and 'full composite action' cases were analysed to check that this effect would have no detrimental effect on the bridge superstructure, due to changing failure modes or neutral axis level, for example. In future cases, it is likely that the composite action between the bridge components will be used to provide a more efficient design.

CONCLUSIONS

The following were concluded from the experimental, numerical and theoretical work described in this paper:

- To create a competitive, structurally efficient advanced composite deck profile, an optimised fibre architecture consisting of uni-directional and off-axis fibres was required.
- FE analysis using shell elements with 'average' properties satisfactorily modelled the global behaviour of the ASSET profile and bridge deck.
- To obtain detailed data on the local effects on the ASSET profile, a more complex, layered shell element model is applicable.
- Deflection (a serviceability limit state) governs the design, as opposed to strength (an ultimate limit state).
- The level of composite action between the longitudinal box beams and ASSET profile bridge deck was significant. With further research and data, use can be made of this composite action.
- Advanced composite components provide a lightweight, durable solution for replacement bridge decks and new-build. In addition, the lightweight nature of the material allows for swift and easy installation.

REFERENCES

[1] Gangarao, H., Thippeswamy, H., Shekar, V., and Craigo, C., 'Development of Glass Fibre Reinforced Polymer Composite Bridge Deck', SAMPE Journal, Volume 35, Number 4, July/August 1999, pp12-24.

[2] Davey, S., Van Erp, G., and Marsh, R., 'Fibre Composite Bridge Decks – An Alternative Approach', International Conference ACUN2, Composites in the Transportation Industry, Sydney, Australia, February 2000, pp216-221.

[3] Karbhari, V., and Zhao, L., 'Use of Composites For 21st Century Civil Infrastructure', Computer Methods in Applied Mechanics and Engineering, Volume 185, 2000, pp433-454.

[4] 'Loads For Highway Bridges', Departmental Standard BD 37/88, Department of Transport, Highways and Traffic, 1988.

[5] Canning, L., 'The Structural Analysis and Optimisation of An Advanced Composite/Concrete Beam', PhD Thesis, University of Surrey, 2001.

[6] Åström, B.T., Hallström, S., and Knudsen, E., 'Mechanical Properties of Pultruded Triaxially Reinforced Composites', 6th International Conference on Flow Processes in Composite Materials, Auckland, New Zealand, 4-5 Feb. 2002

7.3 Novel use of FRP composites in building construction.

D C Kendall,
Director, CETEC Consultancy Ltd, Romsey, UK

INTRODUCTION

FRP composite materials have been used successfully for many years in building construction to produce items such as cladding panels, rooflights, canopies etc. These have generally provided a weather-tight, aesthetic finish and formed elements of secondary structure, taking the imposed loads back to a structural frame, most commonly made of steel or concrete. Unfortunately, this methodology of distinct elements of secondary structure and primary structural frames has not fully exploited the structural benefits that can be achieved by using FRP composites.

Other industries, where structural efficiency has been of greater importance have taken a rather different approach. For example, aircraft, boats and racing cars, see Figure 1, are designed as monocoque structures to provide the optimum structural efficiency. Every element of the structure is designed to work as effectively as possible, in these instances to save weight, but the same philosophy can be used to increase structural efficiency to reduce costs, which will be of greater interest in the construction industry.

Figure 1 Typical monocoque structures built in FRP composites
(Photographs courtesy of Lockheed Martin, ACG and Red Bull Sauber AG)

It must be accepted that FRP materials are significantly more expensive than those conventionally used in building construction. However, by changing the structural methodology and producing highly efficient structures, FRP structures can be very competitive. Once through-life costs are considered and other possible benefits of weight reduction, such as reduced supporting structure, easier installation etc are allowed for, the FRP solution will often be less expensive than the conventional one.

ACIC 2002, Thomas Telford, London, 2002

DESIGN METHODOLOGY FOR STRUCTURAL EFFICIENCY

There are several ways in which the efficiency of the structure can be increased and hence the costs reduced. As the raw materials in an FRP structure are relatively expensive, it is important to minimise the material content and produce efficient designs. To produce efficient structural solutions engineers need to be encouraged to optimise the structures to a greater extent than may be commonplace in conventional building structures. It must be accepted that this may entail an increased engineering cost, but this will be small compared to the overall costs and must be less than the amount saved due to optimisation to make the exercise worthwhile. The advent of economic methods of computer modelling, finite element analysis, see Figure 2, and dedicated composites analysis software will help in the role of structural optimisation.

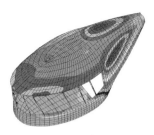

Figure 2 Finite Element Analysis can assist in structural optimisation.

Figure 3 Factory moulding of a large roof structure.

PARTS INTEGRATION

The incorporation of several components into a single part generally has the effect of significantly reducing assembly time and costs. In a construction project this may also have the added benefit of shifting a substantial amount of work from the building site to the factory, see Figure 3, where work can be carried out in a more controlled environment with improved quality control and health & safety and reduced risk of delays due to the weather, conflicts with other contractors on site etc.

Parts integration may be as simple as incorporating stiffening elements, the outer skin and joint features into a single pultruded panel as shown in Figure 4.

Figure 4 Pultruded FRP panel with integral stiffeners & joint details.

This can be taken further by including primary structural members, internal linings, insulation, services etc into a single factory made component as shown in Figure 5.

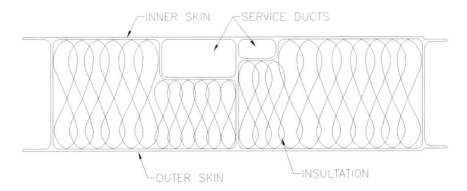

Figure 5 FRP panel forming primary structure with integral service ducts, insulation etc.

To improve the structural efficiency, it is desirable for the skin and frames to work as a single component. The outer skin will be more fully utilised if it contributes to the global strength of the primary structure in addition to withstanding the local imposed loads. In a conventional structure the panels forming the outer skin have to withstand the local imposed loads and normally transfer these to the separate primary frame structure, which will then have to carry these loads independently to the supporting foundations.

There are limits to the extent to which the skin will contribute to the strength of widely spaced frames due to problems with the stability of the skins, shear lag and the resulting loads through attachments. However, these are all issues that can be readily addressed and predicted with modern structural analysis techniques, but not by using existing building design codes.

Structural Sandwich Panels

FRP laminates can be combined with lightweight core materials such as structural foams (PU, PVC, phenolic etc), end-gain balsa wood or honeycombs (aluminium, Nomex®, thermoplastic etc) to produce very lightweight panels. Figure 6 shows how with only a marginal increase in weight, the strength and particularly the stiffness of the panel can be dramatically increased, by switching from a solid laminate to a sandwich construction. It should be noted however, that the core materials and manufacturing methods can be expensive and good quality sandwich structures can be more difficult to manufacture than solid laminate structures. Significant effort, both in manufacturing techniques and subsequent non-destructive testing is therefore required to ensure the structural integrity of the panel, particularly adequate skin to core bonds.

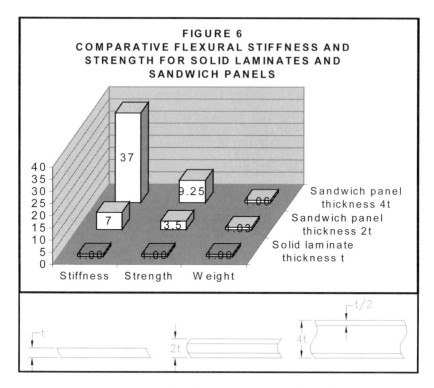

Figure 6: Comparative flexural stiffness and strength for solid laminates and sandwich panels

Sandwich structures will provide significantly higher thermal insulation than a solid laminate, which may be advantageous for building applications. Sandwich structures can be produced as either flat or curved panels and may in some cases be aesthetically superior to a stiffened solid laminate. Stiffeners laminated to a solid laminate can produce visible distortions on the outer face as the resin cures and shrinks. There are methods for minimising such effects, but by using sandwich constructions it is often possible to omit all secondary stiffening.

Geometric Optimisation
One of the great advantages of using FRP composites is the ability to mould complex shapes. This enables buildings and structures to be produced with unusual forms, including concave and convex, single or double curvature sections, which can produce stunning aesthetic effects, but can also be utilised to produce much more efficient structural forms.

One of the most efficient structural forms is the sphere, part of which has been used on numerous occasions to produce structures such as the FRP radomes shown in Figure 7. This photograph also demonstrates how the lightweight nature of the

Figure 7 FRP radomes.   **Figure 8**   FRP spherical buildings.

structure has enabled the radome to be assembled on the ground and then lifted into position with a modest crane. This structure consists of a series of FRP sandwich panels, which are bolted together to form a continuous monocoque structure, completely free of any internal framing.

Based on the technology developed for producing radomes, these highly efficient spherical structures may also be used to produce buildings as shown in Figure 8.

Adding curvature to the structure, particularly double curvature, provides natural stability and resistance to buckling, enabling the amount of material to be significantly reduced compared to a flat structure, resulting in a more economic solution. Such curved structures tend to resist imposed loads with membrane action rather than in pure bending and will utilise a much greater proportion of the material in resisting such loading.

In addition to producing smooth geometric shapes such as the spherical structures shown above, it is possible to mould FRP into much more complex shapes, such as the domed roof shown in Figure 9. Once the tooling costs are amortised, this complex corrugated domed structure, is approximately half the cost of a more conventional structure, also produced in FRP, consisting of separate beams and panels, as shown in Figure 10.

Figure 9   FRP monocoque roof.   **Figure 10**   FRP beam & panel roof.

There will often be an increase in tooling costs associated with producing complex double curvature shell structures and this must be offset against the reduction in material and manufacturing costs. The overall economic balance will therefore be highly dependent on the complexity of the tooling and the number of mouldings to be produced from each mould. Careful design and definition of the geometry will often enable an increased utilisation of the tools and therefore further cost reductions.

LOW-COST TOOLING

Several methods have been developed to produce low-cost temporary moulds to produce a very low number of FRP mouldings economically.

Figure 11 Glasgow Science Centre   **Figure 12**   FRP cabin during manufacture.
Observation Tower.

Techniques for producing one-off FRP structures without any moulds are also established and commonplace in industries such as boat building. Such methods were recently used to produce the FRP structure for the viewing cabin installed on top of a 100m high rotating tower at the Glasgow Science Centre, as shown in Figure 11.

The cabin was produced by laser-cutting a series of plywood frames, which were assembled to form an internal skeleton, over which a structural PVC foam core was laid as shown in Figure 12. The outer skin of the FRP sandwich was then laminated

Figure 13   FRP cabin during manufacture   **Figure 14**   Home Planet Zone

onto the foam core. Once this was completed and cured, the structure was inverted and the plywood frames removed, exposing the inside of the foam core. The inner skin was then laminated against the foam core, completing the integrity of the FRP sandwich as shown in Figure 13.

CASE HISTORIES
Home Planet Zone Building
The building housing the Home Planet Zone within the Millennium Dome was originally conceived by Fletcher Priest architects as a relatively straightforward steel framed structure clad with aluminium panels. However, the architect's desire was to create a much more organic flowing form to the building, which they realised would be impractical to produce using conventional building techniques and materials.

CETEC developed several designs using FRP composites to produce a highly efficient and aesthetically pleasing structure as shown in Figure 14.

The building consists of a monocoque structure enabling the internal steel frame to be omitted completely. The visible exterior skin of the building provides the outer cladding, waterproofing membrane and primary structure of the building, integrated into a single factory-moulded component. For transportation the mouldings were made in manageable sizes, but were quickly bolted together on site to provide the finished structure.

This method of construction provided a stunning building, which is structurally highly efficient, economic and very quick to erect on site. An added bonus is that the building can be dismantled and moved to a new site if required.

MEETING ROOM PODS
A series of novel building structures have been conceived by Alsop Architects to provide additional meeting rooms within the atria of a refurbished office at Victoria House, London, owned and developed by Garbe (UK) Ltd. These consist of various pod structures of complex curved forms, providing both fully enclosed and partially open meeting areas. The larger pods are 9m high providing two meeting rooms in each pod supported on slender steel legs from the atria walls. Whilst these are

complex structures, their cost is justified by the extra usable space they create. They also add significant further value by providing a talking point and certain "wow factor" to attract tenants and clients to the building. See Figure 15.

Figure 15: The "wow" factor

The original concept was to produce the pods in timber, but initial studies carried out by CETEC, refer to Figure 16, showed that a more efficient structure could be produced using FRP. The pods are to be produced in temporary female moulds to provide a high quality moulded exterior finish. They will be manufactured in very large panels to minimise the number of joints and will consist of a monocoque FRP sandwich structure with minimal internal framing. Laminate reinforcements include E-glass and fire retardant polyester resin to produce a Class O fire rating in accordance with the Building Regulations.

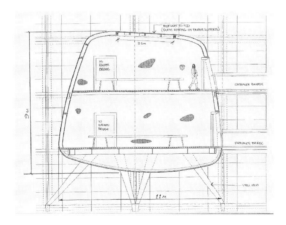

Figure 16: Concept study

A significant advantage of the selected method of manufacture is that the bulk of the manufacturing work will be carried out off-site in factory-controlled conditions. Each pod will be trial assembled at the factory and then delivered to site in the minimum number of sections, enabling rapid site assembly with minimal disruption to other site activities.

FUTURE MODULAR CONSTRUCTION
The FRP monocoque type structures described above will be ideal for adaptation into a true modular building philosophy. This could entail producing a building such as that described above for the Home Planet Zone in several large sections, which are fully finished in the factory, both internally and externally. Each section would include internal linings, finishes, doors, windows and services.

On site, all that would be required would be a rapid assembly operation to join the sections together and connect the services between modules. In essence this is how aeroplanes, ships and cars are produced by finally assembling a number of major sub-assemblies. However, for this to happen in the building industry there will need to be significant cultural changes in the way in which such projects are designed, tendered and managed.

CONCLUSIONS
It has been shown that FRP composites can be used to produce dramatic building structures, economically and with several other benefits over conventional materials. However, to produce such economic solutions the design methodology used may need to change and effort needs to be spent on the optimisation of such structures.

Existing design codes do not adequately cover either the FRP materials or the optimum structural solutions and this issue needs to be addressed to increase the acceptance of FRP materials. However, this should not prevent the use of FRP at the present time, as there is a wealth of data available and case histories to prove the adequacy of the materials.

7.4 Notched strength of woven fabric-based CFRP composites

H Belmonte[1], S L Ogin[1], P A Smith[1] and R Lewin[2]
[1]School of Engineering, University of Surrey, Guildford, UK
[2]Rolls-Royce plc, PO Box 31, Derby DE24 8BJ, UK

INTRODUCTION

In using composites in engineering applications, it is the design of structural details that is often a critical issue. Attachments in particular present a problem. Hence, over the years an extensive literature has evolved concerned with developing methods to predict the notched strength of laminated composites. Most of the early work focused on continuous (non-woven) systems, as reviewed in *e.g.* [1], while more recently there has been a growing interest in the behaviour of woven fabric based composites [2-4]. Ideally, a predictive model for notched strength should capture in some way the physics of the damage development prior to fracture.

In recent work [5, 6] we have developed a model for predicting the notched strength of woven fabric glass fibre composites, which is based on the experimentally observed stable growth of a damage zone at the notch tip, prior to catastrophic fracture. The model requires as input parameters the unnotched strength and toughness of the composite; the latter is measured from a fracture mechanics test using a single-edge-notch geometry. The present work is concerned with the application of this approach to a range of laminates based on woven fabric CFRP. Data are presented for the base-line mechanical properties and the strength as a function of notch size. The fracture model developed in previous work is shown to describe the notched strength data satisfactorily. The underlying assumption of the model, that damage proceeds stably from the notch tip, has been verified by identifying tow fractures near the notch tip in laminates unloaded prior to catastrophic fracture.

EXPERIMENTAL

Plain weave and satin weave CFRP laminates of two different lay-ups (cross-ply and quasi-isotropic) and three different thicknesses have been investigated – a total of twelve systems. Standard rectangular coupons (25 mm by 230 mm) were prepared for base-line mechanical property (stiffness, strength) and single edge notch investigations.

Circular central holes of 2.5, 5 and 10 mm in diameter, were drilled in the 25 mm wide coupons using low helix, high speed steel drill bits normally used with thermosetting polymers. The drill bits used had acute point angles to reduce end pressure, minimise matrix burning and material break away when the drill bit exited the coupons. This ensured that no excess damage was introduced around the notch edge.

All coupons were tested on an Instron 5500R at a constant cross-head displacement rate of 0.5 mm/min using a 75 mm gauge length extensometer to monitor the strain. The ASTM standard D 3039 was followed wherever possible in determining the in-plane tensile properties of interest here, which were the Young's modulus and strength of the unnotched coupons and the strength of the notched coupons. Young's modulus values were determined by taking a linear regression of stress against strain between 0.05% and 0.30% strain, which was within the linear portion of the stress-strain curve.

The critical strain energy release rate, G_c, was measured using single edge notch (SEN) specimens, following ASTM standard E 399-90 where possible. The edge notches were machined using a jewellers saw and sharpened prior to testing with a fresh scalpel blade. The compliance calibration (compliance against crack length) was carried out by recording the load-extension profiles for incremental increases of sawn-in cracks of length 0 mm to 15 mm, again using a 75 mm gauge length extensometer. The samples were loaded to 0.3 % strain and then unloaded; this cycle was repeated three times for each sample and crack size enabling three compliance measurements to be made for each crack length so that the average compliance could be determined. A fourth order polynomial was fitted to the data. The toughness measurements were obtained from fracturing specimens at three crack lengths (4, 8 and 12 mm) and the maximum load was used as the fracture load. The toughness of the laminates were then found using the expression:

$$G_c = \frac{P_{max}^2}{2B} \frac{dC}{da} \qquad (1)$$

where B is thickness of laminate, P_{max} is the maximum load at failure, C is the compliance and a is the crack length.

With regard to characterizing the damage at the notch tip, this was not straightforward, compared to previous work on (transparent) GFRP, due to the opaque nature of the CFRP material. In validating the analytical model, a key point is whether or not stable tow fracture proceeds from the notch tip before the maximum load. In order to investigate this, notched coupons were loaded to just below their failure loads so that a critical damage zone formed, and then unloading rapidly. The coupon regions near the hole, which contain the damage, were sectioned and the matrix was burnt off to expose the reinforcement. The damaged reinforcements were gold coated and viewed in the Scanning Electron Microscope (SEM) using secondary electron imaging mode. Each layer of the reinforcement was examined separately, using the depth-of-focus of the SEM to determine the extent of fibre damage.

RESULTS

The mechanical data will be discussed in detail in an associated paper. In this paper we will focus on the four layer quasi-isotropic $(0/90/\pm45)_s$ woven fabric reinforced (five harness satin weave) laminate, designated 5Q4. The laminate was 1.53 mm thick with a fibre volume fraction of 36 %, determined by acid digestion.

The base-line properties of the laminate were: Young's modulus, $E = 34.1 \pm 0.3$ GPa, unnotched strength, $\sigma_o = 375 \pm 22$ MPa and a strain-to-failure of 1.10 ± 0.1 %. The values for the strength of the notched specimens are presented in Table 1 and plotted in Figure 1. The reproducibility of the results for a given notch size and coupon width was reasonable.

Table 1 Notched Strengths of 5Q4 laminate

Notch Size – d (in mm)	d/W Ratio	Strength - $\square_{max}$ (in MPa)
2.5	0.1	242.4 ± 8.3
5	0.2	202.1 ± 4.7
10	0.4	154.9 ± 6.2

Three coupons were used to determine the compliance characteristics for each material. Figure 2 shows the measured compliance calibration curves and the average compliance against crack length curve obtained from these data.

The toughness measurements were carried out on 25 mm wide coupons for three crack lengths (4, 8 and 12 mm). The mean toughness value was 19.2 kJ/m². The fracture toughness, K_c, was calculated from the critical strain energy release rate, Young's modulus, E, and Poisson's ratio, v, by the relation:

$$K_c = \frac{\sqrt{EG_c}}{(1-v^2)} \qquad (2)$$

Sections were cut from samples unloaded just before failure from the four-layer quasi-isotropic material so that the damage close to the hole could be examined in more detail. Figure 3 shows the damage visible on the surface of the sample that was unloaded before failure occurred. The fabric layers were separated giving four layers, the first layer being (0/90) orientated, the second (+45/-45), the third (-45/+45) and the final layer being (90/0) orientated. Each layer was examined in the SEM. Some damage to the fibre tows in one of the (+45/-45) layers was seen, but the most important feature was the fractures in the $0°$ tows in the 0/90 layers, see Figure 4. Hence a physically based failure criterion needs to take this into account.

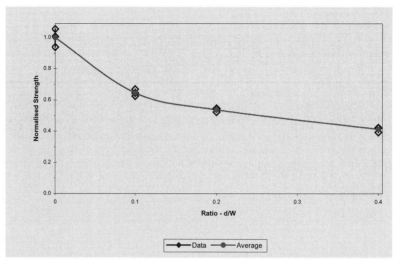

Figure 1 Normalised strength against ratio of notch size to coupon width.

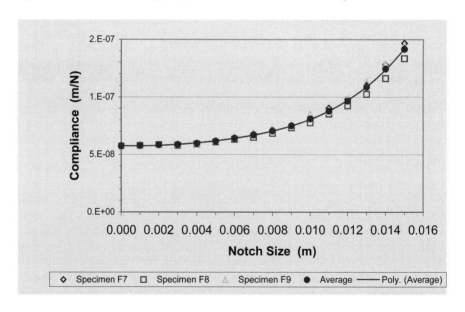

Figure 2 Graph of compliance against crack length for single edge notch specimens

Figure 3 Photograph showing visible surface damage to a specimen loaded to a stress of 198 MPa without catastrophic fracture. Photograph shows front and back faces of the laminate at one side of the hole.

(a)   (b)

Figure 4 Secondary electron micrograph showing damage adjacent to a 5 mm circular centre notch in the first (0/90) layer showing tow fractures on both sides of the hole.

MODELLING
Model Principles
The models considered in the present study were the Whitney-Nuismer point and average stress criteria and a critical damage growth (CDG) model applied previously to notched woven fabric composites [5, 6]; the latter is summarized briefly below.

The analysis is based upon the stable growth and subsequent catastrophic failure of a damage zone at the edge of the circular hole. The applied laminate stress, σ, required to grow a damage zone or crack to a length, c, from the edge of the hole (of radius R) is defined using an average stress argument and is given by:

$$\sigma = \frac{1}{c} \int_{R}^{R+c} Y_1 \sigma_y \, dx \qquad (3)$$

In equation (3) σ_y is the stress adjacent to the notch edge and Y_1 is a finite width correction factor. A fracture mechanics approach is then used to predict the point at which the damage zone becomes unstable and catastrophic fracture occurs. The stress required for catastrophic failure is given by:

$$K_c = \sigma_a \sqrt{\pi c} F_0 Y_2 \qquad (4)$$

In equation (4) K_c is the critical stress intensity factor and σ_a is the remote applied stress. The parameter F_0 is a correction factor for cracks growing from a circular hole, while Y_2 is a finite width correction factor for a crack emanating from a hole.

Fracture is assumed to occur when the stress required to advance the damage zone is equal to the stress required for catastrophic crack growth. Thus the model uses a competing mechanisms approach that gives both a prediction of notched strength and the critical damage zone size. The data required for the prediction of notched strength in this model are fracture toughness, K_c, and unnotched laminate strength.

Comparison with Experiment
The point stress criterion (PSC) characteristic distance, d_0, and average stress criterion (ASC) characteristic distance, a_0, were calculated from the average notched strength at any given notch size. It was noted that the characteristic distances for both the PSC and ASC increased slightly with increasing notch size; this behaviour has been documented in other studies [1]. The average value of each characteristic distance was then used to make notched strength predictions. Data are shown in Table 4 for the systems shown in Figures 2 and 3. The unnotched strength and fracture toughness were used in the critical damage growth (CDG) model to predict the strength of the notched coupons. Results are shown in Table 3 and Table 4.

Table 3. Notched Strength Predictions According to the Point and Average Stress Criteria

Notch Size R (mm)	Experimental Strength - σ_n (MPa)	d_0 (mm)	PSC σ_n (MPa)	Error (%)	a_0 (mm)	ASC σ_n (MPa)	Error (%)
1.25	242		273	-12.7		257	-5.81
2.5	202	0.86	207	-2.4	2.23	206	-1.90
5	155		141	8.4		146	5.78

Table 4. CDG Model Notched Strength Predictions

Notch Size R (MPa)	Experimental Strength - σ_n (MPa)	Damage Zone Size c_0 (mm)	CDG σ_n Prediction (MPa)	Error (%)
1.25	242	2.18	255.0	-5.20
2.5	202	2.28	207.1	-2.50
5	155	2.60	151.7	2.10

DISCUSSION
The main result to emerge from the present work is that the damage growth and fracture model developed in previous work on GFRP woven fabric laminates has been shown to be applicable to CFRP laminates based on woven fabric reinforcement. In particular the damage growth studies suggest that tow fractures develop at the stress concentration before catastrophic fracture providing the experimental justification for the model used. Agreement between theory and experiment has been demonstrated in the present work for one thickness of quasi-isotropic laminate based on five harness satin reinforcement. Other results that we have obtained (to be published) show that the approach applies also to a wider range of woven fabric CFRP laminates.

CONCLUDING REMARKS
A range of mechanical property data (stiffness, unnotched strength, notched strength and toughness) has been obtained for a quasi-isotropic CFRP laminate based on woven fabric reinforcement. Fracture from circular holes has been studied. It has been shown that tow fracture occurs at the hole prior to catastrophic fracture. The notched strength data can be adequately described by the well-known Whitney-Nuismer point and average stress criteria models, or by a damage and fracture model which is more physically-based.

ACKNOWLEDGEMENTS
The authors would like to acknowledge support from the EPSRC and Rolls-Royce plc and to thank our colleague, Mr Reg Whattingham, for invaluable technical assistance.

REFERENCES
[1] J.Awerbuch and M.S.Madhukar, J. Reinforced Plastics and Composites, 4 (1985), pp.3-159.

[2] N.K.Naik and P.S.Shembekar, Comp. Sci. & Tech., 44 (1992), pp.1-12.

[3] J.K.Kim, D.S.Kim and N.Takeda, J. Composite Materials, 29 (1995), pp.982-998.

[4] A.Afaghi-Khatibi, L.Ye and Y.-W.Mai, J. Composite Materials, 30 (1996), pp.142-163.

[5] C.I.C.Manger, S.L.Ogin, P.A.Smith and R.Lewin, "Damage Zone Development and Fracture in Notched Woven Glass/Epoxy Composites", Proceedings of ECCM-9, 9th European Conference on Composite Materials, Brighton, UK, 4 - 7 June 2000.

[6] H.Belmonte, C.I.C.Manger, S.L.Ogin, P.A.Smith and R.Lewin, Comp. Sci. & Tech., 61 (2001), pp.585-597.

7.5 A fatigue design methodology for GRP composites in offshore underwater applications

P A Frieze,
P A F A Consulting Engineers Limited, Hampton, Middlesex, UK

INTRODUCTION

The Norsk Hydro Troll C field was exploited using a floating production facility with a large number of risers. The risers are supported by two 80 m high underwater jackets, in 340 m water depth. Each riser has its own support tray atop a jacket [1]. To reduce tray weight and installation time/cost, glass reinforced plastic (GRP) was used for construction. The trays were classified to NPD [2] as safety critical and inaccessible (for inspection and maintenance) and so attracted a fatigue life safety factor of 10. This meant 250-year fatigue design lives for the 25 year service life.

At the time, no formal offshore fatigue requirements existed for offshore structures. A two-staged approach to develop appropriate fatigue design criteria eventuated. The stages are described but concentrate on the first in which basic GRP fatigue failure mechanisms drive the process. The second stage exploits reasonably extensive data determined for another application, and is briefly discussed.

STAGE 1 - BASIC PHENOMENA

Background

According to Read and Shenoi [3], the Palmgrem-Miner rule is a reasonable basis for predicting the life of composite materials although it can yield unconservative estimates of their fatigue lives. Curtis [4] defines the ideal life estimation technique as one that enables the prediction of fatigue life and residual strength from a minimum amount of data, such as static strength and fatigue data on a small number of laminates. The process would then enable fatigue life prediction to be made for various laminate lay-ups and loading modes through extrapolation.

Stress Range

Generally for steel structures, only the dynamic stress range is of importance for fatigue. However, for GRP, recognition is required of the long-term level of static loading because this also damages laminates. Such damage acts as stress-raisers and contributes to fatigue development. Thus, not only is stress variation important but also the average long-term stress. This is catered for by the use of the stress ratio R, the ratio of minimum to maximum stress; a minus sign denotes compression. Thus, R = -1 represents equal compression and tension. R = 0.1 defines tensile (or compressive) loading with a minimum value near zero.

The approach adopted here exploits the ultimate limit state (ULS) results to define the relevant fatigue limit state (FLS) stress. In the present set-up, forces at one end of the tray are essentially static (lower catenary) whilst those on the other end are highly variable (upper catenary UC). Variations in the UC tensions have been determined for eight levels of sea states covering eight wave directions. These are presented in [5]: the two most onerous weather directions are repeated here in Table 1 and relate to 25 years of exposure. From Table 1, the range of riser tensions is Max. Tension minus Min. Tension. The long-term loading is the average of the Max. and Min. Tensions corresponding to Case 8. This is 63.5 kN for 180° and 70.5 kN for 225°. However, for all wave directions, 60 kN is a more appropriate. As permanent loading leads to damage, a first estimate of the tension range relevant for fatigue loading is to ignore Min. Tension values and assume the stress range is Max. Tension minus zero. The resulting equation to determine this stress range from the ULS results is

FLS stress range = ULS stress x (Max. Tension/Factored ULS Riser Tension)   (1a)

Equation (1a) applies to tray locations dominated by riser forces. In other regions, such as supports, these will stress as a result of overall tray action. The stresses thus alternate as the UC tension oscillates about the long-term average. From Table 1, the larger difference between the mean and Max. Tension or Min. Tension is in relation to Max. Tension. In this case, the stress amplitude is

FLS stress amp. = ULS stress x [(Max. tens.–mean)/(Factored ULS Riser Tens.)]   (1b)

Table 1. Riser tensions and numbers of cycles [5,6]

Case	Weather towards 180°		Weather towards 225°		No. of waves
	Max. Tens.	Min. Tens.	Max. Tens.	Min. Tens.	
1	193	4	180	16	140
2	128	21	127	34	1,681
3	107	31	112	40	22,253
4	89	42	95	51	215,807
5	79	49	84	59	961,688
6	75	52	77	65	4,792,695
7	72	55	73	68	18,058,912
8	72	55	71	70	85,559,648
				Total	109,612,824

This stress amplitude is tensile. For alternating loading this could be used as input to a $R = -1$ assessment. However, since the mean is tensile, and that the stresses should reflect this, this stress will be doubled (to give the range) and then assumed to be all tensile. In the assessment, this value was adopted, or the value from equation (1a), whichever was the larger. Part of the reason for assuming only tensile stresses rather than alternating is that, as demonstrated below, the corresponding fatigue design curve ($R = 0.1$) is considerably more onerous than the $R = -1.0$ curve.

For each laminate 'hot spot' the long-term distribution of relevant stress ranges is established. Fatigue life is then calculated using the Palmgrem-Miner's summation:

$$\frac{n_1}{N_1} + \frac{n_1}{N_1} + \frac{n_1}{N_1} + \ldots\ldots\ldots + \frac{n_1}{N_1} \leq 1.0 = D \qquad (2)$$

Fatigue life = 25/D $\qquad\qquad$ (3)

where n_1, n_2 etc. are the numbers of cycles in the design spectrum (Table 1) and N_1, N_2 are the corresponding maximum number of cycles obtained from the relevant S-N curve. As S-N or, as will be seen later, ε-N, curves are based on the ultimate strength of the material, it is necessary to check fatigue life for longitudinal and transverse tension and compression for all laminates, using a different curve in each case. Alternatively the ε-N curve can be used based on the appropriate choice of R. The fatigue life of the structure will be the smallest of these values.

S-N Curve

Outside the aerospace industry, composites are generally designed for static loading. Available data are limited because it is difficult to establish high cycle behaviour because of the amount of time involved. Thus fatigue tests are normally conducted up to 10^6 cycles or, at most, up to 10^7. Here, for a life of 250 years, the number of cycles is 10 x 1.096 x 10^8 = 1.096 x 10^9 (Table 1), ie, two orders of magnitude longer than most test data. Care is required when extrapolating beyond 10^7 cycles. A further issue here is that the loading (at supports) is reversible so that any test data that are to be exploited should relate to fully reversed loading, ie, R = -1 loading.

For large cycles, is there an endurance limit. The issue is complicated as most tests are conducted under constant amplitude loading whereas real sea conditions demand variable amplitude loading. Variable loading can eliminate the endurance limit or change the slope of the S-N curve beyond the endurance limit. At worst, no change of slope occurs for cycling beyond the endurance limit.

An important loading aspect is the angle relative to the ply direction. A summary of the effects of off-axis loading and ply make-up, uni-directional (UD), cross-plies, or angle-plies, is found in [7]. The tray laminates mainly comprise quadraxial plies interlaced with UD plies. Based on UD laminate behaviour loaded parallel to the fibres (0°), [7] identifies three fatigue failure mechanisms:

- composite fracture strain ε_c corresponding to fibre breakage and resulting interfacial debonding
- matrix cracking and interfacial shear failure
- fatigue limit of the resin ε_m that represents the lower bound of fatigue failure.

For UD laminates loaded at 90°, transverse fibre debonding (ε_{db}) occurs. Between 0 and 90°, fatigue failure mechanisms vary from ε_m (0°) to ε_{db} (90°). The transition occurs rapidly with angle as seen in Figure 1 (Figure 13 of [7]).

For cross-ply (ie, biaxial) laminates loaded parallel to one fibre, debonding initiates failure but, because the orthogonal plies arrest the debonding cracks, delamination (ε_{dl}) controls failure. The strain to cause delamination is smaller than the fatigue limit of resin but is greater than the debonding strain. Table 2 lists typical values. In Figure 1, the fatigue limits for UD laminates subject to off-axis loading correspond to the ε_m and ε_{db} strain values.

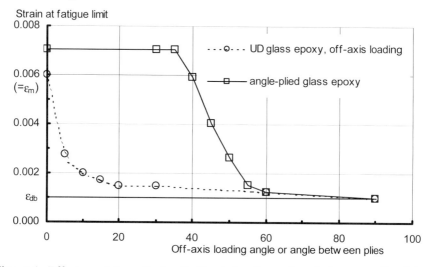

Figure 1. Effects on fatigue limits of off-axis loading and angle between plies [7]

Table 2. Typical values for glass-epoxy failure strain [7]

Failure Mechanism	Strain
Resin fatigue limit ε_m	0.0060
Delamination ε_{dl}	0.0043
Transverse fibre debonding ε_{db}	0.0010

As for off-axis loaded UD laminates, the delamination strain in biaxial laminates is also a function of loading axis. It is also a function of the angle between the fibre directions as demonstrated by the fatigue behaviour of cross-ply laminates. Tests conducted on symmetric cross-ply laminates with fibre orientations between ±30 and ±60° are illustrated in Figure 1 [7]. They demonstrate no apparent fatigue degradation for orientations less than ±35° with a relatively rapid reduction in fatigue strength to the debonding limit for orientations approaching ±60°. Thus the delamination strain for quadraxial laminates should correspond to the ±45° cross-ply results, ie, $\varepsilon \approx 0.004$ This is confirmed [7] where the delamination limit for two laminates composed of combinations of 0, +45, -45, 90° plies is $\varepsilon_{dl} = 0.0046$.

This observation of the variation of delamination limit with loading axis and ply angle explains the result presented in Figure 10 [8]. Here the fatigue behaviour of a cross-ply laminate subject to loading at 45° to the warp was noted to be significantly different from the fatigue behaviour of the same material loaded at 0°. Where loading is at 45°, the knee in the curve occurs at around 10^4 cycles whereas at 0° and more generally, it occurs at 10^5 cycles. The trays are reinforced with plies at no more than ±22.5°, so it is reasonable to assume that the knee occurs at 10^5 cycles.

On the basis of the above, the S-N curve for the E-glass/polyester laminates was determined as a straight-line fit with the stresses plotted as linear and the number of cycles to failure as log-linear (to the base 10). The equation for this is

$$S = S_{max} - m \log N \qquad (4)$$

where S is stress, S_{max} is the intercept for N = 1, ie log N = 0, and m is the slope of the curve. A bilinear curve is required where an endurance limit exists.

S_{max} and m are established from test data. The mean curve is found as a fit the data. To account for the spread of data, the design S-N curve is the mean minus two standard deviations. When insufficient data exist by which to determine the scatter, a lower bound serves the same purpose. Here a generic approach to determine the lower bound curve is adopted and then demonstrated to provide a lower bound through comparison with fatigue test data.

Mean S-N Curve
Test data demonstrate an N = 1 FLS strength in excess of the ULS. To generalise the S-N curve, this increase is ignored and S_{max} is taken as the ULS in the direction of loading. Thus, for N = 1 and log N = 0, S = $S_{ult.}$ To determine the endurance limit, use is made of the observation in relation to test data for an isophthalic polyester laminate that at around 10^5 cycles, the mean stress approximates 0.4 of the stress at N = 1. It follows that if S = 0.4 S_{ult} when N = 10^5, m = 0.12 S_{ult}.

$$S = S_{ult} - 0.12\ S_{ult} \log N \qquad (5)$$

For the lower curve, delamination governs long fatigue life [7]. For epoxy-based glass laminates with quadraxial reinforcement, the delamination strain is 0.0046 [7]. Reducing the delamination limit where polyester resins are involved suggests 0.004 is appropriate. However, [8] provides no indication of whether a corresponding limit on cycles exists. For this, use is made of the results in [9] covering curve-fits to fatigue data for a range of polyester resins including isophthalic and flexural fatigue test data for the same range of resins when reinforced with alternating plies of glass mat and woven roving. The loading applied was R = -1.

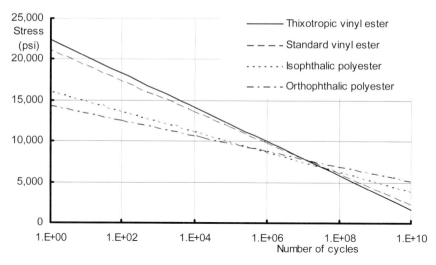

Figure 2. Curve-fit of ASTM D671 fatigue data for various unsaturated polyester resins (R = -1) [9]

Figure 2 presents polyester resin fatigue curves, extrapolated following a curve fit to 10^{10} cycles: the limit in [9] was 10^7 cycles. At 10^{10} cycles, the fatigue strength of the isophthalic resin is 3,855 psi ($\equiv$ 26.6 MPa): the corresponding strain is 0.00714. In Figure 3 (Figure 7b of [9]), a comparison is presented between flexural test results for reinforced isophthalic polyester [9] and the corresponding unreinforced resin curve from Figure 7a. The close correspondence between the resin-only-based curve and the test results is clear which supports a limiting strain of 0.004 at 10^{10} cycles as appropriate for the mean curve. Solving the generalised S-N equation for this condition for a typical tensile modulus of 17,100 MPa and assuming again that S = 0.4 S_{ult} at N = 10^5 cycles

$$S = 0.517\ S_{ult} - 0.0234\ S_{ult}\ \log N \tag{6}$$

Design (Lower Bound) S-N Curve
A lower bound is found by assuming that at N = 1 S is limited to 0.9 S_{ult}. For the lower curve, a larger safety margin is provided by adopting a limiting strain of 0.002, ie, half the value adopted for the mean curve. The equation for the curve is

$$S = 0.9\ Sult - 0.12\ Sult\ \log N \qquad \leq 105\ \text{cycles} \tag{7}$$
$$S = 0.459\ Sult - 0.0317\ Sult\ \log N \quad > 105\ \text{cycles} \tag{8}$$

A typical comparison of predictions given by the design S-N curve {equations (7) and (8)} is seen in Figure 3. Other comparisons have been made. Those based on orthophthalic resin laminates demonstrate, as shown in Figure 2, that provided the data relate to specimens loaded to 106 cycles or less, smaller fatigue lives than those achieved by isophthalic resin laminates will be realised.

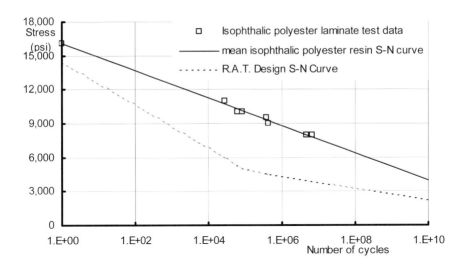

Figure 3. Comparison of isophthalic polyester laminate fatigue data with proposed design curve [9]

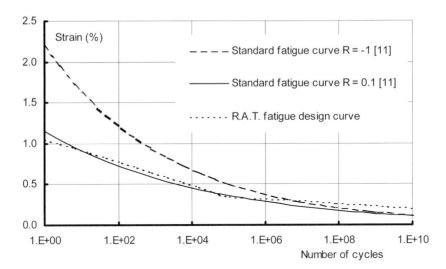

Figure 4. Comparison of fatigue design curves

STAGE 2 - ε-N CURVE

The chosen ε-N fatigue curve is independent of laminate ULS strength and so applies to all laminates. The R = 0.1 design curve is [11]

$$\log \varepsilon = 0.063 - 0.101 \log N \qquad (9)$$

where ε is % and given in amplitude terms. A curve for R = -1 is also provided [11]. Figure 4 presents a comparison of all three curves from which it can be seen that the derived S-N curve lies very close to the R = 0.1 ε-N curve except for cycles in excess of 10^6. Further, as noted earlier, the R = 0.1 ε-N curve can be seen to lead to considerably shorter fatigue lives than the corresponding R = 1 ε-N curve except for lives in excess of 10^8 cycles when little difference is likely.

REFERENCES

[1]   Hill, P S and Cuthill, J, "Advanced Composite Riser Arch Trays - The First Major Use of Fibre Reinforced Plastics for Primary Structures in the Offshore Industry", SAMPE Europe 22nd International Conference, Paris, March 2001.

[2]   NPD "Regulations for the Design of Load-bearing Structures Intended for Exploitation of Petroleum Resources", Norwegian Petroleum Directorate 1997.

[3]   Read, P J C L and Shenoi, R A, "A Review of Fatigue Damage Modelling in the Context of Marine FRP Laminates", University of Southampton, Southampton, May 1994.

[4]   Curtis, P T, "The Fatigue Behaviour of Fibrous Composite Materials", Journal of Strain Analysis Vol 24 No. 4 1989.

[5]   Rockwater, private communication.

[6]   Rockwater, private communication.

[7]   Talreja, R, "Fatigue of Composite Materials: Damage Mechanisms and Fatigue Life Diagrams", Proc. R. Soc. London., A378, 461-475, 1981.

[8]   Boller, K H, "Fatigue Characteristics of RP Laminates Subjected to Axial Loading", Modern Plastics, p145, June 1964.

[9]   McCabe, R T, Burrell, P and de la Rosa, R, "Cycle Test Evaluation of Various Polyester Types and a Mathematical Model for Projecting Flexural Fatigue Endurance", 41st Annual Conf., Reinforced Plastics/Composites Institute, The Society of the Plastics Industry, Inc., January 27-31, 1986, Session 7-D.

[10]   Owen, M J and Smith, T R, "Some Fatigue Properties of Chopped-Strand-Mat/Polyester-Resin Laminates", Plastics & Polymers, February, 33-44, 1968.

[11]   Echtermeyer, A T et al, "Comparison of Fatigue Curves for Glass Composite Laminates", Design of Composite Structures against Fatigue - Application to Wind Turbine Blades, Ed. R M Mayer, MEP, Ch 14, p 209, 1996.

Session 8 : **Case studies**

8.1 The design, construction and in-service performance of the all-composite Aberfeldy footbridge

J Cadei and T Stratford
Maunsell Limited, Beckenham, UK

INTRODUCTION

The world's longest span advanced composite bridge was opened on 3rd October 1992. It crosses the River Tay in Scotland, where it connects the two halves of Aberfeldy golf course. The bridge combines a variety of innovative advanced composite technologies, including a pultruded glass-fibre-reinforced-polymer (GFRP) deck and aramid cables [1]. The bridge was fabricated on site with minimal heavy equipment, causing significantly less disruption than a conventional steel or concrete structure, and offered a cost-effective solution to the golf club [2].

This paper reviews the design, construction and in-service performance of the bridge, as we approach the tenth anniversary of its construction.

DESIGN

The Aberfeldy Bridge is a symmetrical cable stayed footbridge, with A-frame towers (Figure 1). The deck has an overall width of 2.12 m, and a total length of 113m, made up of a 63m main span over the river, a two back spans of 25m. The deck is stiffened by 4 fans of 5 cables, anchored to the ground via short aluminium columns under the back spans. Figure 2 shows the bridge in elevation, and indicates the composite materials used in its construction.

The bridge contains 14.5 tonnes of composite material. The GFRP deck, towers and parapets account for the majority of the composite in the structure. These are combined with aramid stay cables.

Design loads

The bridge was designed to carry live loading of 5.6kN/m ($3.52kN/m^2$) in accordance with UK Highways Agency Departmental Standard BD 37/88. The dead weight of the bridge is significantly less, at only 2.0 kN/m including 1.0kN/m ballast. Wind and temperature design loads were also to BD37/88.

ACIC 2002, Thomas Telford, London, 2002

Figure 1: The Aberfeldy bridge (October 2000)

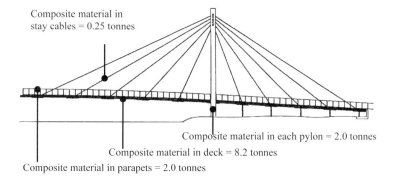

Figure 2: Elevation of the bridge, showing material quantities.

Design criteria for dynamic behaviour

Due to the bridge's high live to dead load ratio and slender proportions control of dynamic effects was a critical design issue. To this end, ballast in the form of concrete filling of the cells of the central deck panel was specified. Its purpose was to prevent uplift under wind, improve the transverse mass distribution of the deck, improve safety against flutter instability, and improve footfall behaviour. The bridge was designed to meet the aerodynamic stability rules subsequently incorporated in BD 49/93. Given the intended use of the bridge in a private golf course, the bridge was not designed to comply with the allowable footfall response specified in BD 37/88.

Deck and towers

The deck and towers of the bridge were wholly fabricated from the Advanced Composite Construction System (ACCS). This comprises a small number of modular component types which are pultruded from E-glass fibre and isophthalic polyester resin and are connectable by a combination of bonding and toggle type mechanical connectors.

A Limit-State design methodology was used, with factors of safety based on previous Reliability theory based development work. The thin-walled composite components are generally governed by allowable strain at the Serviceability Limit State, or buckling at the Ultimate Limit State. Critical buckling modes include both global buckling and local buckling of the walls of the sections.

The components were joined using a combination of bonding and mechanical interlock (Figure 3). Over 2 km of mechanical toggles were used, which draw the bonding surfaces together to a predetermined bondline thickness, to give a high quality joint that can be readily and accurately assembled under site conditions.

Figure 4 shows the structural form of the bridge deck. The deck is formed by 600 mm wide longitudinal panels and is stiffened by edge beams and cross beams formed from 80 x 80 mm '3-way connector' components. Primary cross beams are provided at 5.73 m intervals at cable anchorage locations, and secondary cross beams at 955 mm intervals. A rubber surfacing membrane (formerly a conveyor belt) protects the deck.

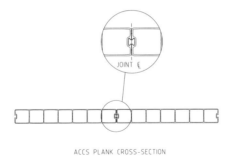

ACCS PLANK CROSS-SECTION

Figure 3: Interlock of ACCS components

Parapets
The parapets were constructed from non-ACCS pultruded GFRP sections. The parapet posts pass through holes in the bridge deck, and are anchored to the deck by a combination of a bonded boundary angle above the deck and dowels below the deck. Top and bottom rails span between the posts, with closely-spaced CHS vertical members between the rails. Dundee University performed load tests to verify the strength of the parapets.

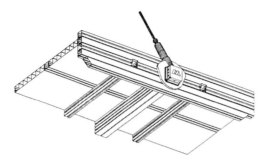

Figure 4: The arrangement of the bridge deck [6].

Cables
The bridge deck is supported by Parafil rope stays. These lightweight cables comprise a core of parallel Kevlar-49 fibres, sheathed in low density polyethylene.

Details at the end of the cables (where high stress concentrations occur) were fabricated from conventional construction materials. The cable terminations and the deck connections are in aluminium, while the tower connections are in galvanised steel. Testing was undertaken by Dundee University to check the strength of these connections.

Foundations
An important advantage of lightweight composite construction is that the Aberfeldy Bridge could be founded on simple pad footings. The main criterion dictating the size of the foundations was the weight required to resist uplift, rather than the contact area to control settlement.

CONSTRUCTION
Fabrication and erection was carried out by a small team of students from Dundee University (with support from the consultant and an engineering management contractor), and took just 8 weeks. No expensive construction plant or major temporary works were required, with minimal disruption and no damage to the golf course.

Fabrication
The tower legs were fabricated by bonding under factory conditions off-site. The deck, on the other hand, was supplied to site in its component parts, and fabricated on one side of the river on the axis of the bridge. Fabrication was carried out under cover of a tented structure, which protected the deck from the elements during the 24 hours required for the adhesive to cure.

Erection

The method of erection used for the Aberfeldy Bridge is unique amongst cable-stayed structures, and was rendered possible by the use of lightweight materials [3].

The tower legs were assembled on the ground and hinged to their concrete footings. They were then raised using an all-terrain forklift to an angle of about 30° above the horizontal, whence they were lifted to the vertical position by means of simple Tirfor hand winches, and held in position by guy ropes (Figure 5). The cable stays and primary cross-beams were then installed, and a temporary cable net created to hold the cross beams in position across the river. The cross-beam and cable net formed a framework over which the deck could be pulled into position by incremental launching, using a winch on the far side of the river.

Figure 5: Erection of the North Tower.

DECK STRENGTHENING FOR INCREASED LOADINGS

The bridge was overloaded when it was crossed by a small tractor towing a trailer of sand. As a result of the high concentrated wheel loads, cracks formed in the top surface of the GFRP deck, parallel to the webs of the cellular sections.

To remedy the situation, the deck was strengthened during the spring of 1997. GFRP pultruded plates were bonded to the top of the deck (beneath the rubber surfacing), with rivets providing fixity while the adhesive cured. At the same time, CFRP pre-preg sheets were applied to the deck edge beams on either side of the primary transverse beams, to handle the increased cable reactions (Figure 6). The strengthening added some 0.17 kN/m to the weight of the structure.

IN-SERVICE PERFORMANCE

Because of its innovative character, the in-service behaviour Aberfeldy Bridge has received considerable interest. A team led by Bill Harvey, then of the University of Dundee, undertook a monitoring programme immediately following the opening of the bridge. At later stages two other research teams carried out field tests to determine the bridge's dynamic performance.   The bridge has also received a number of condition surveys.

Figure 6: CFRP strengthening of the edge beam

Dynamic performance

Two experimental investigations into the dynamics of structures under footfall loading have been conducted on the bridge, in 1995 [4] and 2000 [5]. The latter investigation was prompted by recent experience with the Millennium bridge in London, and investigated the possibility of lock-in of pedestrian loading with the lateral vibration of the bridge. (The results showed that lock-in could occur, but were inconclusive).

Details of the first few natural modes of vibration determined by these studies are shown in Figure 7, and listed in Table 1. The damping ratio is referred to the critical damping level.

During the tests conducted in 1995 [4], a peak acceleration of 0.22g was measured when a person deliberately walked with a pace coinciding with the first fundamental natural frequency of the bridge. This acceleration is considerably higher than the maximum acceleration criterion in BD37/88.

Between the two investigations the deck was strengthened to cater for the increased loading from golf buggies. As a result of the bonding of GFRP plates to the upper surface of the deck [5], the mass of the deck, and the stiffness of the deck (in particular, the lateral stiffness) were increased. As a consequence, the natural frequencies measured during the tests in 2000 were marginally lower, and the dynamic stability of the bridge improved.

The design of the bridge ensured that the first torsional and vertical natural frequencies are well separated, to maximise the critical wind speed for aerodynamic flutter. This was achieved by incorporating ballast along the centre line of the bridge.

The damping ratios measured at Aberfeldy (Table 1) are not significantly different to the value of 0.75% determined for the deck in isolation during a LINK research programme into the ACCS system (with the exception of the low damping ratio for the first vertical mode suggested by the study in 2000). This suggests that the other bridge components, principally cables, parapets, and surfacing, do not make a significant contribution to damping. The parallel-lay aramid fibres in the cables dissipate far less energy than spiral-wound cables, and the thin rubber surfacing contributes little to damping. Although the parapets were fitted with frictional sliding joints to improve damping, these do not appear to have been effective, due to looseness in the sliding joints and in the connections between the parapet posts and the deck. Pavic et al [5] noted that the damping increased with the number of people on the bridge.

Table 1: Dynamic response of Aberfeldy bridge, from studies in 1995, and 2000.

Mode*	Frequency (Hz)		Damping ratio for empty structure (%)	
	1995	2000	1995	2000
H1	1.00	0.98	-	1.0
V1	1.59	1.52	0.84	0.4
V2	1.92	1.86	0.94	0.7
V3	2.59	2.49	1.20	0.7
H2	2.81	2.73	-	1.2
V4	3.14	3.01	-	0.8
T1	3.44	3.48	-	5.5
V5	3.63	3.50	-	0.6
V6	4.00	3.91	-	0.9
T2	4.31	4.29	-	3.2
V7	4.60	4.40	-	0.8
V8	5.10	4.93	-	1.8

* H = horizontal, V = vertical, T = torsional

Weathering
Within its first year of service, the structure withstood hurricane force winds, unprecedented snowfall, and flooding to above deck level in the back spans [6]. These moderately extreme adverse conditions did not cause damage, and the primary structure of the bridge continues to perform very well.

However, from a surface weathering point of view, some details have fared less well, although in no case has weathering adversely affected the overall structural safety of the bridge.

Superficial weathering effects
There has been considerable erosion of the resin-rich surface layer of the parapet components, in some instances exposing the fibres. The parapets were fabricated

from non-ACCS GFRP sections, which were pultruded to a different specification from the ACCS components in the primary bridge structure. These were the only suitable sections available within the cost and time constraints at the time. This shows how important the detailed specification of resin and manufacture of these materials is in practice. The worst affected areas are the upper surface of the handrails and the lower region of the posts. The handrail erosion is likely to be the combined effect of environmental weathering and abrasion, since resin loss is less widespread on the lower rail. The handrail is inevitably subjected to a large amount of wear, both due to its normal function, and to people who climb over the parapet to jump into the river below. Another region where surface resin loss has been observed is at the base of the posts, many of which are badly scuffed.

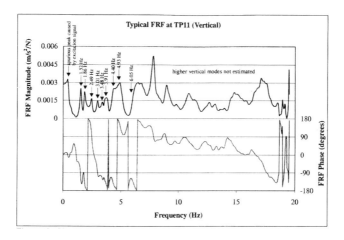

Figure 7: The frequency response of Aberfeldy bridge, from the 2000 study [5]

There is very little loss of surface resin over the remainder of the structure. The ACCS components incorporate a protective surface veil that guarantees a resin rich surface of constant thickness. One exception, however, is a substandard ACCS panel that was used as a plaque, which has suffered significant resin loss, and delamination cracking. This panel is not subjected to any load, and thus degradation is purely driven by the environment.

Performance of the parapets
Flexing of the deck is accompanied by movement of the parapets (Figure 7). This was recognised during design, and the parapet rails were connected to posts using sleeved connection to a FRP plate pinned to the posts so as to allow some relative movement. However, the connections have generally worked loose, and in some cases the rails no longer remain in their sockets. Furthermore, many of the post to deck connections are loose. These effects are the result of movement cycles of the deck and could be avoided by giving the parapet connections greater   movement capacity or by reducing deck displacements by means of a stiffer cable system.

In some cases, the parapet posts have suffered impact damage near their base, leading to delamination. Golf buggy use was not envisaged in the original design, and the parapet was not designed for impact. Thus, damage would also have occurred even if the posts had been made from conventional materials. Such damage could be mitigated by the addition of protective kick-boards at the base of the railings.

Mould growth
Both the parapets and the primary structure have been affected by mould and moss growth, especially on north-facing surfaces and in the gutter areas where due to lack of maintenance free drainage is impeded and water is trapped (Figure 8). This is due in part to standing surface moisture retained by dirt and debris, and in part by moisture absorption of the composite, which, in contrast to the aluminium components, which are free from mould, absorbs up to 1.5% by weight of moisture. Similar effects can be observed on traditional masonry structures in the area, due to the damp climate. The mould growth observed has little impact on structural strength, however, being primarily a maintenance and aesthetic issue. Due to the modular construction system, the edge beam inevitably contains grooves and indentations, and these also act as moisture traps, providing a foothold for minor mould growth.

Figure 8: Looking north along the bridge, October 2000.

A strategy towards managing mould growth under similar exposure conditions might include selection of appropriate colours for the composites to minimise the visual impact of mould growth, the use of mould-inhibiting additives in the resin system, detailing to avoid water traps, and a maintenance regime to ensure that drainage paths are kept clear of leaves and other debris.

CONCLUSIONS

The primary structure of the advanced composite Aberfeldy bridge has proved very successful in demonstrating the feasibility of advanced composite materials for constructing long span bridge structures cost-effectively. It has been readily strengthened to cater for additional loading imposed by golf buggies by the addition of carbon fibre composite to the upper surface of the deck. The dynamic behaviour of the bridge is relatively lively, on account of its ultra-lightweight construction and its below expected damping value, but the dynamic response is considered acceptable for the bridge's intended use. Studies have demonstrated that the dynamic response can be readily controlled by calibrating the magnitude and distribution of the mass of the deck, and tailoring the damping of the parapets, cables, and surfacing.

The GFRP parapets have proved less durable than the primary structure, due to movement of the joints associated with flexing of the deck and the different specification of the parapet components. The surface resin has been eroded from the top handrail (exposing the glass fibres), and some of the posts have suffered impact damage from golf buggies. Both these problems can be readily mitigated.

The structural use of advanced composite materials continues to grow, and systems are now available which are capable of carrying full highway loading [7]. Ten years after its construction, the Aberfeldy footbridge remains an innovative demonstration of the structural use of composites, and continues to attract world-wide attention.

REFERENCES:
[1] Harvey W.J., "A reinforced plastic footbridge, Aberfeldy, UK", Structural Engineering International, 4/93, pp. 229-232

[2] Lee D.J. (1993) "Project Linksleader: The first major cable-stayed GRP bridge." FIP symposium, "Modern prestressing techniques and their applications" Kyoto, Japan: FIP. Vol.2 pp. 671-678

[3] Burgoyne C.J., Head P.R., "Aberfeldy bridge – an advanced textile reinforced footbridge", TechTextil symposium, Frankfurt, Germany, June 1993.

[4] Pimentel R.L., Waldron P., Harvey W.J., "Assessment of the dynamic behaviour of Aberfeldy GRP plastic cable-stayed footbridge", Seminar on

Analysis and Testing of Bridges, pp. 38-40, Institution of Structural Engineers, London, April 1995.

[5] Pavic A., Reynolds P., Cooper P., Harvey W.J., "Dynamic testing and analysis of Aberfeldy footbridge", Final report, Ref. CCC/00/79A, The University of Sheffield, Department of Civil and Structural Engineering, Vibration Engineering Section, 31 July 2000.

[6] Head P.R. "Cost effective advanced composite structures designed for life in the infrastructure", BPF congress, 1994.

[7] Daly A.F. "Developments in modular FRP decking for highway bridges", Bridges 2001, Aston University, Oct 2001.

8.2 The use of advanced composite laminates as structural reinforcement in a historic building

H N Garden[1] and E G Shahidi[2]
[1]Taylor Woodrow Construction, Southall, UK
[2]The Advanced Composites Group Ltd, Heanor, UK

ABSTRACT
This Paper describes the development and implementation of a novel system for restoring the capacity of two principal steel beams to above their original levels. The beams had suffered a loss of flange and web section of approximately 30% due to electrochemical corrosion. They support floors on which commercial businesses operate as tenants within the Grade II listed building. The beams also support Terra Cotta cladding units, which comprise the architectural façade of the structure. Since the corrosion damage led to a loss of section, the beams had to be re-evaluated for their structural suitability as part of the asset management programme.

It was necessary to enhance the beams to 30% above their original flexural capacity to allow for future changes of floor loading. The options were to either replace or strengthen the beams. Replacement was dismissed due to the long lead time for new beams, which were curved in plan. The use of advanced composite materials was shown by Taylor Woodrow to be a viable alternative, which satisfied cost and programme constraints. Glass and high modulus carbon fibre epoxy prepregs were used for this application.

The CFRP material provided a rapidly installed reinforcement system, allowing the work to be undertaken during a strict period of just two weeks. This short time ensured there was no delay to the main contract, which included extensive cathodic protection services, designed and supervised by Taylor Woodrow, to the steel frame of the building.

The laminates were installed using vacuum bag technology, using a low temperature moulding (LTM) prepreg, in which the material was held under vacuum pressure against the beams' profile, and the local temperature was raised to promote full cure of the epoxy resin matrix. The resin achieved cure at 60°C.

INTRODUCTION
The Use of Composites for Strengthening
The rehabilitation of structures has been achieved for many years using externally bonded steel plates. Disadvantages of this method include transporting, handling

and installing heavy plates and corrosion of the plates. The use of composite materials overcomes these problems and provides equally satisfactory solutions. For members of unique profiles, steel is not a convenient rehabilitation material since it does not conform to the parent shape. However, moulded materials, in which the member to be strengthened is the mould, are appropriate. Polymeric composites offer particular advantages, including high strength- and stiffness-to-weight ratios as well as excellent environmental durability. In dynamic circumstances, they show outstanding fatigue behaviour.

It is both environmentally and economically necessary to apply rapid, effective and simple strengthening techniques for the upgrading of existing civil infrastructure. This is certainly the case for historic structures, in which modifications must be sympathetic to the original appearance of the structure due to heritage preservation requirements. Bodies such as English Heritage stipulate strict limits on the modifications that may be made to listed buildings.

Composite materials, also known as fibre reinforced polymer (FRP) materials, comprise strong and stiff reinforcing fibres embedded in a polymer matrix resin. The main material types finding uses as reinforcing fibres in FRPs are glass, carbon and aramid. FRPs are generally more expensive than mild steel but the majority of the cost of a strengthening scheme is labour; the easy handling of FRPs reduces labour costs considerably.

The use of bonded composite reinforcement, to carry only live or live and dead loads in concrete structures, has been reviewed in several previous publications by the first author and many others. An extensive reference list will be found in Garden [1]. For steel structures, FRPs have gained more recent recognition.

HISTORY OF THE BUILDING
The Boots Building, High Street, Nottingham, shown in Figure 1, is a skeletal-framed building clad in Terra Cotta. The building was constructed in two phases in 1903 and 1921, the latter phase being an extension of the first. The Boots building is now a Grade II Listed structure. The early section of the building has a frame consisting of cast iron columns and wrought iron beams, while the 1921 section has a steel frame. All Terra Cotta is built tightly about the structural frame. The 1921 section of the building was showing physical signs of distress, typical defects including the opening of mortar joints, the cracking of Terra Cotta blocks and the displacement of some blocks.

Previous investigation work identified that deterioration of the 1920s Terra Cotta façade was the result of expansive corrosion of the underlying steel frame. This investigation enabled Taylor Woodrow to examine the construction details and perform a number of preliminary tests to determine the suitability of cathodic protection as an appropriate method of controlling corrosion of the steel frame and preventing further loss of section. In addition to this, Taylor Woodrow proposed the use of advanced composite materials to rehabilitate two principal steel beams at the second and third floor levels.

Figure 1: Elevation view of the Boots Building, showing the curved corner section

The Problem to be Overcome

One of the beams in need of strengthening is shown in Figure 2. The corrosion product can be seen on the flanges and the web. The beams were curved in plan to connect two straight members around the corner of the building. The detail of the connection between the curved and straight beams is shown in Figure 3. The purpose of the strengthening scheme was to raise the flexural capacity of the beams to above their original, uncorroded level so that an anticipated increase in floor loading could be accommodated.

DESIGN AND INSTALLATION CONSIDERATIONS

The Site Environment

The activities on site comprised the beam strengthening, cathodic protection installations for the steel frame, and related refurbishment activities inside the building. The external works were undertaken from scaffolding access. It was important that the beam strengthening be achieved with minimal disruption to other work on the scaffolding. The use of easily and rapidly installed composite laminates satisfied this criterion.

Figure 2.   Corroded steel beam in need of restoration

Figure 3. Bolted connection detail by which the curved and straight beams are joined. The grey appearance is due to the corrosion product having been grit blasted off to reveal the bare steel beneath.

Design of the Reinforcement

The purpose of the rehabilitation work was to restore the flexural and torsional capacities of the beam sections. The composite material acted as additional material bonded to the steel to provide the increase in capacity. The beams were analysed using the QSE program to confirm that the quantity and type of composite designed would achieve the desired increase in beam capacity. A combination of glass woven fabric and high modulus carbon / LTM prepreg was used to achieve the stiffness and strength requirements.

The thickness of the composite strengthening layer comprised unidirectional, 0° and $\pm 45^{\circ}$ fibre orientations. The unidirectional fibres were aligned along the direction of the beams' length for flexural strengthening. The $\pm 45^{\circ}$ fibre glass orientations were used to improve shear and torsional stiffness, necessary to resist loading generated by the curvature of the beams. The high modulus carbon layers provided high flexural stiffness with a minimum of material thickness.

The bonding of pre-cured carbon strips (ie. pultrusion) was ruled out for this application by the curvature of the beams in plan. The wet lay-up method was rejected because the laminating area was restricted and the risk assessments showed that there were concerns about handling wet resins in a congested area on the site. The use of liquid resin infusion was considered but was rejected on the grounds of practicality in this particular application. The use of LTM prepreg technology provided the required level of flexibility. The materials were easy to work with and offered a practical solution. The distortion of the laminate was prevented by the tack of the resin in the prepreg layers. The use of low temperature curing enabled a high quality repair without specialist equipment on site, as explained in the following sections.

INSTALLATION PROCEDURE AND CONTROL OF WORKS

Access to the Faces to be Strengthened

A number of the Terra Cotta units were removed as part of the cathodic protection and related refurbishment works. This provided access to all faces of the steel beams. The local environment of each beam was protected within a tented enclosure. This was to isolate the area from surrounding activities and ensure that the bonding surfaces of the beams were kept free of dust. The enclosure also maintained a warm environment for the purpose of curing the resin in the composite materials.

Surface Preparation of the Steel and Composite Material

The steel surfaces were grit blasted to reveal the bare steel beneath the corrosion product, as shown in Figure 3. The cleaned surface was solvent-washed to remove grease other contaminants, after which the beams were kept dry using silica gel packs placed on the beams, with the beams enclosed in polythene. An epoxy adhesive was painted on the cleaned and dry steel surfaces, as shown in Figure 4. The purpose of this was to act as a bonding aid. This epoxy adhesive was an ambient-cure material – ie. required no elevated temperature to set. It provided a good bonding surface for the subsequent composite layers.

Epoxy Adhesive Curing Régime and Installation Procedure
The composite reinforcement was in the form of continuous fibres in woven and unidirectional format, with resin already pre-impregnated throughout the fibre arrangement. Prior to installation, the resin is prevented from premature curing by keeping the prepreg in a chilled environment (eg. a freezer). The prepreg layers are flexible when taken from the freezer, so they can be moulded around any given profile, as shown in Figure 5. When in place, the layers are compacted under a vacuum bag and heated, which causes the resin to flow throughout the fibres and encourages full cure development of the resin.   Also, the vacuum promotes expulsion of air from the resin, ensuring low voidage. This is important because the resin transmits the mechanical properties of the fibres better if there is a minimal void ratio. The cured resin is solid, so the prepreg remains bonded to the mould (ie. beams in this case) in the formed shape. The tented enclosure on site helped to maintain the elevated temperature in the vicinity of the beams.

Figure 4. Painting of epoxy adhesive bonding aid onto steel beam surfaces before installation of the prepreg composite materials

The vacuum bag was placed around the beams, as shown in Figure 6. This ensured the material conformed to the profile of the beams, which was particularly important in the corner radii between the webs and flanges.

The completed strengthening is shown in Figure 7. The thickness of the composite material added to the steel section was approximately 5mm, which did not affect the positions of the Terra Cotta units placed back on the structure. Therefore, the requirements of the heritage authorities were met.

CONCLUSIONS AND CLOSING COMMENTS
The low temperature moulding prepreg solution offered a unique opportunity to address a time- and cost-sensitive rehabilitation problem, with additional constraints of limited access.

Figure 5. Placement of a carbon prepreg layer around the flanges and web of a steel beam

The prepreg method offers the ability to specify fibres of various moduli (typically 400 to 1000 GPpa) and strengths, in addition to having good control of the resin content. The quality and content of the resin is critical to ensure translation of the fibre properties into the final laminate with maximum efficiency. The fibre volume fraction was high (65%) and the vacuum method ensured a low resin void content (below 4%).

The quality of the resulting laminate brought about a reliable and cost-efficient solution in this application. Other material processing options would not have been suited to the unique problems associated with this structure.

Figure 6. A vacuum bag used to apply pressure on the composite material.

Figure 7. The completed strengthening to one of the beams.

The key points to be taken from this example are:

- Steel and cast iron structures can easily be rehabilitated using composite materials, irrespective of the geometry and complexity of shape.

- Section properties such as stiffness and strength for steel and cast iron members can be increased significantly using high and ultra-high modulus carbon fibres with the LTM prepreg technology.

- The flexibility of being able to take advantage of fibres and fabric reinforcements with a range of physical properties, and the ability to tailor the properties to achieve the desired stiffness and strength, were easily exploited in the repair.

REFERENCES
[1] Garden, H.N. (1997), 'The strengthening of reinforced concrete members using bonded polymeric composite materials', *PhD Thesis*, University of Surrey.

8.3 FRP bridge retrofit applications in New York State

O Hag-Elsafi, S Alampalli, J Kunin and R Lund,
NYSDOT Albany, New York, USA

ABSTRACT
Three projects demonstrating application of bonded Fiber Reinforced Polymer (FRP) materials for bridge retrofit in New York State are briefly discussed in this paper. The projects include a reinforced concrete bridge capbeam, a deteriorated precast prestressed concrete box beam, and a reinforced concrete T-beam bridge.

INTRODUCTION
Sometimes, concrete bridges are classified as deficient due to loss of steel or prestressing strands as a result of corrosion. Many of these bridges are often replaced or posted for lower loads for being unable to carry current legal loads. Using FRP materials to retrofit suspected deficient members of these bridges is a viable alternative to replacement or load posting to keep many of these bridges in service. Realizing this potential, New York State Department of Transportation (NYSDOT) has recently used these materials in a number of retrofit projects to investigate this issue. A brief discussion of three of these projects is presented in References [1-4]. They include retrofit of a reinforced concrete capbeam, a deteriorated prestressed concrete box-beam, and a reinforced concrete T-beam bridge.

BRIDGE CAPBEAM RETROFIT
In this application, FRP composite plates were used to strengthen the capbeam of Pier 3 of East Church Street Bridge in Chemung County, New York. Addition of a concrete wearing surface and a median barrier to the bridge superstructure substantially increased dead load, leading to deficiencies in moment and shear capacities of the capbeam structure. As a result, the concrete beam suffered flexural and shear cracking, and was considered for strengthening using bonded FRP carbon-glass hybrid plates (Figure 1).

Load tests were performed before and after the plates were installed, to investigate effectiveness of the strengthening system [1]. The load test results, before and after retrofit, for the negative moment region are shown in Figure 2. These results demonstrate linear behavior during the testing and prove consistency of the test data, reflected by the relatively small scatter of recorded strains about the best-fit line. Comparing the before and after retrofit results in Figure 2, it can be concluded that the FRP plates system reduced service load stresses in the steel reinforcing bars by about 10 percent in the negative moment region.

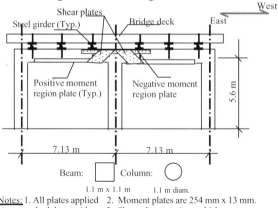

Figure 1. Bridge capbeam retrofit using FRP plates.

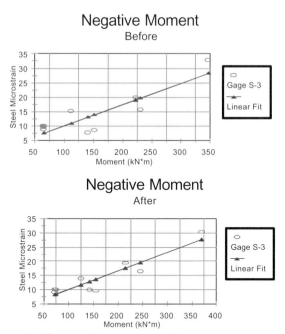

Figure 2. Strain results for negative moment region gauges

PRESTRESSED CONCRETE BOX-BEAM RETROFIT

The objectives of this project were to investigate the feasibility of retrofitting deteriorated prestressed box-beams using FRP materials, and to develop design and construction guidelines for the FRP strengthening retrofit system [3]. Testing and analytical programs were initiated to accomplish set project objectives. The testing program included full-scale load testing of three 70-ft long beams at the Federal Highway Administration's (FHWA) Turner Fairbank Highway Research Center in Virginia. The first two beams were unstrengthend and the third was strengthened using 11 carbon fiber sheets bonded to the beam's bottom flange (Figure 3). The first beam was pre-cracked with about 25% of the strands broken, the second was intact with no visible signs of deterioration, and the third was badly deteriorated with over 50% of the strands broken. The tested beams were all removed from a decommissioned bridge in New York State. The analytical program included development of two independent Mathcad programs for analysis of strengthened and unstrengthened prestressed box beams. The first Mathcad program was developed to estimate current capacity of a deteriorated beam, and the second for design of the FRP retrofit system and investigation of its effect on beam behavior. Design of the system was based on restoring a beam's flexural capacity to its as-designed ultimate, and limiting laminate strain to a theoretical maximum of 1.5% . On this basis, the strengthened beam was expected to fail at a total load (Load P in Figure 3, applied on two load cells) of 60 kip. However, because of unsymmetrical cracking in the beam during the testing, the laminates debonded prematurely on one side of the beam at about a 30 kip load. The results for the FRP gauges mounted at the midspan in Figure 4 indicate that the FRP laminate strain at failure was about 4750 microstrain. This may be compared with that obtained at 30 kip load predicted using the Mathcad program in Figure 5 (4739 microstrain). As a result of this test, a fourth beam will be retrofitted with a mechanically-anchored laminates system and tested early next year. Although the FRP system debonded prematurely during the testing, yet, the system increased ultimate moment capacity of the retrofitted beam by about 50% and reduced its midspan deflection by over 60%.

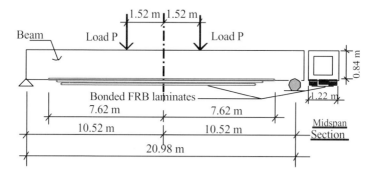

Figure 3. Retrofitted Beam 3

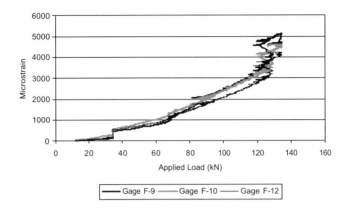

Figure 4. Experimentally obtained laminate strains

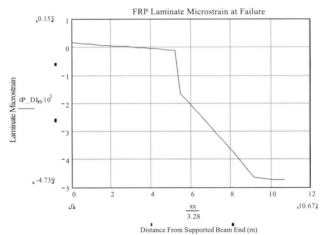

Figure 5. Analytically obtained laminate strains along Beam 3

T-BEAM BRIDGE RETROFIT

Composite carbon laminates in a wet lay-up process were used in this project to retrofit a seventy-year old reinforced-concrete T-beam bridge in South Troy, Rensselaer County, New York. Leakage at the end joints of this single-span structure led to substantial moisture and salt infiltration in the bridge superstructure, causing freeze-thaw cracking and concrete delamination at some locations. A concern about integrity of the steel reinforcing and overall safety of the bridge prompted the FRP retrofitting. Load tests were conducted before and after the laminates were installed to evaluate effectiveness of the strengthening retrofit system and its effect on the bridge structure. Design for shear and flexure was based on the retrofit FRP system compensating for an assumed 15% loss of steel rebar area, due to corrosion [4]. The designed retrofit system is shown in Figure 6. Shear force verus shear stress results in Figure 7, obtained from rosette gauges mounted at critical locations on the web of the most stressed beam, show a moderate increase in shear stress for the retrofitted beam. This may be attributed to the confinement of the pre-existing freeze-thaw cracks provided by the U-jacketed laminates system used for the shear retrofit and its effect on shear resistance mechanism. The figure also reflects the linear behavior exhibited during both the before and after retrofit load tests.

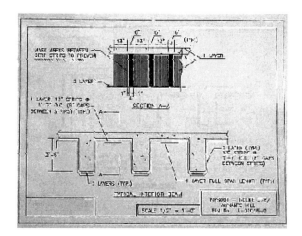

Figure 6. FRP laminates retrofit system

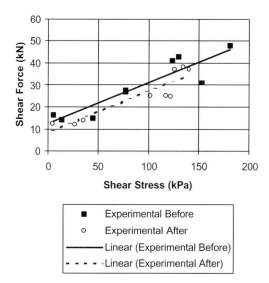

Figure 7. Shear force versus shear stress at critical locations

CONCLUSIONS

FRP materials may have a potential for retrofitting deficient concrete bridge members. Although the FRP retrofit system only moderately reduced stress levels in the capbeam project, ultimate capacity of the retrofitted beam is expected to increase significantly. Use of the FRP laminates system in the deteriorated prestressed beam application demonstrated significant increase in ultimate moment capacity and overall improvement in structural behavior of the retrofitted beam. Application of the FRP system in the T-Beam bridge project resulted in slight increase in the beam shear stress, which may be attributed to the influence of the system on the beam's shear transfer mechanism.

ACKNOWLEDGMENTS

The authors acknowledge the contribution of George Schongar, Harry Greenberg, Joseph Savoie, Jerry O'Connor, and Timothy Conway of NYSDOT, Glenn Washer and Paul Fuchs of FHWA, Dr. Fassil Beshah and Richard Walther of Wiss, Janney, Elstner Associates, Inc., Hardcore Composites, and Mitsubishi Chem. Corp. Prestressed concrete box-beam retrofit studies were conducted in cooperation with the FHWA.

REFERENCES

[1] Hag-Elsafi,O., Lund, R., and Alampalli, S. "Strengthening of Church Street Bridge Capbeam Using FRP Plates," Research Report, Transportation R&D Bureau, NYSDOT, Albany, NY (in preparation).

[2] Hag-Elsafi,O., Lund, R., and Alampalli, S. "Strengthening of a Bridge Capbeam Using Bonded Composite Plates," Structures Congress 2000, Philadelphia, Pennsylvania, 2000.

[3] Hag-Elsafi, O. and Alampalli, S., "Strengthening Prestressed-Concrete Beams Using FRP Laminates," Structural Materials Technology, An NDT Conference, Edited by Sreenivas Alampalli, Atlantic City,NJ, 2000.

[4] Hag-Elsafi, O., Kunin, J., Alampalli, S., and Conway, T. "Strengthening of Route 378 Bridge Over Wynantskill Creek In New York Using FRP Laminates," Special Report 135, Transportation Research and Bureau, NYSDOT, Albany, NY, March 2001.

8.4 Repair and refurbishment of a Prill Tower in Qatar

S Luke, R Karuna, A Jarman, and S Godman
Mouchel Consulting, UK

INTRODUCTION

Carbon Fibre Reinforced Polymer (CFRP) plates are being considered as a more viable and cost effective method for strengthening concrete structures. Over 200 structures including bridges, buildings, industrial plants, chimneys, towers etc have been strengthened using CFRP plate bonding. The benefits of the technique have become apparent to infrastructure owners, designers, contractors following the success of the ROBUST project, and more recently the publication of the Concrete Society Best Practice Technical Guidance Report, TR55, together with the successful installation of over 200 structures.

The benefits of the technique include, reduced installation times, reduced plant, equipment and scaffolding on site, due to the lightness of the plates compared to steel, minimal alterations to the existing structure, improved durability, fatigue and creep. This all leads to lower installation and whole life costs.

The structures in which the technology comes into it's own are those where it is imperative that business interruption or disruption to train or traffic movement is kept to an absolute minimum. Infrastructure owners such as Railtrack, the Highways Agency, Local Authorities, and owners of industrial plants, such as the Prill Tower in Qatar have begun to appreciate the benefits of this technology in minimising disruption to their business critical infrastructure. The technique is also being exported to other parts of the world such as the Middle East and South East Asia.

QAFCO PRILL TOWER
Mouchel was commissioned by the Qatar Fertiliser Company (QAFCO) in the Qatar to carry out a detailed assessment of a 30 year old, 65m tall tower which produces 400.000 tonnes of Urea Fertiliser Prills per year. The brief also included specifying remedial measures with a view to minimising business interruption. The tower circumference is around 36m (diameter 12m), with a 3m stair tower, and was constructed using the slip form technique.

Figure 1: Internal section of tower showing exposed rebar

Inspection and detailed finite element (FE) analysis

A detailed inspection and site survey of the facility was carried out and this included concrete cube tests and tests to establish the levels of carbonation and chlorination in the concrete. The inspection revealed significant vertical and horizontal cracking in the concrete in many regions on the tower, together with loss of section on the inside of the tower in various locations. Internal and external surface temperature measurements were also carried out at various key locations on the tower. Figure 1 shows the internal section of the tower with the rebar exposed with the concrete and the concrete coating spalled off in many locations.

Detailed FE analysis was carried out again to mimic the current condition of the tower, and to understand the behaviour of the tower. The analysis was carried out under dead, live, wind and thermal loading. The analysis also included the effects of the loss of section and the cracking pattern observed in the inspection. The analysis results clearly showed that the vertical bending capacity was just adequate, whereas the horizontal (hoop) bending was significantly under capacity. It was deduced that the original design must have excluded or underestimated the effects of the thermal gradient, which was in the region of 31°C.

Options for Strengthening

A number of options were proposed for strengthening the tower and these included:

1. CFRP plate bonding in the hoop direction
2. Steel plate bonding (jacketing) in the hoop direction
3. Addition of concrete rings equally spaced along the height of the tower

Mouchel recommended the CFRP option as this would:
- have the least disruption to the operations of the tower
- The CFRP option was estimated at 25 days

- The steel or concrete options would require at least 90 days
- not require strengthening works to the foundations
- The concrete or steel options needed extensive strengthening of the foundations
- Required the least amount of scaffolding and lifting plant to carry out the works

The CFRP strengthening option required 2 No 160 mm x 1.8mm thick plates per m throughout the height of the tower. The steel option required 2 No 300 mm x 7mm thick plates per m.

There were client concerns at the use of CFRP on this type of structure in the Middle East, but recent strengthening projects on West Burton power station in the UK and a number of silos and chimneys in Sweden and Russia together with the cost and programme implications of the other options were instrumental in obtaining client approval.

The strengthening contract
Due to the nature, complexity and sensitivity of the project, the plate bonding aspect of the project was tendered to specialist sub-contractors in the UK and Europe. Balvac, Concrete Repairs, Makers and Skanska (Sweden) were invited to tender for the project. The contract was won by Balvac, who sent a team of 4 to Qatar to carry out the works, with assistance from Appollo, the local contractors who won the main contract. In effect Balvac were acting as specialist sub-contractors to Appollo. The CFRP plates and specially formulated adhesive for the Middle East were supplied by MBT FEB. The 1.8mm thick plates were supplied in 250m rolls, see Figure 2 below.

Figure 3 shows the tower fully scaffolded up. It was recommended that this extensive type of scaffolding be used to ensure that works could be carried out at various lifts on the tower during the shutdown period, and that all the mixing and application of the adhesive, etc could be carried out on the tower to facilitate the installation process.

The scaffolding was also completely shrouded as shown to provide protection from the direct sunlight and to maintain a suitable temperature and humidity regime for effective bonding. The outside temperature was between $38 - 45^{\circ}C$ during the works, and the temperature inside the scaffolding was $5 - 6^{\circ}C$ below the ambient.

The plates were supplied with peel-ply on one face and cut to length on site using a guillotine. The tower diameter is around 36m, with a 3m stair tower. The plates were installed in 11m lengths (3 plates per band) with lap plates installed for continuity. The plates were then anchored to the stair tower as shown in Figure 4. Figures 5, 6, 7 show various stages of the plate bonding process.

Figure 2: CFRP plate rolls   **Figure 3**: Fully scaffolded-up tower

Figures 4 & 5: Application of adhesive to substrate, slots in stair well for anchoring
plates

Figures 6 & 7: Installed plates with temporary supports, and with supports removed once adhesive cured

As the tower was not completely circular in shape, it was necessary to support the plates for 6 – 12 hours after bonding them on to ensure they remained in intimate contact with the concrete substrate. Steel flats with timber backing plates were attached to the tower using 70mm long screws drilled into the concrete at 300 mm – 500 mm centres, as shown in Figures 6 and 7. The plates were then tap tested 24 hours after bonding to check for voids. The number of voids found was very small indeed, only about 2 – 3% of all the plates had any voids in them. Plates with voids greater than a 35 mm x 35 mm area were drilled with two small 3 mm holes and filled with a low viscosity resin, see Figure 8, the offending area was then lapped with a "band-aid" plate.

The shutdown period for the tower was limited to 25 days. Any delays to this period would have incurred significant liquidated damages. The first 3 days of the contract were quite challenging, with 3, 6 and 8 plates done respectively as opposed to the plan of 15 plates per day. However, once the huge learning curve was overcome the operatives were easily able to bond 22 plates per day. In the end the contract was completed 5 days early, saving the client significant sums in lost production.

Over 3,500m of CFRP plates were bonded to the outside face of the tower in 20 days. The whole tower was then painted to give an additional level of protection and enhance its durability, see Figures 9 and 10.

Remedial works were also carried out to the internal face of the tower, with the old concrete broken out and re-concreted to the upper 10 m of the tower. Further concrete repair works on the inside will be carried out during the next yearly shutdown of the tower.

Figure 8: Sealing of voids using resin injection

Figures 9 & 10:   The refurbished Prill Tower

CONCLUSIONS

The QAFCO Prill Tower clearly shows the cost-effective and innovative way in which CFRP plates have been used strengthen a concrete tower. It required minimal business interruption. The strengthening of the Prill Tower enabled the shutdown period to be reduced from 25 days to 20 days. Alternative methods would have taken in excess of 3 months and would have required strengthening of the foundations. This will result in enhanced durability and therefore low maintenance and minimal whole life costs

A partnership approach was adopted, whereby client, consultant and contractor worked very closely to develop and implement an effective and durable solution. The strengthening managed to preserve the nature of the structure, with minimal changes made to its overall appearance.

CFRP plate bonding will always be a cost-effective solution to strengthening structures provided the design and installation is carried out by specialists, experienced in the technology. A sound understanding of the principles involved in the design, specification and installation is required to ensure the final solution is robust and durable.

8.5 In-service performance of GRP composites in buildings

S Halliwell,
BRE, Watford, UK

INTRODUCTION

Fibre reinforced plastics (FRP) have a number of advantages when compared to traditional construction materials such as steel and concrete. FRP materials have been utilised in small quantities in the building and construction industry for decades. However, because of the need to repair and retrofit rapidly deteriorating infrastructure in recent years, the potential market for using FRP materials for repair (and for a wider range of applications) is now being realised to a much greater extent. There have been many successful applications using FRP materials in the construction industry, such as bridges, piers, building panels, walkways, pipelines, and offshore structures.

One of the most significant perceived obstacles preventing the extensive use of such materials is a lack of long-term durability and performance data comparable to the data available for more traditional construction materials like steel and concrete. Although there have been numerous studies in the areas of creep, stress corrosion, fatigue and environmental fatigue, chemical and physical ageing, and natural weathering of FRPs in the past five decades, most of these are not aimed at applications for the construction industry. The expected service life of a structure is much longer in infrastructure applications. For instance, bridges are designed to last for over 120 years, and buildings in the region of 60 years. This paper highlights some of the many examples of GRP structures demonstrating good long-term performance as well as the lessons learned from past misapplications.

The work reported in this paper forms part of a much larger site investigation project covering many types of application of composites in construction. It is funded through DTI.

OUTDOOR USE

Outdoor use means exposure to several influences simultaneously; ultraviolet light, fluctuating temperatures, wind and moisture. If the component is buried in the soil, there may be microbiological activity, but this will not usually be a problem with the materials used in conventional reinforced plastics. Urban situations have the added complication of pollution.

ACIC 2002, Thomas Telford, London, 2002

The effects of outdoor use on structural reinforced plastics such as glass/polyester or carbon/epoxy laminates are confined to the surface and do not often involve a serious threat to their structural integrity. The problems are mainly cosmetic [1]. Resins vary a great deal in their ability to withstand outdoor use for long periods. Those with poor performance can sometimes be completely transformed by trace additives, so the problem becomes one of using the right grade of resin.

Changes in appearance occur because of:
1. Changes in the matrix
2. Debonding
3. Light induced changes in additives such as pigments or fire retardants.
Yellowing is usually the first change to appear in the matrix itself, caused by the radiation of shortest wavelength in sunlight. Yellowing is not reflected in corresponding decreases in strength.

CASE HISTORIES [2-4]
Cladding
Houses and apartments
A number of houses and apartment blocks were constructed in Harrow, north of London, in 1996, featuring 'resiform' preformed load-bearing panels made from a mixture of polyester resin, sand and alum, reinforced with chopped strand glass mat. Behind this was a mineral wool insulation layer, covered with plasterboard, and integral metal fixing points were provided. The surface was textured in a number of variations. When inspected in 1996, they were in exceptionally good condition, although some had been painted. Local authority housing officials who recently refurbished one of the sites, expressed their satisfaction with the performance of the panels.

Naval Base, Cornwall
The Naval Base in Torpoint, Cornwall is a Royal Navy training establishment where a series of buildings have been replaced, over several years during the mid 1970s, by new buildings which have smooth surface white GRP single skin cladding on first and second floors (Figure 1). The main structure is made of steel universal beams encased in in-situ concrete and steel columns clad in concrete blockwork.

The first storey is clad in 267mm brick cavity walling. The upper storeys are clad in GRP panels. Inspection in 1997 found that the GRP had been overcoated with Decadex. This was for several reasons. The surface of the GRP had chalked and there was evidence of crazing of the gelcoat. The north facing panels were affected by algae so needed cleaning by a power wash. No bowing or distortion of the panels was evident. The construction retained heat well. There were no reports of leakage problems. The major problem with the panels was in the design around the window areas where an air gap was produced between the gelcoat and chopped strand mat, leading to eventual damage of the GRP. There was no evidence of delamination of the GRP before coating. The GRP was painted with two coats of Decadex which has proved effective but gives a rough surface and so accumulates dirt.

Figure 1: Naval Base, Cornwall

Figure 2: Uneven colour fading between GRP panels on motorway service station

Office block
An office block in Stevenage clad completely in GRP which has a Class 0 fire rating was erected in 1974. The cladding on this block of offices and flats consists of, mostly glazed, white single skin GRP panels. In 1997 the building had stood empty for over 2 years due to change in ownership and is due to be refurbished in the near future, the main problem being leakage from joints and lack of insulation in the building. The panels had been painted a pale cream colour by the previous owners.

Motorway Service Station
A motorway service station in Buckinghamshire was given a 'face-lift' in 1973 by using single skin GRP panels over the existing fabric of the buildings. The buff-coloured panels are in various sizes and are approximately 4mm thick. The panels were inspected in 1997 and the following observations were made. There was uneven colour fading between panels and also on some individual panels (Figure 2).

Crazing was evident on many panels. Areas on some panels, particularly at corners and edges, were devoid of gelcoat exposing fibre. Some edges also showed cracking. Little repair work appeared to have been carried out on the GRP.

Structural applications
GRP Church Spire
A 50ft (17m) GRP church spire was erected in Smethwick, near Birmingham, in 1961 and when the church was demolished 35 years later, the spire was taken to the Avoncroft historic buildings museum at nearby Bromsgrove. A crude orthophthalic gelcoat, to which phthalocyanin green pigment paste had been added, was used on the exterior. On recent (1996) close-up examination of the spire, the gelcoat was seen to have entirely disappeared leaving a fibrous surface of chopped strand glass mat embedded in green resin. The approximately 3mm thick laminate however appeared to be structurally sound and good for many years, since the glass and a certain amount of ingrained dirt were protecting the underlying structure and the steep angle allowed all water to drain off immediately. The intention at that time, however, was to recoat the surface with an isophalic flowcoat (gelcoat plus was to

overcome air inhibition of cure) in the original colour and re-erect the spire on a brick base at the museum.

Classroom extension
A GRP structural shell forms a single classroom extension (built in 1973) to a primary school in Preston. The building is the shape of a modified icosahedron and is made up of 35 self-supporting tetrahedral panels (Figure 3).

Figure 3: Classroom Extension, Preston   **Figure 4:** Surface cracks in gelcoat

Most of the panels have a solid GRP skin but 7 panels contain non-opening triangular windows and 5 contain circular apertures for ventilation fans. The panels have a 50mm phenolic foam backing, which also provides the interior lining of the building. On inspection in 1997, discolouration of the GRP panels was found to have been a major problem, although this appears to be dirt accumulation rather than the GRP itself, and so the building has been cleaned several times with a power hose. There was evidence of chalking on panel surfaces and some surface cracks were evident (figure 4). No leakage problems were noted between the panels. There are leakage problems around the windows despite the fact that they have been resealed.

International telephone exchange
Mondial House, London is an international telephone exchange built in 1974 which has a very large area of GRP cladding panels, having GRP/Polyurethane/GRP sandwich construction. The panels also add structural integrity to the building. The building is a 45m high reinforced concrete 'ziggurat' structure, with each floor except the second being slightly smaller than the floor beneath. Each floor is clad in GRP (Figure 5). The panels have a white fluted edge and are reinforced by means of top hat sections of GRP ribbing. The panels also have complicated side ribbing which allows for a degree of flexibility of support and fixing. The panels are fixed to the building at two points at each level by means of clamps. The clamps fit into a specially formed groove and are fixed to the main bracket by a single stud. The bracket allows for movement and adjustment both laterally and at right angles to the building.   The joints between panels are designed to accommodate thermal movement between panel and fixing tolerances, and also to prevent the ingress of water and air into voids formed by the panels. Inspection in 1997 found the white GRP to be in good condition, some chalking was evident and there had been gasket movement between some joints. No repairs had been done. The building had been

cleaned several times. The maintenance house, known as the 'chocolate box' is the only part of the building to be constructed in brown GRP. This had faded and chalked and there was some evidence of crazing. Fibres were showing through in some areas.

Modular structures
GRP modular stores building
A modular GRP stores building was constructed at Wollaston in 1968, to demonstrate the suitability of this material for construction purposes (figure 6).

Figure 5: Mondial House **Figure 6**: Modular stores building at Scott Bader site, Wollaston

The walls are of a folded plate design made with filled, FP resin and a green (chrome oxide) pigmented FR gelcoat; the roof utilised a translucent FR resin and a translucent gelcoat. When examined 28 years later, the green gelcoat on the outer walls had some grime and algae growth, but when cleaned was found to be smooth, uniformly green but completely matt (it was originally semi-gloss). There had been some impact damage from low level traffic. Vertical, corrugated infill-panels using a bright green pigment were also in good condition in respect of both colour and surface uniformity, again apart from some vehicular damage at edges. However, on the translucent roof, only traces of the original FR gelcoat remained and it had badly discoloured. Some areas where the gelcoat had been lost were smooth, but others had patches of exposed reinforcement. However, the structure was mechanically sound and light transmission inside the building was still sufficient to avoid the need for artificial lighting during the day. Fire-retardant gelcoats would not now be recommended for exterior exposure and today's materials would undoubtedly have a much improved service life.

Transmitter Cubicles, South Wales
Transmitter Cubicles in South Wales comprising several complete GRP structures constructed in early 1974, are being used to house electrical equipment (Figure 7). The walls and roof of the cubicles inspected are all GRP/expanded steel/PU foam/GRP sandwich construction. The door, also GRP, has steel frame surround

and is fixed into a steel frame in the cubicle. The floors are fixed down by bolts that are embedded in the concrete and pass through holes in the floor.   The cubicles were designed to withstand winds of up to 140mph.   No provision had been made for differential   expansion between the concrete raft and the cubicle floor, except that the bolt holes are considerably oversize and this has proved sufficient to accommodate any thermal movement. The cubicles achieve a Class II fire rating. Three sites were inspected in 1997.

Site 1: Erected 1973. Some crazing type cracks on the back wall and evidence of repairs over crazed areas. Gel coat peeling in one small area. No discolouration, some dirt accumulation and lots of lichen growth. No leakage problems.

Site 2: Erected 1973.   South face is badly crazed (Figure 8) and shows patchy discolouration. Gel coat thinned in places revealing brown resin underneath. No leakage problems.

Site 3: Erected 1973. Painted over. Reports of leaks – not obvious from where.

Multi-storey structure
The world's first advanced composite multi-storey building constructed using the Advanced Composite Construction System (ACCS) was commissioned by the client as an alternative to conventional accommodation for the Government's Agent team on the Second Severn Crossing project (Figure 9). This has since been converted into the Severn Bridges Visitors centre. Good thermal insulation and fire-resistance are key features of the building design. Panels are insulated to a high standard and double-glazed windows have been fitted. In critical areas, panels meet Class 0 fire regulation specifications with a fire resistance of at least 30 minutes when tested in accordance with BS 476: Part 21. The ACCS provides rapid, cost effective, high quality construction, even in locations where access is difficult. The key advantages from using composites in this application are:

Figure 7: Transmitter Cubicle,
South Wales

Figure 8: Transmitter Cubicles –
surface crazing

- Monocoque structure, no additional framework required
- Lightweight modular components

- Designed for easy erection
- Thermally insulated and fire resistant.

The system consists of high quality interlocking pultruded glass reinforced panels which are bonded together to form a membrane structure in which walls, floors and roof provide complete structural integrity without additional framework.

Figure 9: Severn Bridges Visitors Centre

Figure 10: Sharjah International Airport Terminal Building

After 8 years service the GFRP was in excellent condition and had undergone minimal maintenance being jet-washed just once. Some dirt accumulation was evident. The structure is in an exposed position with no shelter.

Roofing applications
GRP dome on international airport
Some GRP dome structures supported by metal frames were constructed in the mid-1970s for the Sharjah International Airport terminal Building (Figure 10), United Arab Emirates and these were recently inspected. The cream isopathic gelcoat was complete but appeared dull when viewed close-up and a few blisters were seen. However, the laminate structure was sound with no signs of delamination.

School roof
The roofs of the different parts of a major extension (built in 1974) to a secondary school in Bethnal Green, London are made from moulded GRP sections (Figure 12).

All the units contain integral translucent GRP window panels. The ends of each span were reinforced with an integral steel plate enclosed within the GRP skin. Stiffening ribs run under the trough of each span for its full length and similar ribs were also moulded between the window panels. The majority of the roof panels were taken down in 1995 and the rest was due for demolition in September 1997. The GRP had become embrittled and was considered unsafe. No large cracks were evident, but some smaller cracks in the gelcoat could be seen. There was no apparent leakage between the GRP panels, leakage only occurred at the joints where the roofs were connected to the brick walls. There was little discolouration. The translucent GRP window panels had darkened and the gelcoat was peeling in several places leaving exposed fibres (figure 12). The window panels still let in some light to the room.

Figure 11: School Roof, London

Figure 12: Deterioration of an unpigmented polyester/glass laminate

REPAIR AND MAINTENANCE OF FIBRE REINFORCED PLASTICS STRUCTURES

Panels facing prevailing winds are generally 'self-cleaning' whilst those in the shadow will become contaminated and require periodic pressure/chemical cleaning.

Mechanical damage, during erection or in service, can be patched [5, 6] and an attempt made to match the original but site conditions are not usually favourable for proper curing of the resin and a patch is likely to show after a period of weathering. It is better to replace a complete panel or member if this is practicable.

Restoration of degraded surfaces can be difficult. In the extreme case of the glass fibres becoming badly exposed, they must be scrubbed off completely before any new surface treatment is applied.

CONCLUSIONS

GRP composites already have a market in the construction industry. With the development of other fibres this market can be expanded. Current developments include the use of fibre composites in listed building restoration and pultruded composites in various structural applications. New markets need to be explored in order to fully exploit the potential benefits fibre composites can offer the construction industry.

The work reported here will lead to the development of a document illustrating the long-term performance of composites in construction applications. The final document is intended to consist of reference text and several case history examples.

REFERENCES

[1]   Chester R J and Baker A A. Environmental durability of F/A-18 graphite/epoxy composite. Polym Compos, 1996, **4**(5), 315-323.

[2] Halliwell S M. Reinforced Plastics Cladding Panels. Proc Composites and Plastics in Construction, BRE, UK, November 1999, Paper 20.

[3] Layton J. Weathering. In: Reinforced plastics durability (Ed G Pritchard), Cambridge, Woodhead Publishing, 1999. Chapter 6.

[4] Halliwell S M. Survey of Glass reinforced plastics cladding panels. DETR report, London, DETR, 1998.

[5] Baker A A and Jones R. Bonded repair of aircraft structure. The Hague, Martinus Nijhoff, 1988.

[6] Streiffert B. Glass-fibre boat manual: practical repairs, maintenance and improvements. London, MacDonald Queen Anne Press, 1989.

BIBLIOGRAPHY

1. Halliwell S. Polymer Composites in Construction. BRE report BR 405, ISBN 1 86081 429 8, 2000.

2. Halliwell S. Advanced Polymer Composites in Construction. BRE Information paper IP 99/7.

3. BRE digest 442: Architectural use of polymer composites.

4. Crump L S. Evaluating the durability of gel coats using outdoor and accelerated weathering techniques: a correlation study. Proc 51st Annual Conference, Composites Institute, SPI, February 1996. Paper 22-B.

5. Wooton A B. Accelerated Weathering specifications used in the polymer industry. Polymer Testing '96, RAPRA, Shawbury, UK, September 1996, Paper 12.

8.6 Application of carbon fibre composites at covered ways 12& 58 and bridge El

D G Church and T M D Silva
Infraco Sub-Surface Limited, UK

ABSTRACT

Covered Ways 12 and 58 are parallel covered ways, each carrying the two tracks of the District Line and Circle Line respectively. They were constructed in 1867 and support the highway, Kelso Place a twin ended cul-de-sac, and predominantly residential Properties over the London Underground's Railway. The two covered ways are each approximately 110m long with interconnecting cross passages. The roof comprises brick jack arches spanning onto cast iron beams, 45 in each covered way. The cast iron beams are 'hog backed' in profile and the 775 deep jack-arches are concrete backed and waterproofed with asphalt.

The cast iron beams were assessed and found to be over stressed. A load restriction of 13 tonnes was imposed for vehicles over the roadway but the cast iron beams were also significantly over stressed due to the building loads.

The covered ways were strengthened using 'aerofoil' type steel beams inserted between the cast iron beams into the jack arches

The proposed staging of the work would have induced differential settlements causing further increases in tensile stress leading to a sudden failure of a beam during the works.
To mitigate premature failure in the bottom flange, the cast iron beams were strengthened using two strips of carbon fibre composite (CFC) bonded onto the bottom flange of each cast iron beam in both covered ways. The specified structural design life being 2 years with no de-bonding for a further 40 years.

Overbridge EL31 is situated in the London Borough of Tower Hamlets, comprises cast iron beams and carries the Lower Road over the Railway. Two of the cast iron beams failed BE 4 loading assessment. This bridge is due for reconstruction in five years time and as a temporary measure the two beams were strengthened using carbon fibre composites.

ACIC 2002, Thomas Telford, London, 2002

INTRODUCTION
Covered Ways 12 & 58
Covered Ways 12 and 58 are parallel covered ways, see Figure 1, each carrying the two tracks of the District Line and Circle Line respectively. They were constructed in 1867 and support the highway, Kelso Place a twin ended cul-de-sac, and predominantly residential properties over the London Underground's Railway.

The two covered ways are each approximately 110m long with interconnecting cross passages. The roof comprises brick jack arches spanning onto cast iron beams, 45 in each covered way. The cast iron beams are 'hog backed' in profile and the 775 deep jack-arches are concrete backed and waterproofed with asphalt.

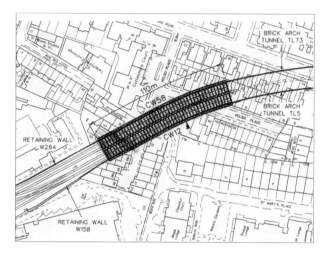

Figure 1:   Location of Covered Ways

PROBLEM
Understrength Cast Iron Beams
The cast iron beams were assessed for the highway and building loadings and found to be over stressed. A 13 Tonnes loading restriction was imposed for vehicles over the roadway but the cast iron beams were also significantly over stressed due to the building loads.

The foundations of the abutments and the central pier were stable with no obvious sign of distress. However assessment indicated that the factors of safety were below the required values given in the London Underground's Engineering Standards for resistance against sliding and bearing failure.

Track
The track comprised ballasted trackform with timber sleepers and mostly limestone ballast. The ballast was degraded and a full replacement of the track formation had

not been carried out since the Covered Ways were constructed. The track formation was considered life expired.

The vertical clearances, between the highest rail levels and the soffit of the cast iron beams were below the agreed clearance of 3962mm, see Figure 2.

Figure 2: View of covered way showing cast iron beams

Noise & Vibration
There had been a history of noise caused by the passage of trains through the covered ways, which gave rise to complaints from residents living above the District and Circle Line tracks south of High Street Kensington station. In recent years the complaints have grown stronger with representation being made to the Local Authority (Royal Borough of Kensington and Chelsea).

SOLUTION
The principal reason for the works was to strengthen Covered Ways 12 and 58 in order to meet London Underground's statutory obligations under the section 117 of the 1968 Transport Act and to prevent sudden failure of the cast iron beams.

In order to improve the existing structure gauge clearances the tracks were lowered.

To minimise the overall construction depth and required track lowering, the new superstructure comprised 'aerofoil' type semi-circular steel beams erected within the jack arches, see Figure 3.

The space between the jack arch intrados and the curved top plate of the beam was pressure grouted to reduce the loading on the existing cast iron beam and ensure that the cast iron and steel sections shared additional loads.

Figure 3: Cross section of 'Aerofoil' Beam

Figure 4: Fabricated beam on site

FOUNDATIONS AND SUBSTRUCTURE

The abutment recesses were infilled with concrete to provide support for the new steel beams. A single row of vertical piles was constructed in front of the abutments to provide stability during the excavation works for track lowering, and to increase the vertical bearing capacity. Raking piles were provided through the centre pier to increase the bearing capacity.

Track

The tracks were lowered by up to 250mm to allow for the required minimum vertical clearance 3962mm plus a further 50mm for track maintenance. A further allowance for deflection and settlement of the structure had been incorporated into the steel beam design.

New Track Form was installed comprising a minimum of 230mm granite ballast below the underside of the concrete sleepers and BS113A flat bottom rail.

Noise and vibration was mitigated using a ballast mat placed between a sand blanket and the ballast. To provide a transition between conventional trackform and the ballast mat, a stiffer transition ballast mat was provided at the end of each overall length of ballast mat and the mat was wrapped up the abutment wall at the ends of the sleepers.

STRENGTHENING OF CAST IRON BEAMS USING CARBON FIBRE COMPOSITES

The staging of the works would have induced differential settlements between the abutments and centre pier. The cast iron beams, whilst nominally simply supported, are built into the walls at the supports. Whilst they cannot be considered to be fully fixed neither are they pinned. The weight of the buildings above added to the degree of potential fixity at the supports.

As the cast iron beams were already overstressed, the differential settlements could have induced further increases in tensile stresses leading to a sudden failure of a beam during the works.

Whilst previous failures have not proved to be catastrophic, due to the relatively shallow depths in some locations, the beams would not have remained in place after failure and could displace vertically infringing the load gauge. This was of particular concern during works to CW58, which could have initiated a failure in the still operational CW12.

In the event of a failure in CW12 during the closure of CW58, the impact on the already modified District and Circle line services would have been catastrophic.

Due to the limited clearances, the ability to provide temporary works was extremely limited. The temporary use of carbon fibre composites on the bottom flange of the cast iron beams to mitigate premature failure in the bottom flange was therefore proposed.

The proposal, as shown in Figures 5, 6 and 7, was to bond two strips of carbon fibre composite (CFC) onto the bottom flange of each cast iron beam in both covered ways prior to the closure of CW58 (the first Covered way to be closed and strengthened). The requirement was for a design life of 2 years with no debonding for a further 40 years.

The strengthening works were carried out during separate long closures of each Covered Way, with enabling works carried out during Engineering Hours.

Figure 5: View of CFC strengthened beams and view of end beams

Figure 6: Plates being ready for installation and plates being attached to beam

Figure 7: CFC plates being erected

DESIGN CRITERIA
Strength requirement
The 90 cast iron beams in both covered ways were strengthened by the application of carbon fibre composites such that tensile failure of the lower flange of the cast iron beam did not result in beam failure.

Beams situated below the three storey buildings were strengthened to be capable of carrying a working load of 185kN/metre run of beam. All other beams were strengthened to be capable of carrying a working load of 167kN/metre run of beam

The carbon fibre composite were designed to carry all the tensile load after cast iron cracking and were designed to prevent compressive stresses exceeding 154N/mm2 from developing in the top flange of the beam at Service Limit State.

Design life, application and Operating Conditions
The strengthening was to be effective for a minimum of two years, but the actual life was to be the maximum achievable by means of appropriate controls relating to materials and workmanship, and taking full account of the local environment. Electrolytic insulation was to be effective for 120 years.

Ambient temperatures were to be close to normal surface temperatures near the covered way entrance and the system was designed such that it was capable of successful application at these temperatures.

Base metal temperature extremes between –15°C to +53°C at headwall and –5°C to +30°C beyond 20 metres from the headwall. Humidity extremes between 35% and 65% were assumed.

Fire safety of materials
Carbon fibre composites were to comply with the material requirements of The Fire Precautions (Sub-surface Railway Stations) Regulations 1989. These specify requirements limiting flammability, rate of heat release on ignition and rate of spread of flame over surface. In addition Carbon fibre composites were to comply with LUL Engineering Standard E 1042 A2 "Fire safety performance of materials used in the Underground"

CHEMICAL AND ELECTROLYTIC RESISTANCE
The CFC material were to remain structurally effective for the specified design life of 2 years and was not to debond from the cast iron surface in the next 40 years in the presence of the chemicals released from above the railway and within the railway.

Electrical safety
The structural integrity was not to be destroyed by accidental transmission of traction current or shorting to earth if insulation fails on adjacent power cables, which may carry current at voltages up to 22kV.

Radiation degradation
The composite was to retain structural integrity for its design life when subject to solar radiation (at headwall) and surveying lasers.

Fatigue & creep

The composite strengthening was to retain structural integrity when subject to fatigue loading and thermal cycling for its design life. Road fatigue loading was to be as specified in BS 5400: Part 10.

Over the structural design life, creep of the composite strengthening must not cause an increase in deflection of more than 10mm from the initial cracked section deflected profile.

Structure gauge clearances

The total depth of the carbon fibre composite and any temporary clamping arrangements was not to exceed 35mm. The vertical clearance above the highest running rail was not to exceed 3,835mm.

CARBON FIBRE STRENGTHENING OF BRIDGE EL31 AT SURREY QUAYS

Introduction

El 31 is an overbridge carrying the A200, Lower Road, in a Northwest/ Southeast direction over the East London Line Railway. The bridge is located at Surrey Quays station on the London Underground Limited's East London Line, see Figure 8. At the bridge the road divides into three notional lanes carrying one way traffic. The bridge has a footpath on either side of the road. Several services cross the railway within the carriageway, footpaths and attached to the south parapet. There is a service bridge, EL30, supported by the substructure of EL31 carrying several large diameter pipes. The railway is in a cutting and consists of twin tracks, with junction work forming a crossover to the north of the structure.

Structure

The superstructure consists of several different types of construction. Masonry jack arches spanning on to cast iron beams form the extremities of each span. The central section is of early riveted trough decking. The bridge has three simply supported spans supported on crosshead girders, in turn supported by cast iron columns.

Figure 8: View of Surrey Quays Station platforms and cast iron beams and columns

Problem

The bridge was assessed in 1985 and was found to be of insufficient strength in certain areas to support the BD 21/84 standard loading. Most of the beams were found to be capable of supporting various reduced loads from BD21/84. The sections that failed to support the BD21/84 Standard Loading were further assessed to the superseded document BE4, and most were found to be capable of carrying the full Construction & Use Vehicle Loading as specified in this document. Two cast iron girders and two secondary beams in the northern end of span were found to be unable to support the BE4 loading. This is the legal minimum loading situation that the bridge must support

It was concluded that, for the short-term, the strengthening of the two cast iron beams that fail the BE4 assessment with carbon fibre composites was the most suitable option considered. The strengthening design made use of High Modulus Carbon Fibre Composites (CFC) that were epoxy bonded to the bottom flanges of the two beams, see Figure 9.

As an added precaution Trief type kerbs were installed to help prevent vehicles from mounting the footpath over the beams. However, the carbon fibre strengthening was designed to carry the full accidental wheel loading on the footpath.

Analysis and Design

The Beams were analysed as simply supported and restrained against buckling by the jack arches. A Fracture Mechanics Analysis was used where the cast iron was overstressed due to dead loading.

The CFC was designed in accordance with the recommendations given in the Joint Industry Project Manual. The use of the Carbon Fibre Composites was submitted to London Underground's Building Control Group for approval for use in the underground.

The original design was carried out by LUL based on parameters and strength properties of fibres given by Devonport Management Ltd (DML) and the final design was independently checked by David Morris of Brown & Root Ltd

Carbon fibre composite (CFC) plates were used to increase the flexural strength of the two cast iron beams by attaching the plates with epoxy resin to the lower flanges. The repair works on the beams were carried out during engineering hours as the beams were located over the tracks. The road was partially closed in the area of the beams several hours to allow the epoxy adhesive to bond in an unstressed state.

Method of construction

Scaffolding was erected to allow the lower flanges of the beams to be blast cleaned. Carbon fibre composite plates were then carried via the station, attached by epoxy adhesive to the bottom flange of the beams and allowed to set while the bridge was in an unloaded state. The epoxy was designed to have sufficient strength to carry the weight of the CFC plates when applied but temporary magnetic clamps were used during initial positioning.

The CFC strengthening plates were repainted to match the existing beams. Regular inspection is in place be to ensure that the carbon fibre composite plates do not debond from the cast iron beam.

Figure 9: Views of strengthened beams

8.7 The design and testing of an FRP composite highway deck

A F Daly[1] and W G Duckett[2]
[1]*TRL Limited, Crowthorne, UK*
[2]*Maunsell Ltd, Beckenham, UK*

INTRODUCTION
Use of fibre-reinforced plastic
The use of Fibre Reinforced Polymer (FRP) as an alternative to the more traditional civil engineering structural materials such as steel and concrete is a technology which is developing rapidly in the construction industry. These materials have considerable advantages in terms of weight, strength and durability and they have been used for a number of decades in high-tech applications in the aerospace and automobile industries. Although uptake in civil engineering has been slow, FRP has already been used in a number of bridges around the world. It can be expected that, as the production technology develops and definitive design guidelines become more generally available, these innovative materials will be used more widely to provide cost-effective and durable alternatives to steel and concrete.

There is considerable scope for the use of lightweight FRP decks to carry highway traffic. Potential applications are new design, replacement of under-strength decks in existing bridges, and the provision of temporary running surfaces. A range of commercial FRP materials are already available and they are beginning to be widely used in bridges.

In spite of the obvious advantages, however, there is still a lack of information on the behaviour of FRP components in bridge applications. This has impeded the development of generally accepted guidelines for the design and application of bridge schemes. In response to this, the UK Highways Agency commissioned the Transport Research Laboratory (TRL) to carry out a programme of research into the performance of FRP bridge decks. The overall objective of this research is to produce generic guidelines for the design of FRP decks that are capable of carrying the large wheel loads imposed by Heavy Goods Vehicles.

Outline of the research project
The main requirement of the project is to examine the use of FRP as the main structural element in new and replacement bridge decks and to devise guidelines to be used by designers particularly in relation to fatigue.

ACIC 2002, Thomas Telford, London, 2002

The objectives of the project are:

- to obtain evidence that FRP decks can provide adequate service performance in terms of the level of routine maintenance and the life to major refurbishment or replacement;

- to provide data on which to base generic fatigue requirements for technical approval of FRP deck systems, i.e., to define the evidence needed and develop a method of providing assurance of durability;

- to study the performance requirements for the connections of the FRP running surface to the supporting deck;

- to develop a standard giving generic requirements for technical approval of FRP deck systems.

To meet these objectives, laboratory tests on an FRP deck are being undertaken to investigate its performance under highway traffic loading. A desk study will address other issues such as design requirements, resistance to environmental effects, repair and maintenance. The project started in February 2000 and is expected to be completed by the end of 2002.

While the primary aim of the project is to produce generic fatigue requirements for FRP decks the project necessarily involves the design and fabrication of an FRP deck system capable of carrying heavy traffic loading. The design commission was carried out by Maunsell Ltd who developed the Advanced Composite Construction System (ACCS). This consists of standard Glass FRP cellular components and connectors that can be assembled together in a wide variety of configurations. The system of construction has been used as the basis for the Linksleader footbridge at Aberfeldy in Scotland and a lift bridge at Bonds Mill in Gloucestershire.

DESCRIPTION OF TEST PANELS
Form of construction
The ACCS bridge system is based on a cellular glass pultruded FRP plank, which is 600mm wide, 80mm deep and has a wall thickness of between 2.5 to 4.0mm, as shown in Figure 1. The components have undercut grooves on the sides used to connect them together, either side by side or at right angles, using a square 3-way connector piece and various other interlocking profiles as shown in the figure.

The components are assembled by placing a bead of epoxy adhesive along grooves moulded in their sides, bringing the faces to be bonded together and sliding the bone-shaped rod or "toggle" into place. The planks can be assembled together in different configurations depending on the structural requirements.

Analysis showed that the ACCS components are capable of carrying the load induced by Construction and Use vehicles when assembled in a box or T-beam form. However, they are not satisfactory in withstanding the concentrated stresses induced by heavy wheel loads. A limited amount of dynamic testing on individual plank specimens resulted in fatigue failure at about 40,000 cycles. Thus, in order to provide a workable bridge solution, a heavier roadway panel was required. This is described in detail in a later section.

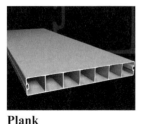
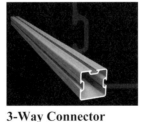
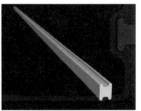

Plank   **3-Way Connector**   **Toggle Connector**
Figure 1. ACCS Components

Assembly of sub-structure
The sub-structure used for the test consists entirely of ACCS components assembled using epoxy adhesive. There are four vertical ACCS planks forming webs and three ACCS planks forming the top flange. The overall width of the sub-structure is 2.12m. For the short span to be used in the test programme, it was sufficient to construct it in the form of a T-beam. For longer spans, the cross-section could be built up into, for example, a box section.

The ACCS planks weigh about 11 kg/m, light enough to be manhandled into place without the use of a crane. All components of the bridge were bonded together using Vantico Araldite 2015 adhesive. This is a toughened epoxy paste ideal for bonding glass FRP components.

The sub-structure spans 4.0m, which is the maximum span that could be fitted into the loading rig. It is supported on 15mm rubber bearings under the webs. End diaphragms are installed transversely at each support to provide lateral restraint to the webs and to assist in distributing the reaction loads. The diaphragms consist of 600mm lengths cut from an ACCS plank. The diaphragms are fixed in place using lengths of 50mm FRP angle glued to the top flanges and webs. Figure 2 shows the sub-structure after fabrication and before being installed in the test rig. For convenience, it was assembled upside-down, with the top flange constructed first, followed by the webs, and finally the diaphragms. This assembly took four days to complete. Figure 2 shows the hardwood inserts that are fitted into the first cell of the ACCS web over the bearings to prevent local distortion and premature buckling of the FRP section.

ROADWAY PANELS
Maunsell Ltd carried out the design of the Roadway Panel capable of carrying HGV wheel loads. The panel was designed to be overlaid on the sub-structure as previously described with the Roadway Panels running transverse to the direction of traffic flow, so that they are spanning onto the webs of the sub-structure. The FRP deck was designed to withstand the highway loading specified in HA document

Figure 2. ACCS sub-structure (upside-down)

BD 37 with 45 units of HB. For fatigue, the loading specified in BS 5400: Part 10 (British Standards Institution 1980) was used. The load partial safety factors for the serviceability and the ultimate limit state were as specified in BD 37.

The Roadway Panel, shown in Figure 3, has a similar layout to the ACCS planks with vertical webs. They are designed to predominately span in their longitudinal direction as they are relatively flexible in their transverse direction due to the shear deformation of the Vierendeel section. This design concept minimises the stresses that have to be carried across the longitudinal joints between panels.

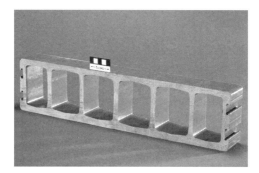

Figure 3. Roadway Panels

When compared with the ACCS planks the wall thickness of the Roadway Panel has been increased to 12-15mm to withstand the high local stresses imposed by wheel loading. The depth has also been increased to 115mm. Because of the heavier component walls, the width of the panel was decreased to 500mm to reduce the weight of the component to about 37kg/m. This facilitated the pultrusion of the units and eases the final assembly of the deck. To form the running surface, the units are joined together using a pair of the ACCS toggles. For this project, the Roadway Panels were bonded uniformly over the surface to provide an efficient connection between the running surface and the sub-structure.

The design was undertaken in accordance with a coherent Limit State methodology consistent with standards for conventional structural materials. The partial material factors of safety for ultimate limit states were based on factors developed by Maunsell from first principles on the basis of reliability theory in connection with previous projects approved by the HA.

The partial material factor γ_m is built up as follows:

$$\gamma_m = \gamma_v \gamma_e \gamma_t \tag{1}$$

where the three sub factors γ_v, γ_e and γ_t relate to material variability, environmental conditions, and age (damage) respectively. For ultimate limit states the following values were adopted:

Material direction	γ_v	γ_e	γ_t	γ_m
σ_1 parallel to fibre	1.3	1.15	1.50	2.25
σ_2 transverse to fibre	1.5	1.2	1.67	3.00
τ_{12} shear	1.5	1.2	1.67	3.00

The material variability factors are based on the coefficient of variation of the material strength in the relevant direction and are calibrated to achieve an equivalent reliability index to that built into BS 5400 partial material factors.

Failure analysis of each individual ply was considered using the Tsai-Hill failure criterion:

$$\left(\frac{\sigma_1}{\overline{\sigma}_1}\right)^2 - \left(\frac{\sigma_1 \sigma_2}{\overline{\sigma}_1^2}\right) + \left(\frac{\sigma_2}{\overline{\sigma}_2}\right)^2 + \left(\frac{\tau_{12}}{\overline{\tau}_{12}}\right)^2 \leq 1.0 \tag{2}$$

where (σ_1, σ_2, σ_3) are the stresses due to factored loads relative to the principal material directions and the barred values are the corresponding material strength parameters reduced by partial material factors.

For preliminary design the following fatigue life equation was used:

$$\frac{\sigma_{max}}{\sigma_U} = 1 - k \log\left(N_L\right) \tag{3}$$

where σ_{max} is the maximum stress in a stress cycle, σ_U is the corresponding ultimate stress under the same loading pattern, N_L is the number of cycles to failure and k is an experimental coefficient whose value depends on the R ratio ($\sigma_{max}/\sigma_{min}$). For preliminary design k was assumed to be approximately 0.1. The research carried out during this project should give further information which will clarify the S-N curve for typical laminates in road decks.

Manufacture of FRP material

The FRP material was fabricated by Strongwell Ltd, to a specification devised by Maunsell Ltd. It was shipped to the UK in July 2001. The material produced is sufficient to produce two full-scale deck panels for testing using a rolling wheel load as well as for the smaller scale tests which will be carried out after the panel tests.

DESCRIPTION OF TESTS

Rolling wheel tests

The first series of tests consisted of rolling wheel loading tests on a full-scale deck using the Trafficking Test Facility (TTF) at TRL. The main features of the TTF are shown in Figures 4 and 5. The rig was originally built to carry out dynamic testing of bridge joints. It was modified to simulate the traffic loading on a full-scale bridge. For these tests a wheel load of 35kN was applied to the FRP road deck: this is substantially higher than the 20 kN wheel load specified for the standard fatigue vehicle in BS 5400.

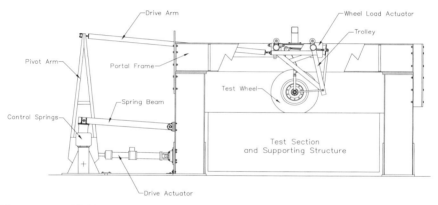

Figure 4. Trafficking Test Facility at TRL

Figure 5. Test Wheel

Two different panel configurations were utilised. The first used ACCS planks to form the running surface. Although these components are not designed to withstand the stress concentrations arising from directly imposed wheel loads, testing in this way provided valuable information on the likely locations and modes of failure. The second configuration used the newly designed Roadway Panels as the running surface. It was predicted that the Roadway Panel would not fail during the rolling wheel tests. However, information was obtained on the state of strain in this panel which will be used in the next phase of testing. This information will also be used predict the state of stress in the component and, in conjunction with finite element analysis, to determine the expected fatigue life. The tests will also confirm the validity of the FRP design and the overall robustness of the FRP deck.

Test results to date
The first test with the ACCS planks forming the running surface was carried out at a significantly reduced wheel load of 6 to 8 kN. The first signs of failure were recorded after 5,000 cycles and the test was stopped after 80,000 cycles. The failure occurred at the junction of the webs and the top skin of the ACCS Plank. Cracking was visible in the top surface of the plank adjacent to the webs of the ACCS planks and is believed to be due to hogging moments at this location. It should be borne in mind that these panels had not been designed to carry the concentrated wheel loads and this type of failure was predicted.

The second test with the Roadway Panels was stopped after 4.56 million cycles. After this period of testing there was no visible sign of any deterioration in the performance of the Roadway Panels and it was concluded that if this was a real life structure there would be no cause for concern with the roadway panel. However, some of the strain gauges indicated local failures. It was not clear whether this was a failure of the gauge or an indication of delamination of the outer ply. Further investigation and coring of the test sample will be undertaken at these locations. The influence lines for the various strain gauges have been plotted and by inspection the critical stress range occurs at the top of the internal webs where a reversal of stress occurs. The "R" ratio $(\sigma_{max}/\sigma_{min})$. for this location is 0.67. A plot of this influence line is shown in Figure 6.

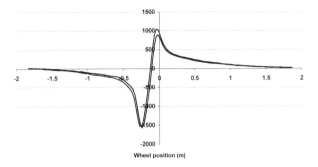

Figure 6. Influence line showing microstrain

Further work is required to identify a full S-N curve for the laminates in the roadway panel for a significant number of cycles. However, given the S-N curve assumed in the design 4.56 million cycles can be extrapolated to in excess of 37 million fatigue vehicles.

ON-GOING WORK
The next phase of testing
The number of cycles in the TTF is limited by the cost and rate of testing. It is therefore planned to use the strain data generated by the rolling wheel tests to design specimens and loading arrangements suitable for a conventional servo-hydraulic testing machine to be carried out in the next phase of the project. The small-scale specimens will consist of full-scale sections cut from the standard FRP panels. Strain gauges installed at the same locations relative to the critical detail as on the rolling wheel tests will be used to set up the loading to give the correct stress distribution. Tests can then be carried out at up to 5Hz, to long endurances, and at several stress ranges to obtain points on an S-N curve.

The objective of the tests is to determine whether appropriate testing procedures using small-scale specimens can be devised. These should be capable of determining the fatigue performance of the detailing of the FRP panels at critical points as identified by the full-scale testing and analysis.

Development of design guidelines
The experience gained in the design and fabrication of the trial panel will enable design guidelines to be drafted. Several approaches are possible: the guidelines could be based on performance requirements, or on imposed limits on the stress in the FRP. The exact form and content of the guidelines will depend on the output and conclusions of the testing programme.

Future work
It is possible that at some stage in the future a trial structure will be built using the bridge system developed for this research. The testing to date has shown that the system used here results in a structure with an acceptable service life. A typical application might be for the re-decking of a concrete bridge which has failed assessment either because of deterioration or because of increased design loading or a complete FRP replacement bridge. The construction of a full-scale trial structure is the logical next step.

ACKNOWLEDGEMENT
The project described in this paper is being funded by the UK Highways Agency. The work is being carried out in the Infrastructure Division of TRL in collaboration with Maunsell Ltd. Any views expressed are those of the authors and not necessarily those of the Highways Agency. The paper is published by permission of the Chief Executive of TRL.

Index of authors